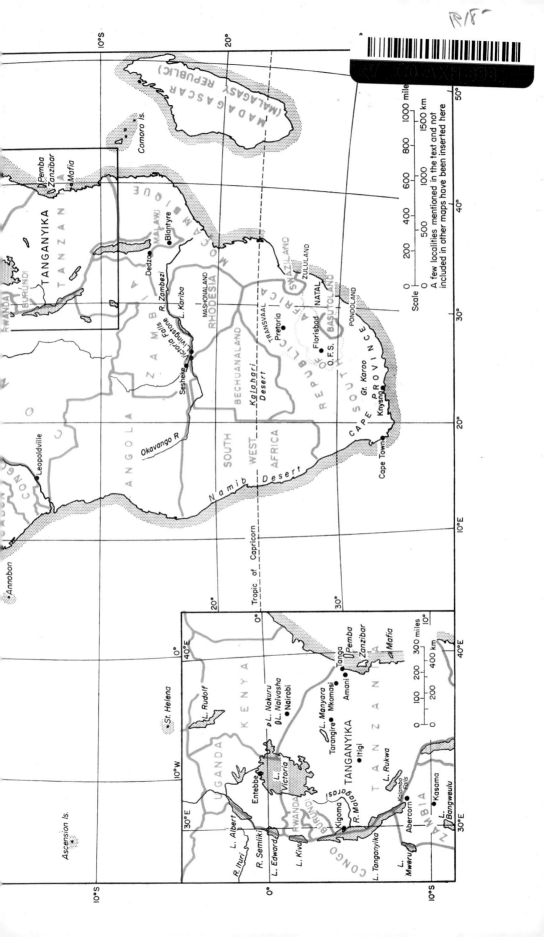

R18

TANGANYIKA

TANZAN

BURUNDI
RWANDA

Pemba
Zanzibar
Mafia

MADAGASCAR
(MALAGASY REPUBLIC)

Comoro Is.

MOZAMBIQUE

Dedza
Blantyre

MALAWI

Kariba
L.
R. Zambezi
Victoria Falls
Livingstone
Sesheke

ZAMBIA

ANGOLA

Okavango R

CONGO

Leopoldville

GABON

Annobon

SOUTH WEST AFRICA

Namib Desert

Tropic of Capricorn

Kalahari Desert

BECHUANALAND

MASHONALAND
RHODESIA

MOZAMBIQUE

SWAZILAND
ZULULAND

TRANSVAAL
Pretoria
O.F.S.
Florisbad
NATAL
BASUTOLAND

REPUBLIC OF SOUTH AFRICA

PONDOLAND

CAPE PROVINCE

Gt. Karoo
Knysna

Cape Town

St. Helena

Ascension Is.

Scale

| 0 | 200 | 400 | 600 | 800 | 1000 mile |

| 0 | 500 | 1000 | 1500 km |

A few localities mentioned in the text and not
included in other maps have been inserted here

10°S
20°
0°
10°S

50°
40°
30°
20°
10°E
0°
10°W

UGANDA

L. Rudolf

KENYA

Nairobi
L. Nakuru
L. Naivasha

Entebbe
L. Victoria
L. Albert
L. Edward
R. Semliki
R. Ihuri

RWANDA
BURUNDI
CONGO

L. Kivu
Kigoma

L. Tanganyika

L. Mweru
Abercorn

Kalambo
Falls
Kasama
L. Bangweulu

ZAMBIA

TANGANYIKA
TANZANIA

R. Malagarasi

Tarangire
L. Manyara
Mkomasi
Amani
Itigi
L. Rukwa

Tanga
Pemba
Zanzibar

Mafia

| 0 | 100 | 200 | 300 miles |

| 0 | 200 | 400 km |

30°E
40°E
10°
0°
10°S

20°

30°

FRIENDS OF WILMINGTON COLLEGE, INC.

THE BIRD FAUNAS
OF AFRICA
AND ITS ISLANDS

THE BIRD FAUNAS
OF AFRICA
AND ITS ISLANDS

R. E. MOREAU

Edward Grey Institute of Field Ornithology

Oxford, England

1966

ACADEMIC PRESS

NEW YORK · LONDON

ACADEMIC PRESS INC. (LONDON) LTD
Berkeley Square House
Berkeley Square
London, W.1.

U.S. Edition published by
ACADEMIC PRESS INC.
111 Fifth Avenue
New York, New York 10003

Library of Congress Catalog Card Number: 65–27736

PRINTED IN GREAT BRITAIN BY
W. & J. MACKAY & CO LTD, CHATHAM, KENT

Foreword

This book owes its original inspiration to the insistence of my friends, Dr. David Lack and Professor Charles Sibley. They pointed out that since I had been studying and writing about various aspects of African ornithology for forty years, it would be appropriate for me to attempt some sort of synthesis. In the result, this book has taken a form quite different from what they or I originally expected; and the reason is that in the last few years advances in various fields of knowledge, especially in the interpretation of geological material and its dating, have given us an entirely new background to the current biological scene in Africa and a new conception of the forces moulding the latest stages of evolution there. They have in fact made possible a dynamic zoogeography of the continent which, without those advances in other fields, would have been unthinkable. At the same time the interdependence of ecology and geographical range has more and more been emphasized. If I were allowed an eighteenth-century latitude of sub-title for this book it would be "an eco-geographical discussion with its roots in the past".

As it stands, its title might suggest that it is addressed only to ornithologists. This is far from being so. However imperfectly I present it, the complex picture provided by the bird faunas of Africa, on an immense canvas, is of the utmost relevance to all those who are concerned with natural history, whether they are amateurs or professional zoologists. It offers many illustrations of biological problems and principles and of the consequences of a succession of sweeping ecological changes. I may add that I have chosen my title partly as a warning that it will not give any intimate studies of particular birds, but will deal rather with groups; and I use the words "bird fauna" with a wide range of meanings—from the bird community in a certain habitat or a circumscribed area of that habitat to the sum total of bird species in a geographical area, however large or small.

In the pages that follow I have here and there departed from what some might regard as due balance, simply because a topic has held special interest for me. I have been influenced by my long periods of residence in two extremely different parts of Africa, first on the edge of the Sahara in Lower Egypt and then in a montane rain forest in northern Tanganyika. Each of these periods aroused certain enthusiasms in me and gave me certain special opportunities; but for dealing with some aspects of my subject my personal equipment is very defective, and here I have leaned heavily on the help of friends. What I have to say is mainly descriptive. Often I have been able to put forward only some preliminary analysis, as when I make comparisons between the wealth of species in different parts of the Ethiopian Region, or compare a local African bird fauna with one of not too dissimilar environment outside the tropics. Here and there, as in discussing the distribution of montane birds, I believe I am able to offer something like real enlightenment.

It will be seen that in different contexts two subjects recur again and again: one is the mutability of the conditions throughout the continent and of its bird fauna, especially during the last 20000 years; the other is the problem of the number of species comprising a given community or inhabiting a given area. As a special aspect of ecological adjustment on a great scale the long-distance seasonal movement of birds within the tropics and, still more, that of Palaearctic migrants into Africa hold a fascination all their own. In the final chapters I discuss the land-bird faunas of the islands off the African coasts, including Madagascar, for the interesting features of their structure and for the varied courses that, on the basis of material derived from Africa, have been taken by evolution in isolation.

There are few people better qualified than I am to realize how superficial much of this book is. I can claim only that I have made what I can of such information as I can find. I cannot emphasize too strongly that we owe most of African ornithology to men, amateurs in every sense of the word, who worked on birds for the fun of it, without organization, subsidy or biological training, often in most unpleasant conditions. Of course, their results have been largely haphazard, but they are immensely better than none at all. The hunters and colonial servants and planters, who used to keep the study going, have been disappearing from the continent with more or less of contumely, in their capacity as unwanted anachronisms. The short-term purveyor of "technical assistance" steps in; but there is, most regrettably, almost no sign of Africans coming into their own as naturalists. Meanwhile endless opportunities exist for expeditions, even if only by a single worker, organized and dispatched from outside, to enter and pick up some of the information that is lacking. Anyone peering around for a thesis in field biology should have no difficulty in getting some ideas from the loose ends that litter these pages. Moreover bases from which work can be organized in comfort and with facilities, undreamed-of only a few years ago, multiply in the national parks and the educational institutions of the African states; while in Cape Town the Percy Fitzpatrick Institute of African Ornithology is the first in the continent to specialize on birds.

I am sure that relatively small parts of Africa will remain much as they are for a long time; but I am also aware that, if things go for the Africans as many of them and of their friends would wish, they will in a few years and over most of the continent have hunted and burned and cultivated and built and procreated themselves out of any semblance of their natural *oikos* as thoroughly as the people of the Lower Yangtse, or the cornlands of the Middle West, or the Cyclades, or the shore-line of Sussex. By the time the Africans are ready to become amateurs of field biology most of them will have to scrabble about in the ruins of their fauna and flora, as everybody else in a "developed" country must do; and instead of studying the grand designs of natural biomes they will have to finick around with the impoverished and lopsided remnants which are the by-products of man's multiple incontinencies.

This is not to say that the phenomena we can study today are wholly "natural". Over most of Africa they are already far from that; and they ceased to be so when, by an indiscriminate use of fire, primitive man became something other than a natural component of the eco-system. In a sense, scientific ornithology was lucky to open its eyes on the African scene when it did. For example, it was just in time to see the last of the small cormorants *Phalacrocorax africanus* of Egypt before the species disappeared from the northernmost 1000 miles or so of the continent. Such an example and some of the relict distributions we see today help us to a salutary sense of the nature of the data, fragmentary and hence liable to be delusive, on which much of our discussion is of necessity based. In attempts at zoogeographical reconstruction the most humble of approaches is the only proper one.

This book has owed a very great deal to the generosity of friends and to other correspondents who have provided me with information. The frequent incidental references to advice and to unpublished material show how much I am indebted to those who are named and I would ask each one of them to accept my thanks. Among ornithologists who have lived in Africa, I must mention especially Mr. John Elgood, late of the University at Ibadan, Nigeria, and Mr. C. W. Benson, late of the Rhodes-Livingstone Museum in Northern Rhodesia. I should like also to take this opportunity of paying a tribute to that best-loved of African ornithologists, Dr. J. P. Chapin, to whose precision of published statement and generosity in sharing his information so many of us were indebted over the years. For assistance in interpreting Pleistocene events I am indebted especially to Dr. J. Desmond Clark of the Department of Anthropology, University of California (formerly of the Rhodes-Livingstone Museum), to the palynologist Dr. E. M. van Zinderen Bakker of Bloemfontein University and to Mr. H. H. Lamb of the Meteorological Office, Bracknell, Berkshire. Mr. L. T. Wigg has helped to guide me in the literature of African vegetation.

In connexion with the photographs, apart from the organizations to which individual acknowledgement is made, I am indebted for generous permission to use their own work to Dr. O. Hedberg, Mr. A. C. Hoyle and above all to Dr. H. Lamprey and my old friend Dr. P. J. Greenway, who first helped me over African botanical matters as early as 1930.

As for the manuscript itself, various chapters have most kindly been examined critically for me by different friends, while the whole work was read at various stages by John Elgood, Mrs. B. P. Hall of the British Museum (Natural History), Dr. David Lack, Professor Charles Sibley and finally by Professor A. J. Cain, who gave it meticulous and con-structive criticism as it was approaching its final version. I would add my thanks for the facilities I have enjoyed—and I mean enjoyed—at the Edward Grey Institute under Dr. Lack's direction.

Finally, I should like to express my thanks to Academic Press for most friendly dealings and to Reproduction Drawings Ltd. for preparing the maps.

8 *October* 1966   R. E. MOREAU

Contents

1

Introduction

Most zoogeographical discussion concerned with Africa has been in terms of "Regions", the whole system of which has been under criticism in recent years. Darlington (1957) has discussed and defended it. He points out that over certain great land areas of the world—and not only those which are bounded on all sides by sea—the animals show strong resemblances and that between these areas strong differences exist. The system of zoological regions "is not and is not intended to be a common pattern, which many different groups of animals fit exactly. It is the best average of all the very different distributions of different animals that zoogeographers have been able to devise." In this system the Palaearctic Region connotes Europe, temperate Asia and also Mediterranean Africa, though here the mammals are anomalous (Chapter 4). There have been arguments about the southern limit of the Palaearctic in the Sahara, but, whatever the precise situation in which it has been drawn on the map of Africa, the rest of the continent to the south has been accepted as the Ethiopian* Region. To this some have added Madagascar and some south-west Arabia. Such additions are matters of opinion; I think it better to regard Arabia as transitional between regions and, as will be shown below, the Saharan fauna is both transitional and peculiar. Certainly it is a convenience, if nothing more, to use "Palaearctic Region" in the sense specified above, and "Ethiopian Region" for Africa south of the Sahara.

In the past the bird fauna of the Ethiopian Region has as a rule been discussed as if it belonged to a different continent, if not a different world, from the 3 million sq. miles to the north of it. Such treatment passes by much that is of interest to the biologist and becomes all the more unrealistic when placed on the background of the immense climatic changes that are now known to have transformed so much of this part of Africa in the last few thousand years (Chapter 3). Darlington (1957) had justification when he wrote of the Himalayas and the Sahara that, as barriers, "in the long run they have been less important than they seem". At the present day the biota on the northern and

---

* It is unfortunate that the name "Ethiopia" has in recent decades acquired a specialized connotation, having been officially appropriated by a single territory in north-eastern Africa. For this territory I shall in this book revert to its older name, "Abyssinia". By contrast, for other territories, the names of which have recently been changed, the new ones have been used, especially Malawi for Nyasaland and Zambia for Northern Rhodesia. But Tanganyika has been retained for the mainland territory of Tanzania. "The Maghreb" has been used to cover the three countries of Morocco, Algeria and Tunisia, often called "Barbary".

southern edges of the Sahara certainly show immense differences (see Chapter 4), but they should be considered in conjunction. With this fully in mind it is none the less a convenient arrangement, in approaching the complicated subject of this book, to start in the north, with Mediterranean Africa and the Sahara, then proceed with the far larger tropical area to the south about which so much more needs to be said.

In general, the discussions in this book rest upon the composition of each bird fauna, of whatever status, in terms of bird families and of the number of species in each. Treatment of these data is influenced throughout by three basic considerations. One is that, as will be seen from the analysis in Chapter 5, in Africa south of the Sahara (as presumably elsewhere in the tropics) a profound dichotomy exists between birds of evergreen forest and birds of other habitats. Owing in part to the fact that the existing evergreen forests, other than the Guinean blocks, form no more than relatively small islands in a vast area of drier country, the distributional and other problems involved in the two bird faunas are so different that it is necessary to keep the statistics of them separate for nearly all purposes and to discuss them separately.

The second consideration is that, within each category, forest and non-forest, most species can be allocated as typically lowland or typically montane. (The term "montane" is preferred to "highland" because the latter has been used by Chapin (1923, 1932) in a somewhat different sense in the term "East African Highland District".) When I went from Egypt to live on a forested mountain in Tanganyika nothing impressed me more than the fact that a few hundred feet down the slopes, although they were still clothed with forest, the component species had changed greatly and the birds no less. Similarly, discussing the eastern Congo, Chapin (1932) has commented: "one might expect that in and about the mountain forests there would be highland races of many of the tropical lowland birds, but these are relatively few"; and again "it is strange that so few of the lowland forest species are in evidence" in the montane areas. Many birds not belonging to the evergreen forest are similarly affected, though the proportion showing altitudinal restriction is not so great (see Chapter 5).* The phenomenon of a marked change in the bird fauna at a certain altitude is indeed found very widely in the tropics; for example, in South America it has been

* Among African mammals the evolutionary and faunistic situation seems to be very different from that in birds. As quoted by Moreau (1963), R. W. Hayman of the British Museum (Natural History) writes (in litt.): "I cannot find that the montane forest mammal communities are specialized as a whole, though a few species are restricted, or almost entirely restricted, to such habitats. In general, there is no clear-cut division between the montane forest and the lowland forest mammals . . ." Dr R. C. Bigalke (in litt.) tells me that he has arrived at a similar conclusion. This is not the only major unconformity between mammals and birds in Africa for which there is no obvious explanation: as discussed especially in Chapter 4 of this book, the mammals of north-west Africa (Morocco–Tunisia) are predominantly Ethiopian, while the birds are strongly Palaearctic. So are the butterflies; and indeed in point after point these insects show in their faunistics, as described by Carcasson (1964), a remarkably close correspondence with those of birds, from one end of Africa to the other.

remarked on by Chapman (1926) and in south-east Asia by Smythies (1960). Similar generalizations have been made for plants. According to Richards (1952), montane forest (which is very different floristically from lowland) has been recorded as starting at about 1200 m in the Malay Peninsula (even lower on isolated mountains), at 1200–1500 m on wet eastern slopes in South America, at 1650 m on the mountains on the east side of the Congo basin and also in New Guinea. In Honduras Carr (1950) put the transition between 1050 and 1370 m and commented that many animals "rarely cross" the montane boundary. For tropical Africa Keay (1959) has made the generalization that the floristic composition changes at about 1300 m. The change is, of course, more abrupt in some places than in others and more than one author has described a sub-montane zone.

The meteorological correlations of the boundary between the lowland and the montane zones do not seem to have been fully discussed, but the prime factor is generally assumed to be temperature. The rate at which this changes with altitude is to a limited extent affected by topography and exposure to prevailing winds, so that although such general averages as those above can be worked out for the lower edge of the montane zone in different areas, marked divergences occur very locally; in particular, the boundary descends on steep slopes and those near the sea (see, for example, Fig. 26, p. 82). Actually the location of the local cloud layer is probably of direct importance to the development of montane forest, if not of other montane vegetation, and this climatic feature probably deserves more attention in this connexion than it has received. For example, it is very noticeable how on many days in the year a belt of cloud hangs for hour after hour over the forest girdle of Kilimanjaro, while the peaks, Kibo and Mawenzi, above are as clearly sunlit as the places below (Fig. 1). Through both the shade and the occult precipitation which such a layer of mist and cloud provides some forests are enabled to survive in climatic areas where the length of the dry season would otherwise be fatal to them. Whatever the immediate importance of the cloud layer, its altitude would appear to be intimately connected with temperature, and in this book I shall argue on the assumption that temperature is a satisfactory indicator of the montane boundary. For birds in Africa, Chapin (1932) on his personal experience placed it at about 5000 ft (just over 1500 m), and I believe that this figure requires important modification only on a few mountains that are very isolated or close to the sea, or both.

The important point for the present discussion is that, while the lowland birds tend to have continuous ranges, the montane birds are of necessity confined to isolated stations, which, however, have many species in common. Hence the two categories of lowland and montane species, no less than those of forest and non-forest, need to be discussed separately. It is, of course, true that ecologically the non-forest habitats are far more heterogeneous than the forest (some main divisions are

FIG. 1. An example of a cloud-layer (here lifting and dispersing) that is regularly associated with a belt of mountain forest. Western Kilimanjaro, with the main peak, Kibo, seen from the north. Photo: Len Young.

indicated in Chapter 2), but, although a proportion of the non-forest birds are typically or even exclusively associated with individual vegetation types, they are in general not nearly so habitat-specific as are forest birds. It follows from the foregoing that, whatever inferences can be made about the distribution of vegetation types in Africa in the past (Chapter 3), they give us an indication of the distribution of bird species, but more definitely for forest birds than for any others.

In this connexion it is worth bearing in mind that the nucleus of a population of a plant species can survive in a smaller space than a bird species, and that in general plants have far greater powers of survival than birds. Individually many of them are much longer-lived; some to such an extent that, especially in a favourable spot, they can weather centuries of inimical climate all around them. Moreover, through their seeds, a great many species have powers of survival that enormously exceed the powers of the individual plant; and finally certain vegetation types, particularly forest, tend to be self-perpetuating. By the power of its foliage and its epiphytes to condense moisture from the air, by the beneficial effect of its shading of its own roots, by the utilization of nutrients in its own leaf-fall and by so checking run-off that any rain is fully effective, a patch of forest on a mountainside may be capable of maintaining itself long after the local climate has deteriorated to such an extent that it would be impossible to establish forest anywhere in the neighbourhood. Hence, on the one hand, there may in certain cases be a considerable time-lag before a marked climatic change fully asserts its effect on the vegetation and, if the patch of vegetation is big enough, on its associated birds; on the other hand, plant species of a certain vegetation type may persist locally long after the associated birds have been wiped out or, when conditions for them improve, may be able to recolonize an area, through seeds or isolated survivors, much more rapidly than the birds, which would *ex hypothesi* need to immigrate from other areas. Hence, however future research may help to increase our knowledge of the distribution of vegetation types (and of plant species) in the recent past, they will not necessarily be applicable in local detail to bird species that are normally and often exclusively associated with them.

The third ruling consideration throughout this book is that much of value and significance is lost if a bird fauna is dealt with *en bloc*. It will be seen in subsequent pages that, if each bird fauna is divided into the following groups, which are broadly ecological as well as taxonomic, the comparisons revealed are enlightening in point after point, and that in the various analyses significant differences are obtained. (For explanation of taxonomy followed, see p. 8.)

GROUP A. Those families which are typically dependent on watery habitats: Anatidae (ducks), Anhingidae (darters), Ardeidae (herons), Balaenicipitidae (shoebill), Balearicidae (cranes), Charadriidae (plovers), Ciconiidae (storks), Heliornithidae (finfoots), Jacanidae (lilytrotters),

Laridae (gulls), Pandionidae (osprey), Pelecanidae (pelicans), Phalacrocoracidae (cormorants), Phoenicopteridae (flamingos), Podicipitidae (grebes), Rallidae (rails), Recurvirostridae (avocets, etc.), Scolopacidae (snipes, etc.), Scopidae (hammerhead), Threskiornithidae (ibises, etc.).

GROUP B. Predatory and scavenging birds: Aegypiidae (vultures), Aquilidae (eagles, etc.), Falconidae (falcons), Sagittariidae (Secretary Bird), Strigidae (owls), Tytonidae (barn owls).

GROUP C. Ground birds: Burhinidae (stone curlews), Glareolidae (coursers and pratincoles), Otididae (bustards), Phasianidae (game birds), Pteroclidae (sandgrouse), Struthionidae (ostriches), Turnicidae (button quails).

GROUP D. The remaining non-passerines: Alcedidae (kingfishers), Apodidae (swifts), Bucerotidae (hornbills), Capitonidae (barbets), Caprimulgidae (nightjars), Coliidae (colies), Columbidae (doves), Coraciidae (rollers), Cuculidae (cuckoos), Indicatoridae (honeyguides), Meropidae (bee-eaters), Musophagidae (turacos), Phoeniculidae (wood-hoopoes), Picidae (woodpeckers), Trogonidae (trogons), Upupidae (hoopoes).

GROUP E. The passerines (i.e. "song birds", "perching birds"): Alaudidae (larks), Campephagidae (cuckoo shrikes), Corvidae (crows), Dicruridae (drongos), Emberizidae (buntings), Estrildidae (weaver-finches), Eurylaemidae (broadbills), Fringillidae (finches), Hirundinidae (swallows), Laniidae (shrikes), Motacillidae (wagtails and pipits), Muscicapidae (flycatchers), Nectariniidae (sunbirds), Oriolidae (orioles), Paridae (tits), Pittidae (pittas), Ploceidae (weavers), Promeropidae (sugarbirds), Pycnonotidae (bulbuls), Salpornithidae (salpornids), Sturnidae (starlings), Sylviidae (warblers), Timaliidae (babblers), Turdidae (thrushes), Zosteropidae (white-eyes).

There are, of course, specific abnormalities within groups, such as the dry-country plovers, whose family is included in group A, and cases of inter-group competition. For example, the Marabou *Leptoptilos crumeniferus* as a member of the Ciconiidae is included in group A, but it attends carcasses alongside vultures (group B); and small falcons (group C) and rollers *Coracias* spp. (group D) take similar large insects; but broadly it remains true that the extent to which any one of these groups is represented in a particular bird fauna may be presumed not to affect directly the representation of any other group.

Underlying all the ecological side of the discussion in this book there is the concept of the "niche". Nowadays this is related by most people almost entirely to food utilization, though other factors, especially that of nesting sites, when these are of a specialized nature, cannot be neglected (see discussion in Udvardy, 1959). The concept of the niche is indispensable, but it must be admitted that it is somewhat complicated and imprecise, involving as it does both the resources of the habitat and the use made of them by individual species or groups. Compare these phrases used in the same recent work: "the important niche used by woodpeckers"; "a heterogeneous niche in a somewhat diversified habi-

tat"; [an animal occasionally] "shifts into new niches"; "each species
is selected to fit into a definite niche in the environment"; "the richer
and more diversified the habitat, the more easily . . . minor niche
differences can develop" (Mayr, 1963). For this last phrase read "can
be developed" by animals, as was no doubt intended; and the relation
between diversity of habitat and multiplicity of niche is, of course, in
any case unimpeachable. Compare a treeless steppe and a humid woodland as a habitat for insectivorous birds. The woodland offers food of
diverse forms in different strata, in different species of trees, in different
parts of the same stratum in the same tree. (See in this connexion Gibb,
1954, for examples.) One or more of such divisions as these can be more
or less exclusively appropriated by a particular species of bird, either all
the time or perhaps especially at certain times of the day or at certain
seasons of the year by birds that may travel a long way to do so. It is
regarded as axiomatic that no two species will occupy, except trans-
iently, exactly the same niche in the same area of any given habitat. If
I had to define my concept of the "niche", it would be as "a segment of
the opportunities offered by a habitat, that segment being utilized at
regular intervals or permanently by one particular species or group, to
the exclusion of any other".* When in any habitat all the niches are
fully occupied all the time, then it is "saturated"; in that condition no
new species can enter, either to breed or as a seasonal visitor, except by
displacing existing species. Such a condition will most nearly be reached
in a habitat with the minimum of seasonal change. Habitats that change
in carrying capacity, in the tropics usually as a result of alternating wet
and dry seasons, will have their permanent resident population limited
by the worst time of the year. In the best time seasonal visitors can be
accommodated, whether breeding or non-breeding, and there may be
such superabundance that competition temporarily disappears.

Such considerations underlie especially those chapters of this book
which deal with the bird communities of different vegetation types,
with the migrations of Ethiopian birds within the Region and with the
accommodation of the Palaearctic migrants; and in the descriptions of
the insular bird faunas the island ecology is linked with an indication of
how evolution appears to have met the opportunities offered. For all
these topics the available information is so imperfect that I can hope
to do little more than throw into relief what I think are significant or
unexpected features, sometimes to make a suggestion, and generally to
provide a basis for the further investigation and discussion that I hope
this book may provoke.

On the taxonomic side the study of birds is probably better advanced
at levels below the genus than it is for any other class of animals, but it
is not clear that the same can be said of the higher categories. However,

* A slightly different aspect of the same concept is put forward by Slobodkin (1962) when
he equates the "niche" of a species with the peculiar "job" it is doing in life; and that, I feel,
is helpful.

at levels above the genus we have several recent classifications; Wetmore (1960), Storer (1960) and Mayr and Amadon (1951) have provided comprehensive ones, while Amadon (1957) and Delacour and Vaurie (1957) have dealt with the oscine passerines. I have followed Mayr and Amadon except where for the purpose of this book it was not convenient to do so. Thus, because most of them form obviously separate ecological categories, I have kept the warblers, the flycatchers, the thrushes and the babblers as distinct families instead of sinking them all in the Muscicapidae, and I have kept the wood-hoopoes (Phoeniculidae) distinct from the Upupidae. Among the passerines there are a few small genera of which the familial allocation is still doubtful. There is no need to mention these in detail, except that *Salpornis*, hitherto changed around between the Sittidae and the Certhiidae and tentatively kept in a subfamily of the Sittidae by Mayr and Amadon, has been put in the Salpornithidae, as suggested by Vaurie (1957).

That most subjective of categories, the genus, has come in for some rough handling in recent years. For example, as recently as 1930 Sclater in his "Systema Avium Aethiopicarum", Vol. 2, arranged the passerines of the Ethiopian Region in 339 genera; White (1960, 1961, 1962a, b, 1963) has divided the same birds into no more than 195. Partly for this reason the genus is a category that is mentioned little in the course of this book. On the other hand, much of the argument and discussion turns throughout on numbers of species. Here there is room for considerable disagreement according to the extent to which the polytypic concept is applied, i.e. the extent to which the concept of the species is widened to include distinctive populations that are obviously close relatives, but are isolated geographically, or at any rate do not intergrade. Such cases are particularly numerous in Africa. "Check list" convention has always insisted that every such distinguishable population shall appear unequivocally as either a subspecies* or a full species; in the result, as Mayr (1963) has said: "An outsider would never realize how many interesting cases of evolutionary intermediacy are concealed by the seeming definiteness of the species and subspecies designations." In truth the taxonomist's decision on the status of an isolated population is based on nothing more than his individual assessment of the extent to which the birds concerned look different; and, as we all know, species whose distinctness is proved by their geographical overlap can look very much alike. A further complication is that many congeneric birds which look sufficiently unlike to be generally or universally accepted as distinct "full" species are allopatric, and are, in fact, recognizable as recent exaggerations of isolated subspecies. That is, together they form a superspecies, defined as "a monophyletic group of very closely related

---

* Most of the trinomials in bird nomenclature relate to populations that are regarded as having some distinctive character, however minute, but are not isolated and hence are of minor evolutionary significance and potential. A wholly disproportionate amount of attention has been devoted to describing and naming subspecies of this nature and to disputing the validity of many of them.

and largely or entirely allopatric species" (Mayr *et al.*, 1953). For a better understanding of the mutual relationships of birds, not only within a region but as between regions, we need a comprehensive revision of birds from the point of view of the superspecies, and the preparation of this has begun for Africa by Hall and Moreau. Meanwhile, the lack of it does not invalidate the taxonomic basis of this book. For one thing, I have consistently employed the broadest polytypic concept of species that has been put forward for the birds in question by a reputable taxonomist.* (Incidentally this has had the result of reducing the number of species in the Ethiopian Region from the 1750 admitted by Stresemann (1927–34) to the 1481 arrived at in Chapter 5.) For another thing, in the discussions in this book the exact statistics are not so important as the comparisons between bird faunas; and the statistics of each of these are derived by the same technique, imperfect as it is.

As a background to the whole discussion it is greatly to be wished that we could get some satisfactory idea of the ages of the bird species we are dealing with, but this is far from being attainable. Bird fossils are deficient in most parts of the world and nowhere more thoroughly than in Africa. Most of what few there are have not been examined. Moreover, as Darlington (1959) has said, much of the comparatively scanty literature on bird fossils in general, especially a number of pronouncements of general importance made well back in the nineteenth century, cannot be regarded as reliable.

Various generalizations have been put forward about the age of bird species. In the opinion of Wetmore (1959) "most species of the present avifauna had attained development by the beginning of the Pleistocene and . . . the major changes must have come through extinction of numerous kinds". This implies that most modern species date from the Pliocene, and are considerably more than 1 million years old. The statistical or other basis of this opinion is not known. Storer (1960) does not support quite this degree of general antiquity of the species: "most modern genera were probably in existence by the Pliocene, but many of the species were different from the recent ones". Brodkorb (1960), working on a statistical basis, has concluded that "the average longevity of Pleistocene avian species is approximately one-half million years and the top limit of longevity approximately one million years"; but for statistical and other reasons these figures appear to be altogether unsound. Using similar methods and bases, Fisher (1964) and Fisher and Peterson (1964, p. 45) have calculated that "Bird species probably survived an average of a million and a half years at the end of the Pliocene before they died out. By the end of the Pleistocene their life expectation

---

* It may be useful to explain the nomenclatorial conventions that have been followed in this book. When a familiar generic name has been "sunk" I have, to help recognition of the species concerned, shown such a name in parentheses after the generic name currently recognized. English names of species have been cited when they are in common use and have then been capitalized. For other species the reader has been assisted by appending some general English term such as "a sunbird".

was only about 40 thousand years . . . Now it is only about 16,000 years." These figures also are unacceptable (Moreau, 1966).

In fact, because of the nature of the data there is every sort of diffi- culty in the way of reaching acceptable conclusions in this matter. A major objection is that all the foregoing pronouncements are over- generalized. Because of the much greater prevalence of non-passerine fossils, especially of large birds, any assessments of longevity calculated from fossil data are mainly applicable to those groups and they surely need to be modified for the passerines. Storer, with whom I have had the opportunity of discussing this problem, agrees. A most important consideration is that, out of about 8600 species of birds in the world today, 3500 are non-passerine and 5100 passerine (including 1100 sub- oscines, nearly all South American). The 8600 species have been ar- ranged by Wetmore (1960) in ninety-seven non-passerine and seventy passerine families, by Storer (1960) in 161, and by Mayr and Amadon (1951), who, unlike the other authors, use subfamilies, in ninety-six and thirty-eight families. The fact that with such different techniques the number of families recognized in the non-passerines remains almost unchanged, while the number of passerine families varies greatly, shows how much more imperfectly differentiated the passerines are. These small birds are represented by many closely allied species, some of them difficult to allocate to families, in contrast to the more marked dis- continuities that exist among the remaining 3500 of the world's birds. One possible interpretation of this is that the passerine species are on average much younger than those of other birds (cf. Mayr, 1963: "large birds . . . seem to go further back in the past without changes trans- gressing the specific level"). To Darlington (1957), indeed, the evidence, admittedly inferential, suggests that the oscines radiated and dispersed no longer ago than the mid-Tertiary, perhaps primarily in the Miocene. He rejects the ascriptions made many years ago, and often quoted, of Oligocene fossils to certain modern genera—*Lanius, Motacilla, Sylvia, Passer*—"because it may be doubted whether the fossils are what they are supposed to be". Palaeontologists with whom I have discussed this question fully share these doubts.

In the face of the foregoing it may be doubted whether the generaliza- tion that "mammals show perhaps the fastest transformation of species, birds apparently changing far more slowly" (Mayr, 1963) would have validity if we knew enough about the palaeonotological history of the passerines to give the evidence from this group due weight. It is true that most of the present-day mammal species seem to have arisen long after the beginning of the Pleistocene and may perhaps not average more than half a million years old (for African data one may use Cooke's (1963) list of the mammals of Barbary). But just as Brodkorb's data were in the main from large birds, so are those for mammals from large mammals, and both large birds and large mammals have considerably fewer generations in a given period than small ones—as an extreme case, about seven generations of sparrows to one of albatrosses. While this is only one factor in the speed of evolution, it is surely of some significance in determining the rate at which a population can adapt itself to changing conditions and, other things being equal, one would expect the passerine species to be on the average much younger than the half a million years calculated by Brodkorb on the basis of the large non-passerine birds. If their ages were half that period, the passerine species we see today would have endured in their present form the vicissitudes of the Last Glaciation and of the interglacial before it, but hardly those of the penultimate glaciation.

Meanwhile, ideas about the period of time needed for subspecific differentiation have been changing very remarkably. When about thirty-five years ago (Moreau, 1930) I inferred, from the colour-relationship of certain Egyptian subspecies of birds to soil that could be roughly dated, that subspeciation could take place in birds in not more than about 4000 years, I seemed to be a bold pioneer. In fact, although evidence for subspecific changes within historical times has been accumulating for other classes of animals (Simpson, 1953; Mayr, 1963), it seems that no further inferences were made on birds until the startling work of Johnston and Selander (1964). By investigating series of the House Sparrow *Passer domesticus* in North America, where it was introduced on the east coast in the 1850s, and in Hawaii (1870), they have shown that populations divergent in colour, size and proportions, which would by conventional standards be classed as "good subspecies", have developed in various areas in the course of periods that vary from 110 to thirty generations (calculated on the time since the species is known to have colonized the respective areas).

It is true that, as pointed out by Mayr (1963), such results at the subspecific level cannot necessarily be extrapolated to indicate the formative period needed for full speciation, since for this process the essential factor is that populations should be isolated long enough for genetic incompatibility to be established. Clearly, however, it must now be regarded as a very real possibility that full speciation in birds has often been far more rapid than the quarter of a million years tentatively suggested above, and not only in that group which seems most labile, namely the passerines. For example, Hall (1963a) has argued that in some African francolins, the isolation of which by ecological barriers can be accepted as complete, a certain amount of speciation, within the superspecies, must have taken place since the Last Glaciation began to relax, i.e. within the last 18000 years. Similarly Fisher and Peterson (1964) feel that five sibling species in the group typified by the Herring Gull *Larus argentatus* have probably evolved from their common ancestor in the last 10000–15000 years. The shortest of these periods is about 300 times as long as proved to be necessary for "good" subspeciation in the House Sparrow. Given isolation, it might seem sufficient for genetic incompatibility, and so full specific status, to be achieved, even in a slower-breeding species such as a gull. For birds such a conclusion is, of course, still speculative. But from another class of vertebrates, fish, comes evidence that three new species, which are understood to be generally accepted as such on morphological grounds, have evolved in Lake Nabugabo, Uganda, which has been narrowly separated from Lake Victoria for a period which radiocarbon dating of material at the base of the barrier indicates as about 4000 years (Beadle, 1962).

Final evaluations of all such cases as those of the francolins will only be possible when the genetical incompatibility of the populations forming part of the superspecies has been tested experimentally. Meanwhile it is surely reasonable to bear in mind as a working hypothesis the possibility that in birds speciation can be achieved not only in islands surrounded by water, but also in ecological islands, within no more than a small fraction of the Pleistocene. Acceptance of this as a possibility is, of course, by no means the same thing as insistence that this has been the rule, but it is clearly a factor to be borne in mind. Furthermore, as a corollary of rapid evolution of species we must envisage frequent extinction, while a greater contemporary abundance of species at various times in the past is also a possibility. The utter inadequacy of the fossil record of African birds denies direct evidence on this point: but if mammals can be accepted as giving any guide it may be mentioned that, in the early part of the Middle Pleistocene, East Africa alone held fifty-eight species of large Bovidae and twenty different Suidae (pigs), according to Leakey (1964 and *in litt.*). The contemporary habitat range could not have been greater than it is today, when the whole gamut from forest to desert exists there, and consequently it may be inferred that individual communities were richer in species than they are at present. How far this may have applied to birds we have no means of knowing.

2

The Geography and Environments of Africa

The purpose of this chapter is only to give a general outline. Discussion of various points will be found, as necessary, in later chapters.

GEOGRAPHICAL FEATURES

The continent of Africa has an area of about 11 500 000 sq. miles and extends from 37° N. to nearly 35° S. In the north it is separated from Iberia by the narrow but ancient strait at Gibraltar, less than 10 miles wide, and from Sicily by barely 90 miles, while at its north-east corner it is joined to the Eurasian land-mass by the flat arid isthmus of Suez, under 100 miles from north to south. The trough of the Red Sea, separating Africa from Arabia, is nowhere more than about 200 miles wide, narrowing to 10 miles at the southern end. From here all round the eastern, southern and western coasts the continental shelf is narrow in comparison to the size of the continent, reaching at its widest about 130 miles to the 100-fathom (200 m) line at the south-west corner of West Africa, off Portuguese Guinea. Only a very small proportion of the coastline is formed of cliffs (and practically no sea-birds breed anywhere on the mainland).

The surface relief of Africa shows strong contrasts to that of any of the other continents, particularly the two others in the southern hemisphere. Compared with South America, a much larger proportion is considerably elevated, but there is no great north-and-south cordillera to provide an obvious means of communication from one end to the other of the continent for montane forest or for organisms adapted to life above the timber-line. Compared with Australia, the general level of Africa is much greater, with highlands dominating the centre of the continent rather than the periphery. As will be seen from Fig. 2, a great deal of eastern, central and southern Africa is above 1500 m. What a map on such a scale cannot show is the large number of detached mountains, some volcanoes, some the result of the circumerosion of ancient raised surfaces, which reach above 1500 m in eastern Africa, mostly between Abyssinia and the Zambezi, and form ecological islands of great interest. Apart from the most elevated parts of the Abyssinian plateau, some of which rise well over 4000 m, there are a number in eastern Africa between 3000 and 6000 m, the biggest being Ruwenzori on the eastern edge of the Congo basin, Elgon on the Kenya–Uganda border, Mt Kenya and Kilimanjaro and Meru in northern Tanganyika. The great line of highlands that form the eastern edge of the Congo basin and the western edge of the Great Rift, including the active volcanoes of Kivu, is of special biological importance. Far away to the west and greatly isolated from all the other big mountains, Mt Cameroon, an active volcano at the corner of the Gulf of Guinea, rises to 4070 m, and just to the north of it the Banso-Bamenda highlands provide a stretch of montane environment at about 2000 m. West Africa and indeed most of Africa north of the Equator form a great contrast to the foregoing.

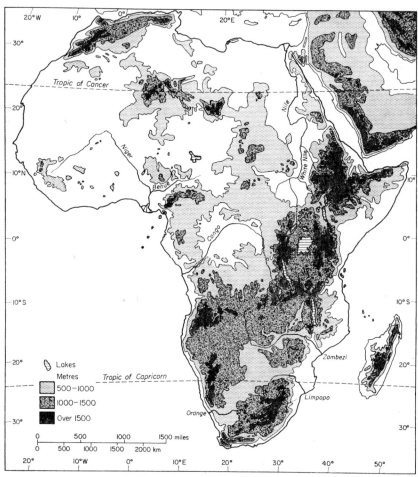

FIG. 2. The surface relief of Africa.

The greater part of it is below 500 m, including even localities 1000 miles from the sea, and in the great area west of the Cameroons only very small patches of ground reach 1500 m. The 1 million sq. miles of the Sudan are nearly all low, and so is much of the Sahara except for a few isolated massifs, especially Ahaggar (2750 m) and Tibesti (3265 m). Then on the extreme north-western tip of the continent, through Morocco and Algeria into Tunisia, another mountain system, this time mainly of parallel ranges, culminates in the Grand Atlas (4071 m).

The great rivers of Africa, most notably the Niger, the Niles, the Congo and the Zambezi, are in themselves of less direct importance to birds than indirect. But for the most part they are subject to wide fluctuations in water level, so that periodically sandbanks are exposed and surrounding lowlands inundated. The sandbanks provide safe breeding-places for a few species, especially pratincoles *Galachrysia* and bee-eaters *Merops*. Some of the shallow floods, in particular the inundation zone of the upper Niger and the "flats" of the Kafue River in Zambia, form water-bird habitats of the greatest importance. Lake Chad and the Okovango swamps on the northern edge of Bechuanaland are intermediate in type, with big areas of permanent swamps and seasonal extensions. These last are relatively small in such areas as Lakes Mweru and Bangweulu in Zambia and above all the sudd of the White Nile, in all of which immense areas are smothered in papyrus, reeds (*Phragmites*) and rushes (*Typha*). As bird habitats these marshes are of altogether more importance than most of the great lakes which show so prominently on the map of Africa. In particular, Lakes Nyasa and Tanganyika are deep and steep-sided, filling rift valleys and providing little for birds, though Tanganyika has a wonderful endemic fish fauna. However, the biggest of them all, Lake Victoria, is in a class by itself not only in its size; though so shallow it fluctuates but little and the rich vegetation on its edges accommodates many birds. In contrast to the lakes and marshes so far named, which are all fresh water, it remains to mention the relatively small alkaline lakes, mostly in eastern Africa, of which Naivasha and Nakuru are examples, the favoured habitats of a few species, notably flamingos (see in this connexion Brown, 1959).

CLIMATE

In determining the distribution of birds, climatic factors are of very slight direct importance compared with their indirect importance, in conjunction with soil factors, in determining the vegetational environment. Hence for the purposes of this book it will not be necessary to describe the African climates at length. Comprehensive basic data are now readily available in the "Climatological Atlas of Africa" edited by Jackson (1961) and in Kendrew (1961).

In Mediterranean Africa and in all Morocco rainfall is confined to the six winter months. East of Tripoli it is very light indeed, even on the coast; and the whole way across the continent from the Atlantic to the Gulf of Suez and the Red Sea these winter rains diminish very rapidly southwards. For 10° south of 30° N., the latitude of Cairo, every rainfall station included in the "Atlas" has a long-term annual average of less than 50 mm, except in the vicinity of the Atlantic. Moreover, these averages are far from telling the whole story, since in Africa there is conformity with the world-wide generalization that the lower the average rainfall the more variable are the annual totals. For example, in the Fezzan, where the rainfall averages only 10 mm a year, successive years are wholly rainless and storms may come at long intervals (Guichard, 1955a). This is the climatic régime that has engendered the Sahara desert, where neither the storm tracks of the temperate zones nor the monsoons of the tropics penetrate with any regularity, even in an attenuated form.

South of this area, by far the greater part of Africa on both sides of the Equator receives its annual rainfall almost entirely during the six months centred on the local summer solstice. The main exceptions to this generalization are (a) that in a narrow belt mostly just north of the Equator—that occupied by the blocks of lowland evergreen forest shown in Fig. 3—the rainfall is well distributed round the year; (b) that in part of eastern Africa, from Zanzibar through the Kenya Highlands to Somaliland, there are two well-defined rainy seasons in the year, the more marked centred round the months of May and June, the "short rains" round October; (c) that in the south-western corner of the Cape Province, for rather over 100 miles north and east of the Cape Peninsula, the climate is "Mediterranean", with rainfall mainly confined to the winter half of the year; (d) that further east along the coast of Cape Province for about 500 miles the rainfall is well distributed round the year, though inland and towards Natal the limitation of rain to summer becomes marked. Within the summer rainfall area the tendency on both sides of the Equator is for the total annual precipitation and also the period of the rains to diminish as the latitude increases. Thus, towards the southern edge of the Sahara in Northern Nigeria at 12° N. the annual rainfall is around 800 mm, at 14° N. in the same longitudes it is 500 mm, at 16° N. 200 mm and at 18° N. 100 mm; meanwhile, the number of months with dependable rain diminishes from five to two. Similar diminutions apply to the western part of southern Africa, but not to the eastern part, which receives rain from the Indian Ocean in the local summer. The foregoing general pattern is, of course, modified by the surface relief, since mountains tend to have a moisture régime of their own. On the coasts another factor contributing to the climate is the nature of the ocean currents. Cold currents, in particular off Somalia and off South West Africa and Angola (the Benguela current), tend to make the adjacent shore rainless though foggy; and similarly a cold upwelling off Ghana is regarded as being mainly responsible for the dry gap between the forest blocks of West Africa, which appears so prominently in Fig. 3. It may be added that except for a very few isolated spots, such as the seaward side of Mt Cameroon with 4000 mm of rain, even the equatorial rainbelt receives only 2000–2500 mm of rain a year, which is much less than parts of Amazonia or the East Indies.

Temperatures in countries of high relief are the despair of cartographers, because local records, especially the minima, are so greatly affected by aspect and also by altitude—to the extent that, as a worldwide average, mean temperatures fall by about 0·5°C, mean maxima by about 0·75°C, for every 100 m of increasing altitude. Using this factor, the "Climatological Atlas" has reduced all the temperature re-cordings of Africa to an altitude of 1500 m above sea-level. This enables isotherms to be drawn on a continental scale, which would otherwise be impossible, and thus provides a picture of the main temperature

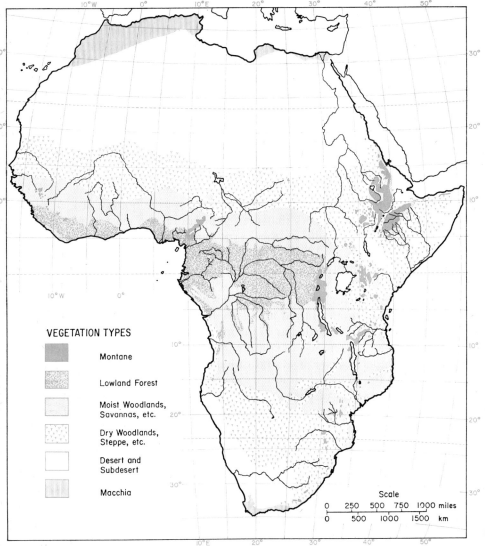

VEGETATION TYPES

Montane

Lowland Forest

Moist Woodlands, Savannas, etc.

Dry Woodlands, Steppe, etc.

Desert and Subdesert

Macchia

Scale

| 0 | 250 | 500 | 750 | 1000 miles |
| 0 | 500 | 1000 | 1500 | km |

FIG. 3. The distribution of the main vegetation types of Africa. (After the map published by L'Association pour l'Étude de la Flore d'Afrique Tropicale; see Keay, 1959.)

zones; but anyone wishing to obtain from the "Atlas" a correct approxi-mation to the monthly mean or monthly maximum temperatures of some restricted locality needs to know its altitude and to apply the necessary corrections. There are, however, important local deviations from the general lapse rates mentioned above, especially in that these tend to be much higher on certain mountains near the sea, and more will be said about this in Chapter 5. Minimum temperatures, moreover, are so greatly affected by the exact siting of the recording instruments that no generally applicable factor can be provided for the reduction of the data, no isotherms of minima can be drawn and the "Atlas" can do no more than insert the spot recordings on the map.

It has always been inexplicable to me that the temperature statistics most often given in publications of all kinds have in the past been daily and monthly means, which are, of course, of little significance in biology or in human economics either. Far more important are the mean maxima of the hottest months and the mean minima of the coldest, and also the incidence of frost, which is of more significance to plants than to animals. The damp equatorial areas have the smallest range of temperature; for example, at Douala in the Cameroons the lowest temperature ever recorded is 65°F=18°C, and the highest 95°F=35°C, while on the average the mean minimum of the coldest month (71°F) is only 15°F lower than the mean maximum of the hottest. It is in the dry areas with clear skies that the temperatures are most extreme, for example at Wadi Halfa on the Egypt–Sudan border, where it practically never rains, 28–127°F (−2° to +53°C), and in the Sahara surface temperatures (as distinct from air temperatures) go up to 177°F (77°C). Such heat is of great biological importance, especially because it connotes extremely high evaporation. This puts a premium on nocturnal activity rather than diurnal and is inimical to animals not adapted to such conditions, in particular to the migrant birds that have to cross the Sahara twice a year. (For a comprehensive discussion of the physiology of birds under stress of heat and drought, see Schmidt-Nielsen, 1964.)

Within the limited latitudinal range of Africa, all within 35° of the Equator, extremes of heat are probably of more importance, at least indirectly, in influencing bird distribution, than are extremes of cold. It may, however, be of interest to note that frost in the innermost tropics is not rare above about 8000 ft and with increasing latitude naturally becomes more widespread. Even as low as 3000 ft at Livingstone in the Zambezi Valley at only 18° from the Equator, very occasional frosts are severe enough to damage the forest department nurseries (W. R. Bainbridge, personal communication). The area of greatest prevalence in the whole continent is that part of the South African plateau from about Pretoria to within 150 miles of the south coast of the Cape Province, in which frost is registered on at least 100 nights a year, and in some places on at least 150. This compares with means of only fifty-nine and seventy-six nights at two stations, not subject to urban influence but otherwise taken at random, in southern England (British Meteorological Office, in litt.). But, by contrast with Europe, in any part of Africa frosts persisting through much of the day are almost unknown except at the very highest altitudes, around glaciers and, outside the tropics, on some mountains of the Maghreb and of South Africa in winter.

The nature of the ocean currents has an important influence on the climate of the margins of southern Africa. The east coast is washed by the warm Mozambique current, the west by the cold Benguela current. This difference contributes to the fact that, as shown in Fig. 3, the coast of South West Africa is desert that is frequently cool and is backed by subdesert steppe, while in the same latitudes from Natal northwards through Mozambique the dominant vegetation is wooded savanna with forest relicts in which tropical species abound. This has an important effect on the ranges of many species of birds, which penetrate much farther south on the east than on the west of the continent.

VEGETATION

No general descriptive account of the vegetation of Africa appears to exist, except for the pioneer sketch in Shantz and Marbut (1923). Phillips (1959) may, however, be consulted for the area south of the Sahara, and there are, of course, numerous local and territorial studies. Special mention should, however, be made of Chapter 5 in Chapin (1932), where he describes and illustrates the vegetation of the former Belgian Congo, with special reference to its ornithology. A noteworthy popular account of the African environments, with special reference to their animals, has been provided by Brown (1965). Church et al. (1964) give useful indications of the extent to which "development" has proceeded in different parts of Africa.

While comprehensive descriptive accounts are lacking, there are two recent vegetation maps that are particularly relevant, namely, that of Monod (1957), which covers the whole continent on a scale of 1 : 34 000 000, and the multicoloured map on a scale of 1 : 10 000 000 published as a result of international collaboration by L'Association pour L'Étude de la Flore d'Afrique Tropicale (AETFAT), with explanatory notes by Keay (1959). This map, extending south from the Tropic of Cancer to the Cape, omits Mediterranean Africa and most of the Sahara. The nomenclature adopted in the two maps is completely different and, as might be expected, there are important differences in the limits of the divisions recognized. It may be added that Rattray (1960) has provided a companion map of grassland types south of the Sahara.

For the present purpose the AETFAT map is the most useful. It recognizes thirty-five categories of vegetation, some composite (as, for example, the one entitled "woodlands, savannas (and steppes):* undifferentiated—relatively moist types"), but from practical necessity

* The original meaning of "savanna", which it retains in part of South America, is "a treeless plain", and a "steppe" is a "level and treeless plain" in Eurasia (O.E.D.); but in African botanical usage both "savanna" and "steppe" imply grasslands of various types that are more or less set with trees. Indeed, closed stands of deciduous trees have actually been referred to as "closed savanna woodland" (Keay, 1959).

delineates only nineteen. These nineteen could not be effectively reduced to any smaller scale or printed with fewer colours than there
used, but although an immense oversimplification is involved, the main
outlines of the African vegetation pattern can usefully be shown by the
six divisions used in Fig. 3, which is based on a simplification made by
Mrs B. P. Hall. Because these things are inclined to be forgotten and
also because what is said applies equally to bird faunal boundaries, it
is worth quoting two sentences from Keay's preamble. "Usually the
boundary [drawn on a map] between say type A and type B represents
an arbitrary line dividing a mosaic of 'type-A-with-patches-of-type-B'
from a mosaic of 'type-B-with-patches-of-type-A' . . . The components
of mosaics and catenas* due to relatively minor variations in relief
cannot usually be shown." How complicated such microgeographical
changes in vegetation can be are shown by such a paper as that of Cole
(1963), with, incidentally, a useful list of "important works" on the
vegetation of Africa.

The importance of vegetation types in bird distribution is very great,
for most bird species are more or less habitat-specific. As already indicated, this is extremely well marked where evergreen forest is concerned
(see statistics in Chapter 5), but it also applies to some extent elsewhere,
so that even in dryer country the presence or absence of certain types of
vegetation can determine the appearance of particular birds or apparently delimit their geographical ranges. Examples of this last have been
given by Benson *et al.* (1962), with special reference to the birds of
brachystegia woodland. In any area each patch of peculiar vegetation
is liable to have its characteristic birds. As an extreme example, in an
otherwise deciduous wooded savanna in Zambia the evergreen thicket
on a single termite hill (which may be as much as 10 ft high and 30 ft
in diameter) can provide the habitat of the shrike *Malaconotus* (=*Chlorophoneus*) *nigrifrons*, which is typically a bird of evergreen forest (Benson
and White, 1957). Similarly, and of more general importance, drainage
lines, not necessarily carrying water all the year, are accompanied by
peculiar biota. In evergreen forest they occasion a break in the canopy,
with consequent sites for species of plants and birds different from those
of the forest interior at the same levels above the ground. In dryer
country, if the drainage line is wide and ill defined it may form a flood-
plain, to be occupied seasonally by a dense mass of heavily seeding grass
beloved especially by weaver birds (Ploceidae). More generally, the
drainage line will be fringed with thicket that is denser and trees that
are taller than any in the surroundings, and again these give harbourage
to different species of birds.

It is worth bearing in mind that the AETFAT map, like nearly all

* The term "catena", now adopted in plant ecology and applied to plant communities,
originated in pedology, for "a succession of soil types from ridge to hollow, characterised by
different degrees of free or impeded drainage . . . a situational succession [of soil types]
determined by topography" (Milne, 1947).

others, is somewhat idealized in another respect. On the one hand it recognizes explicitly that the country round the north and west of Lake Victoria is today "Forest-Savanna Mosaic" and "Moist Savanna", though beyond doubt it has ceased to be dominated by Forest only in recent historical times. On the other hand, it makes no attempt to depict the extent to which, for example, the "Moist Forest" area of Western Nigeria has been given over to cultivation and secondary forest or the "Wooded Steppe (Acacia)" of Usukuma, south of Lake Victoria, has become treeless "cultivation steppe" (as mapped by Gillman, 1949). This question of the human influence on vegetation will be referred to again later.

The essential features of African plant geography are, as shown in Fig. 3, an equatorial belt of lowland evergreen forest and concentric belts of increasingly dry and sparse vegetation that are interrupted only in the neighbourhood of the Gulf of Guinea and reach their extremes in the Sahara and South West Africa. This general pattern is disturbed by the surface relief, one result of which is that the lowland evergreen belt reaches only two-thirds of the way across the continent from the west. As necessary in succeeding chapters, more will be said about the characteristic of the main vegetation types, but it is useful to give a very brief summary at this stage.* For the sake of completeness, before proceeding to the main list, mention should be made also of the coastal mangroves (which on present information are of little importance to African birds; cf. Cawkell, 1964). The swamps have already been dealt with above. It is probably true that something like the present "concentric pattern" of vegetation belts, shown in Fig. 3, has been a feature of Africa since before the Pleistocene (though the component species may not have so long a history); but the limits of each of the vegetation belts have been in constant flux, as discussed in Chapter 3.

In order of decreasing humidity, the vegetation types may be briefly described as follows; more particular reference will be made to certain of these types in subsequent chapters as necessary. It cannot be emphasized too strongly that any classification of the kind offered in the AETFAT map, and still more, of course, in the present summary, involves immense oversimplification.

1. Lowland Evergreen Forest

Lowland Evergreen Forest (="Moist Forest at low and medium altitudes" of AETFAT) is developed only in areas with at least 60 in. of rain well distributed round the year. A valuable description and discussion of this type of vegetation in the tropics generally has been provided by Richards (1952). In its primary condition as a climax community this forest is richer in biomass and in plant species than any other vegetation type in Africa, forming as it does a dense stand of many

* "Tropical vegetation has a fatal tendency to produce rhetorical exuberance in those who describe it. Few writers on the rain forest seem able to resist the temptation of the 'purple passage'." (Richards, 1952). I shall.

FIG. 4. The physiognomy of lowland evergreen forest, primary and secondary. (From Chapin, 1932.)

species of trees, with a continuous canopy often between about 20 m and 30 m above the ground. From this taller trees emerge, but rarely exceed about 60 m in height. A general impression of the profile of the Congo forest and the effects on it of clearing and regeneration is conveyed in Fig. 4. As elsewhere in the forests of the tropics, typically the trees have pale, smooth trunks, which are relatively thin and are unbranched for most of their height, but their branches are loaded with epiphytes, and they support a wealth of lianas, many of them as thick as a man's leg, which reach up into the canopy.

Fig. 5. Lowland evergreen forest in the north-eastern Congo. Photo: H. Lang, by courtesy of the American Museum of Natural History.

Inside primary forest progress on the ground is "scarcely hindered save by old logs and rotting boughs" (Chapin, 1932) and the spindly saplings are not numerous (Fig. 5). A special eco-climate is developed, of which the darkness is the most notable feature, the light intensity being only about 1% of that outside and low enough to check the growth of herbage. Clearly, animals adapted to this eco-climate will encounter special difficulties if they are forced by local destruction of forest to try to survive elsewhere. Conditions are very different from the foregoing where there is any opening in the forest, caused, for example, by the fall of a tree; there ensues a dense tangle of vegetation competing for the light, among which especially quick-growing but short-lived tree species have a temporary advantage. Regenerating forest does not attain nearly the height of climax forest so long as it consists of these secondary species, but some among its components, such as *Musanga* and *Trema*, are exceptionally important sources of food for fruit-eating birds.

FIG. 6. Montane evergreen forest of very rich type at 2000 m on Mt Kenya. Photo: R. S. Troup, by courtesy of Forestry Department, Oxford University.

2. MONTANE VEGETATION

Montane Vegetation (Figs. 6–11) includes Montane Forest, the afroalpine communities (above the timber-line; Haumann, 1955) and

FIG. 7. Montane evergreen forest of poor type at 2000 m in the West Usambara Mountains, Tanganyika. Photo: P. J. Greenway.

FIG. 8. Montane evergreen forest at 2000 m on the Zomba Plateau, Malawi. Photo: R. E. Moreau.

other montane communities, found typically above about 1500 m (see also Chapter 15). The Montane Forest, while generally resembling Lowland Forest, is composed of different species and does not grow so tall and dense. Mostly it is broad-leaved, but in the driest kind of montane forest, the so-called "cedar" tree of East Africa, *Juniperus procera*, becomes the dominant. In East Africa some of the montane forests

FIG. 9. Forest grassland at 1100 m, East Usambara Mountains, Tanganyika, showing the tendency for clearing and firing to leave forest in the valleys. Photo: P. J. Greenway.

include patches or belts of a giant bamboo *Arundinaria alpina*, but, spectacular as this growth is, it is of no ornithological importance. The unforested parts of the montane areas, both above the timber-line and below, are dominated by bushy growth, which in its natural state is largely ericaceous (see Chapter 12) and at the higher altitudes is diversified by giant lobelias and arborescent senecios.

FIG. 10. Afroalpine vegetation on Mt Kenya at over 4000 m, with giant lobelias and the arborescent *Senecio keniodendron*. Photo: from O. Hedberg (1964), by courtesy of the author and of Svenska Växtgeografiska Sällskapet.

FIG. 11. Afroalpine vegetation on Kilimanjaro at 4000 m, with giant heather. Mawenzi peak in background. Photo: P. J. Greenway.

3. WOODED SAVANNA

Wooded Savanna (Figs. 12 and 13) is a term comprehending types 16–22 of the AETFAT map, which together cover a great area of Africa on the periphery of the Lowland Forest, where rainfall exceeds about 35 in. a year, but the dry season is severe.* Unlike the components of Forest, those of the Moist Woodlands and Savannas tolerate the fires to which they are subject in the course of the annual long dry season. The grasses grow in clumps from perennial root-stocks from which fresh growth shoots soon after fire has blackened the ground; and many of the trees have a thick corky bark that protects the interior of the trunk from the blaze. In the most humid areas the grass may form thickets as much as a dozen feet high and, perhaps because of the extremely severe fires nourished by such growth, the accompanying trees are small and sparse, while shrubs are almost absent. Where the rainfall is lower the grass tends to be less rank and the trees taller and predominantly leguminous, casting only a slight shade. North of the Equator the characteristic trees are *Isoberlinia* spp., predominating in a belt from Senegal to Uganda, which curves south past Lake Victoria and, expanding through western Tanganyika, joins the great block of "myombo" (="miombo")—dominated by *Brachystegia* and *Isoberlinia* spp.—that occupies Africa almost from the Atlantic to the Indian Ocean between about 7° and 17° S.

Fig. 12. Brachystegia woodland of rich type near Abercorn, northern Zambia, in dry season (August). Photo: A. C. Hoyle.

* It has been suggested (Walter, 1964) that most of the savanna of this type has been produced by human activity, but if so it is not clear what the original vegetation was.

FIG. 13. Brachystegia woodland under lower rainfall, edge of Zambezi valley, southern Zambia, in dry season (August). Photo: R. E. Moreau.

4. DRY WOODLANDS AND WOODED STEPPE

Dry Woodlands and Wooded Steppe (Figs. 14–19) occupy a belt outside the preceding woodland type and with rainfall between about 12 and 35 in., right across Africa from Senegal to Somaliland, which then extends down through eastern Kenya and central Tanganyika almost to the head of Lake Nyasa; it reappears in the hotter and drier river valleys of the Rhodesias and thence extends, steadily getting drier, south and west over most of the Transvaal, Bechuanaland and South West Africa. The most widespread trees of the Dry Woodlands and Wooded Steppes are *Acacia* spp., all with feathery foliage; but at the lower latitudes, where the rainfall is not so limited, there is considerable admixture of broad-leaved trees, and baobabs *Adansonia* are characteristic. Further from the Equator, in the latitude of Lake Chad in the north and in the Kalahari to the south, but also in the dry belt Somaliland–central Tanganyika, the main associates of the *Acacia* spp. are *Commiphora* spp. A special type of dry woodland is that in the south dominated by *Colophospermum mopane*, another leguminous tree, especially in the middle Zambezi Valley and in the Luangwa Valley of Zambia. In some of these dry woodlands the grasses grow to about 2 ft high, but nowhere are the grass fires so severe as in the moister woodlands of type (3) above.

Fig. 14. Mopane woodland, Luangwa Valley, Zambia, in dry season (August). Photo: R. E. Moreau.

Fig. 15. Mixed, partly mopane, woodland in the Zambezi Valley, showing the greatest degree of defoliation (August). Photo: R. E. Moreau.

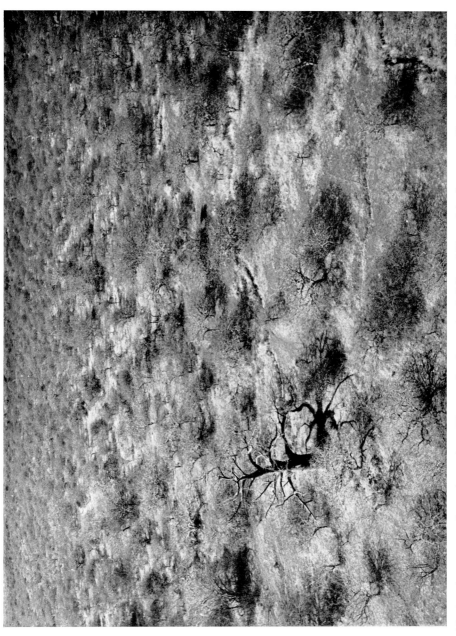

FIG. 16. Acacia steppe, Mkomasi, northern Tanganyika (annual rainfall *ca* 250 mm) in dry season (August). Photo: H. Lamprey.

Fig. 17. Acacia steppe, Tarangire Game Reserve, northern Tanganyika (annual rainfall about 800 mm) in dry season (February), but after exceptionally heavy rains. Photo: H. Lamprey.

Fig. 18. Acacia steppe near Richard-Toll, Senegal, towards end of rainy season (September). Photo: G. Morel, by courtesy of *Terre et Vie*.

Fig. 19. Same spot as Fig. 18 towards end of dry season (May). Photo: G. Morel, by courtesy of *Terre et Vie*.

5. Desert and Subdesert

Desert and Subdesert (Fig. 20) appear peripherally, in the north culminating in the Sahara, in the east in Kenya and the Horn of Africa, and in the south-west in a strip from Angola to the lower Orange River. The main vegetation of the subdesert consists of low perennial plants, often bushy, with a few *Acacia* trees in favourable spots. The grasses are ephemeral; fire is no longer a factor, but the soil is bare for most of the year, since any herbage, on drying, is removed by harvester termites

Fig. 20. Sahara; one of the worst areas, west of Wadi Halfa (annual rainfall under 10 mm). Photo: R. E. Moreau.

and wind. In the true desert plants are solitary, few indeed and far between, with very large areas that are completely barren. Today, as described in Chapter 4, the Sahara, which is about the size of the U.S.A. or Australia, interposes a most formidable ecological barrier, 1000 miles wide, between Mediterranean and tropical Africa. Yet, as shown in Chapter 3, on the evolutionary time-scale conditions in this part of northern Africa were very different indeed only yesterday.

6. MACCHIA

Macchia is a loose term employed to cover the vegetation characteristic of the extremities of the continent, the Maghreb in the north (Morocco–Tripoli) and the western Cape Province. This latter area, including the Karoo, is of great botanical complication and interest, but from the ornithological point of view it forms an insignificant part of the continent as a whole. In the nature of its vegetation it shows convergence with the Maghreb, where also the vegetation is characterized by a wealth of sclerophyllous (leathery-leaved) and aromatic shrubs. Floristically, however, the two areas are quite different, for the vegetation of the Maghreb has nothing of peculiarly Ethiopian affinities and its mountain forests consist especially of oaks (*Quercus*) and pines (*Pinus*), neither of which genera occurs south of the Sahara.

It may be emphasized that the vegetation types dealt with in Sections 1 and 2 above (Lowland Forest and Montane Vegetation), being evergreen, hardly change in appearance from one year's end to the next. Types 3–5 do so very markedly, and they must surely have a corresponding seasonal fluctuation in the amount of food available for birds. Nevertheless, the composition of the bird communities dealt with in Chapter 15, all of which include various species of frugivorous birds and of sunbirds, show that some fruit and some nectar-bearing flowers of one sort or another are perpetually available in vegetation types 3 and 4 as well as 1 and 2. Nothing precise is, however, known about the food resources, and we badly need surveys round the year in sample areas to ascertain the extent to which a sequence of the main kinds of food— insect, fruit, seeds and nectar—is maintained in successive months. This opportunity may be taken to mention the exceptional importance of figs *Ficus* spp. in the economy of fruit-eating birds in every vegetation type from lowland forest to even subdesert steppe (where some of these trees exist along the drainage lines).

Only a small fraction of the African continent has a human density of more than 50 to 1 sq. mile and most of tropical Africa has considerably less than 25, yet over most of its vast surface the influence of man on the vegetation has been great, both directly and indirectly, and overwhelmingly towards impoverishment. For a documentation of the process in Africa generally Harroy (1949) may be consulted, and a map showing the varying degrees to which "danger of erosion" exists throughout the continent has been compiled by Fournier and D'Hoore (1962). The damage started, though doubtless in a small way, when man acquired the use of fire. This was probably well over 50 000 years ago, for between 50 000 and 60 000 years ago he was systematically using hearths at his living-sites at Kalambo, near the south-east corner of Lake Tanganyika (Clark, 1964). It is, of course, true that in prehuman times grass fires must have occurred naturally from time to time, and to this the "pyrophytic" tree species of the African savanna, specially adapted to withstand fires, bear witness; but man's firing would have tended more and more to change the vegetation, because it covered the same ground more frequently and was not associated with thunderstorms, the rain from which may quench the fire that lightning has started. In particular, the intermediate "soft" types of vegetation would be eliminated and stark contrasts with abrupt edges would have been developed between the wooded grasslands ("savanna") that feed a roaring "bush fire" and the evergreen forest too humid to be more than nibbled peripherally by the most ardent blaze.

The next stage of human interference came with the domestication of stock and its consequences, especially in dry years, of excessive trampling, overgrazing and erosion. No doubt stock-keeping spread southwards in Africa from the Neolithic peoples of the Mediterranean and the Sahara; and since in Cyrenaica, for example, domestic stock were first kept less than 7000 years ago (quoted in Bakker, 1964a), these animals were presumably not an important factor in the tropics until after this date. Over Africa as a whole their effects have probably been most felt in what is now the Sahara, and in other dry areas unsuitable for tsetse flies *Glossina* spp., whose carrying of trypanosomiasis has done far more than administrative action to preserve parts of the surface of Africa from the ruination consequent on excessive cattle-keeping. In modern times, and especially since veterinary science became effective against so many ills, no part of Africa has suffered more from overstocking than the Horn and the far south-west. In what was formerly British Somaliland the run-down in the last few years has been catastrophic, and where early travellers described pleasant park-like country (albeit the trees were predominantly acacias), inhabited by herds of game, there is now "a grim wasteland without cover" (Bally, 1964). For South Africa, Acocks (1953) has mapped the degradation of the vegetation that has taken place through the occupation by African and European herdsmen in recent centuries. The semi-arid Karoo is estimated to be half as big again as it was formerly, when, moreover, it included none of that near-desert, having only rarely even a temporary cover of vegetation, to which Acocks shows that one-third of the present Karoo area has been degraded.

Finally has come the impact of agricultural man, which can be accepted as having spread from north to south, but by no means evenly.

On the middle Zambezi cultivators were already present in "the early years of the first millennium A.D." (Clark and Fagan, 1965), but in South Africa they do not seem to have begun to replace the food-gatherers until much later. Most vegetation maps give no hint of the cultivators' depredations. One exception is Gillman's (1949) map of Tanganyika, already mentioned, which shows the thousands of square miles of "actively induced vegetation", respectively "by natives" and "by aliens", the "cultivation steppe" of some authors. Typically the African in the past has cut down at least the smaller trees, burnt the slash, raised a few crops of millet (the only indigenous African food crop) or maize and moved on to do the same thing elsewhere (Fig. 21). In a few years the original vegetation has revived—the trees are not taken out by the roots—and the cycle is resumed. But as the human population increases so does the pressure on the land. There is nowhere for the cultivator to pass on to; and the result is such a baldness of landscape as on the swelling hills of Ruanda and the Transkei (Fig. 22) or on the plains of Usukuma (Tanganyika), devastated in the interests of grain and cotton. Meanwhile alien plantation interests have moved in; for example, with sisal extirpating much of the coastal vegetation of the East African coast and sugar-cane denuding coastal Natal and Zululand. Some permanent crops, such as oil-palms, cacao and bananas, do not, of course, have the same immediately distressing results, partly from their nature, partly because they are grown in such humid climates that the soil is not lost and regeneration is quick and easy. But none the less in, for example, the big area mapped as forest in Southern Nigeria most is, in fact, under crops of one sort and another, and of the surviving forest little if any seems to be truly primary.

More often, forest is locally eliminated. Acocks has inferred that in all probability until recently a continuous belt of "forest and scrub-forest" 50–150 miles wide extended round the entire east and south coast of what is now the Union of South Africa to within 50 miles of Cape Town. Now it has been so generally removed that on his scale of 1 : 7 000 000 Acocks can map few surviving areas. At the opposite end of the Ethiopian Region an example may be cited from Abyssinia, where even a mountain as high as 4000 m (over 13 000 ft) has "every acre cultivated" and is "entirely devoid of trees" (UNESCO report quoted in *Oryx* (1964) **7**, 247–250). From the birds' point of view most of the changes attendant on increased agriculture are indeed disastrous. On the one hand, the destruction of isolated patches of forest, as on mountains or on the scarp of Angola, is irreversible under present conditions and the closely associated bird community is eliminated, for forest birds are highly unadaptable. Further, in savanna the clearing of trees and bushes robs a large number of species of nesting sites and impoverishes the local bird fauna greatly. It is probable that so far there have been few large-scale changes in the geographical ranges of birds as a result of the use or even the misuse of land. But that stage is beginning. It has been stated, for example, that following agricultural expansion in Natal a number of species have disappeared from there, though they are still found in the less-developed country of Zululand (Vincent, quoted in *Ann. Cape Prov. Mus.* (1962) **2**, 154).

FIG. 21. A cultivated area in the Gambia in the dry season (December), after harvest of maize and groundnuts. Large trees have been left untouched; regeneration of the smaller is proceeding from their root-stocks. Photo: R. E. Moreau.

FIG. 22. Cultivation steppe in the Transkei, Eastern Cape Province. Native woody vegetation has been eliminated. Photo: R. E. Moreau.

On the other hand, it must be admitted that "development" is not, from the birds' point of view, altogether without compensation. In its more primitive state, when agriculture is not too intense and remains untidy, with all stages of regenerating "bush" surrounding patches of cultivation, the bird population is probably actually enriched by the resultant mosaic of sub-habitats. Moreover, a by-product of agricultural enterprise and of stock-keeping has been the provision of water-points and dams where none existed before. This has greatly favoured some species, especially water-birds on the big open reservoirs on the South African veld. One passerine species can be cited as having profited from agricultural enterprises and that to a highly undesirable extent. *Quelea quelea*, formerly dependent especially on the seeds of wild grasses, has battened on the new crops, all the more where they are grown in big blocks, as on the wheat lands of eastern and southern Africa or on a rice enterprise such as that on the lower Senegal River (Morel and Bourlière, 1955, 1956).

It is appropriate to mention here also that human installations notably favour some birds. For example, where I lived at Amani, in a forest clearing in northern Tanganyika, we had four species of hirundine and one swift that nested on stone buildings where none had existed twenty years before; and I do not know of any natural sites used by these birds anywhere in the neighbourhood. Gardens of European type, cleared of natural vegetation, are accepted as a refuge by some birds. Also, wires and posts in otherwise bare country serve such birds as shrikes, rollers and bee-eaters as the essential perches from which they can pounce on their prey; and culverts and the cross-arms of telegraph poles are used by a few birds as nest sites. Hence, while it is true that the main tendency of the modernization of the African landscape is, as elsewhere, towards the impoverishment of the bird life, and there will be many shrinkages of range, the evil results are not entirely unmitigated.

A few recent studies make a systematic attempt to determine the population trends of individual species in different parts of Africa in recent years. In the Gambia, which has experienced the results of increased African population and drive for export of groundnuts, with some general impoverishment of habitat in consequence, many species appear to have become scarcer in the last half-century (Cawkell and Moreau, 1963). The Salisbury area of Rhodesia, which Brooke (1963) has studied over a radius of 30 miles, has experienced partial urbanization, extensive cutting of woodland (*Brachystegia*) and ploughing of grassland, but also reduction in fires and a multiplication of dams. One-third of the species in this area have changed significantly in numerical status. In general, birds typical of *Brachystegia* woodland and of grassland have declined, while those dependent on thickets and watery habitats and also "aerial feeders" have increased. Similar effects are likely to be expected wherever much development on European lines takes place and perhaps offers rather more ornithological compensations than the results of the Gambia type of development. Marked changes in the bird population can, however, take place for no obvious reason. In Eritrea, where there appears to have been very little "development" in the course of the last hundred years or so, Smith (1957) found that many species had evidently changed their status over the same period, some gaining strength and some losing. Clearly, then, human interference, though the most potent influence on bird life in Africa today, as it is elsewhere, is not the only one affecting it.

Seasonality

In most of Africa, as in the tropics generally, the seasonal procession of biological events is different from that in higher latitudes and is far more complicated. It is therefore worth while here to give a brief summary of the salient features as they affect the bird life, even though, as part of the ecological background, they are of direct interest to only some of the later chapters in this book.

In the north temperate zone, regardless of when and how the bulk of the precipitation takes place—for example, as winter rainfall in the Mediterranean basin, as all-the-year-round rain in the British Isles, or as snow and rain in the colder half of the year in Russia—the biological year is governed primarily by temperature. When the weather becomes warmer, at the end of winter, vegetation of all kinds grows, insects become plentiful and birds begin to breed. To a great extent and with few specific exceptions, egg-laying and also most of the care of the young in the nest take place between the beginning of March and the middle of July—in fact, mostly within the second quarter of the year. In autumn and winter supplies of various foods decline everywhere in the temperate zone, so that many populations migrate; indeed, from those areas which normally have much snow nearly all of them do. The great point is, however, that in varying degrees the whole vast area of the north temperate zone is affected similarly at the same time, while in their rhythm of life nearly all species of birds, whatever their ecology, conform rather closely to the same time-table.

By contrast, in Africa there is great heterogeneity. First of all, in the country bordering the Mediterranean the rhythm and the restricted breeding season of the Palaearctic prevail, apparently under the control of temperature, though for the most part activity begins and ends earlier in the year, before the full summer drought. South of the Sahara, except at the south-western extremity of the continent, this control evidently does not operate; the dominant influence in the biological year is rainfall, and in many areas (as explained below) the breeding season is indefinite to an extent unknown at higher latitudes. Also, because as a rule the rainy season comes in the summer, it is about six months out of step on the two sides of the Equator. Hence a Palaearctic migrant arriving in the northern autumn encounters a different half of the African biological year, according to whether it crosses the Equator or not. Moreover, an African bird can escape from the local ecological rhythm by migrating across the equatorial belt, as a few species do (Chapter 13).

In the deciduous vegetation types of Africa, i.e. wooded savanna and steppe of various kinds, which cover between them most of the Ethiopian Region, the seasonal contrast in appearance is very great indeed and the biological year is dominated by the alternation of the extremely well-marked wet and dry seasons, the latter of which extends unbrokenly from five to nine months according to the area. In the driest parts, where the ground cover is annual, it grows only when the rains have begun; elsewhere, as in the brachystegia belt, where the grasses are deep-rooted and as a rule are burnt, they begin to show their young green some weeks before the rains start. Similarly, in such areas and also in acacia steppe the trees put on their new leaves before the rains. The result is a flush of insect food, which also begins before the rains, around October (early spring) in most of the southern tropics and April (the equivalent month) or May in the northern tropics. The trees flower and fruit in various months of the year, mainly by species, thus to some extent providing a sequence of foods for birds dependent on these sources. As the rains continue the ground is progressively covered with a dense stand of herbage, which provides seed and the necessary conditions for the breeding of some species, though for a few others it has unfavourable features. Subsequently, as the dry season proceeds the increasing desiccation and the general aspect of the countryside suggest to the observer that the food resources must be diminishing greatly. We do not know how far this may actually be true in any locality; but appearances must be to some extent deceptive, since most species of birds manage to survive the dry season without extensive movements (though there are interesting exceptions, see Chapter 13), and in some dry areas it is even possible for various Palaearctic migrants to be accommodated alongside the resident African species throughout the "worst" months (Chapter 14).

In the absence of overriding control by temperature, and ecological conditions round the year being what they are, it is not surprising that in any particular locality the breeding of the bird fauna as a whole is spread widely over the year, the timing depending on the habits of the different species involved. As a result of a general survey of the somewhat inadequate data available some years ago (Moreau, 1950), I concluded that five categories of birds could usefully be distinguished, with characteristic breeding seasons that fit into the biological year in the same way over most of the continent: (1) the big raptors and scavengers, which lay in the middle of the dry season (perhaps because prey and carrion are more accessible when cover is reduced); (2) some ground-birds, such as nightjars, which tend to lay in the rains in areas where grass fires are not an extensive risk, but elsewhere to lay in the dry season after the grass fires are over; (3) the birds dependent on tall grass, which necessarily wait till well after the rains have started, both for nesting sites and for seeds; (4) the water-birds, which tend to nest when watery habitats have increased in extent as a result of the rains; (5) most other birds (nearly all dependent wholly or mainly on insects), which tend to nest early, beginning with the pre-rains flush of new foliage and of insects.

Since 1950 there have been several important contributions to our knowledge of the breeding seasons in different parts of Africa, especially by Smith (1955) for Eritrea, Hall (1960a) in Angola, Archer and Godman (1937–61) for Somaliland, Morel and Morel (1962) in Senegal, Benson *et al.* (1964) for the Rhodesias and Malawi, Clancey (1964) for Natal, and Ruwet (1964) for Katanga. These new data draw attention incidentally to a number of local and specific peculiarities which are of great interest, but on the whole the mass of new data supports the generalizations quoted above.

The foregoing paragraphs have been concerned with the deciduous vegetation types, which change so strikingly in aspect between the wet and the dry seasons. In contrast to these, the Evergreen Forest, both lowland and montane, and the montane vegetation generally, does not alter its general appearance from one year's end to the next. This might well be taken to imply that conditions were equally favourable for breeding at any time of the year. This may indeed be true in at least some areas of the Lowland Forest close to the Equator, which have an extremely equable climate, but data from individual localities are not sufficient to verify this fully. What is surprising is that in certain montane forests of eastern Africa egg-laying is considerably restricted, to much the same months for most species of birds, irrespective of the diversity of their habits (Moreau, 1936b); and to account for this one has to postulate a much more marked fluctuation in food supply in the course of the year than one would have supposed. To complete the picture it may be added that in the evergreen macchia of the south-western tip of the continent temperature again apparently becomes the main control of the breeding season, for this takes place about the end of the winter rainy season, as the days get warmer, and is thus a counterpart of the arrangement in Mediterranean Africa, but removed by six months of the year.

The foregoing paragraphs give a brief indication, admittedly somewhat oversimplified, of the ecological changes round the year to which the breeding seasons of various types of birds are geared and which presumably control them. In most vegetation types and areas the spread of the breeding season is such that the peak activities and consequently the peak populations of different categories of birds are reached one after another, not simultaneously as at the higher latitudes (including Mediterranean Africa and the Cape). It is, however, true that, since the majority of bird species are wholly or mainly dependent on invertebrate food (Table VI, pp. 90–91), and these tend all to breed at much the same time of the year, before the rains or at their very beginning, there is, notwithstanding the diversification of breeding seasons in the bird fauna as a whole, a marked seasonal increase in the demand for this type of food. However, because clutch sizes average so much smaller in tropical species than in the related species at higher latitudes (Moreau, 1944a), over most of Africa the bird populations and consequently their food requirements will not regularly be subject to such big seasonal increases as at higher latitudes.

3

The Past of Africa

Some years ago I made a pioneer attempt to review the literature dealing with Africa since the beginning of the Tertiary (Moreau, 1952a), in order to depict the background to African bird evolution as well as I could, and with the hope of elucidating some of the distributional problems, especially those posed by the montane biomes, as exemplified by birds. At the time I published my review I felt, as I said, that the hope was not fulfilled. I did not realize that the Pleistocene,* which then seemed to me to be too short, potentially provided the key to the main problems; and in any case the geological evidence for the state of Africa during the Pleistocene was extremely defective. Most of what I wrote in 1952 seems in no need of emendation, but in the last few years, mainly since 1959, there has been a striking and most welcome accession of geological data for this period, more particularly the late Pleistocene, reviewed with reference to the biological implications by Moreau (1963; see also Bakker, 1962a, 1963, 1964a). We now know that ecological changes on an enormous scale have repeatedly taken place, the last of them to affect the whole continent being within the last 20 000 years, and another, which transformed much of northern Africa, in the last 5000 years. What follows is condensed from the reviews quoted, except where otherwise stated.

On the geographical side the basic fact is that since before the Tertiary, for something like 100 million years, the continent of Africa has existed in very much its present size and shape—in fact, it has been one of the most stable land-masses in the world. It has had overland connexions of various kinds to other continents only during the last 20 million years. From the late Miocene to the late Pliocene (which ended 1–2 million years ago) there were connexions eastwards to Arabia in various parts of what are now the Red Sea and the Gulf of Aden, and in the Pliocene a northward bridge in the Tunisia–Sicily area. Since the Pliocene, i.e. throughout the Pleistocene, Africa has again been isolated from the rest of the world, except for the tenuous connexion of the Isthmus of Suez, which was at times no more than 10 miles broad. It may be added that there is no evidence of connexion across the Straits of Gibraltar at any time and that the most recent geological opinion

---

* I follow those geologists who regard the conventional splitting off of the post-glacial few thousand years, commonly called "Recent" or Holocene, as unnecessary and inconvenient; and consequently I use the term Pleistocene as extending to the present day.

also rejects the idea of land connexion between Africa and Madagascar in any time relevant to the evolution of birds.

The present relief of the African surface, which is so dominant a factor in the biogeography, is the product of the very late Tertiary. It seems probable that during Jurassic times (when the class Aves was making its appearance in the world) differential uplift raised masses of hard Palaeozoic rocks to considerable heights, and that these, cut back by circumerosion for something like 100 million years, are represented today by various steep-sided islands of ancient rock, such as the Usambara and Uluguru Mountains of Tanganyika, deeply cleft masses largely of hornblende-biotite-gneiss that are of exceptional biological diversity, considering their modest altitude of some 8000 ft. But apart from these peculiarly resistant rock-masses there is general agreement that during the very long period of tectonic calm extending through the early Tertiary the general surface of Africa was worn down to a gently sloping peneplain, with the result that the greater part of Africa would have been below about 2000 ft (600 m) and only a little of it in the centre of the continent as high as 5000 ft (about 1500 m), i.e. would have exceeded the present lower limit of montane conditions. This is very different from the present state of the continent; and since world temperatures at this period in the Tertiary, and until the Pliocene, were higher than they are now, it would seem that montane conditions as we know them at present could not have existed at this stage of peneplanation. Consequently the typical montane biomes must be the product of subsequent evolution.

The calm was broken about the end of the Miocene, i.e. around 12 million years ago. Vast areas of eastern and southern Africa were uplifted some 4000 ft (1200 m) or more, the Abyssinian and Kenya highlands were largely overlaid with volcanic outpourings, and during the Pliocene the great isolated volcanoes of Africa, especially Kilimanjaro, Mt Kenya, Elgon and Mt Cameroon, began to pile up. Moreover, huge blocks of eastern Africa were tilted, upraising their eastern edges, so that on the one hand they provided north-and-south causeways of montane conditions, as in the Nguru and Uzungwe Mountains in Tanganyika, where probably none had existed before, and on the other hand they produced dry, rain-shadow, effects in their lee. By contrast a great lake filled the centre of the Congo basin. However, it is only in the course of the last million years or so that the most spectacular features of modern Africa became fully developed. Rifting emphasized the troughs now occupied by Lakes Albert, Nyasa, Tanganyika, and some smaller ones. Lake Victoria took on its present aspect, while the Congo lake was drained by a coastal river that captured it and the present Congo system took shape. Many considerable volcanoes, for example Rungwe at the head of Lake Nyasa, Marsabit and Kulal in Kenya, the Virunga group of the eastern Congo, and Emi Kusi (the highest part of Tibesti) in the Sahara were built up during the Pleistocene.

Younger still, the entire Chyulu range in Kenya, some 7000 ft (2100 m) high, may be no more than 40000 years old. But unlike all those mountains named, Ruwenzori, one of the giants of Africa, 17000 ft (over 5000 m), is not a volcano but an uplifted block that has continued to rise during the Pleistocene.

At the time when I wrote in 1952 some highly unrealistic views about the past climatic history of Africa were still current; in particular that the continent had been blanketed with forest until the Miocene and that subsequent desiccation had provided the opportunity for an Asiatic steppe fauna to swarm into it. Actually the evidence pointed to the contrary conclusion that evergreen forest, savanna and arid areas had all co-existed in Africa, in series similar to those of today, though with north-and-south oscillations of the climatic equator through several degrees every few thousand years. Africa has consequently been well fitted to be a centre of evolution for animals adapted to each of these types of environment. Pliocene evidence was notably lacking, owing to the general absence of deposits of that age in tropical Africa, but for the Pleistocene there was sporadic evidence for the occurrence of climatic alternations on a great scale in various parts of the continent. In East Africa, indeed, a whole system of "pluvials"* and "inter-pluvials" had been worked out, but the detailed correlation of the pluvials with the Palaearctic glaciations and the application of the climatic sequence elaborated in Kenya to much of the continent has not so far been justified on the present evidence; the validity of only the last "pluvial", correlated at least in part with the last glaciation, is generally approved (Cooke, 1958; Flint, 1959a, b).†

It has only very recently been realized to what a great extent the alternations of glacial and interglacial stages throughout the Pleistocene affected the lower latitudes of the world. Their influence was twofold: first, directly by means of temperature (and associated changes in evaporation), about which there is no important disagreement; second, indirectly through changes in the atmospheric circulation, which affected the rainfall at lower latitudes in ways that are the subject of opposed opinions. Both the factors exercised an enormous and ever-changing influence on the geographical ranges of the main vegetation types and their associated animals. As will be shown, the ecological picture of the continent of Africa differed greatly from the present only some 5000 years ago, especially north of the Equator, was different

* "Pluvial" has of recent years been a very widely and loosely used term that does not necessarily imply heavy rainfall. A convenient definition is that of Flint (1959a): a "climatic regimen sufficiently long enduring to be clearly recorded in physical or organic evidence, and having an average effective rainfall distinctly greater than that of today in the same area". Conversely an "inter-pluvial" had a rainfall distinctly less than the present.

† The names Kageran, Kamasian, Kanjeran, Gamblian, Makalian and Nakuran, which had originally been devised for "pluvial" and (the last two) "subpluvial" periods in Kenya and Uganda, had been somewhat uncritically applied elsewhere. The Third Pan-African Congress of Pre-history, 1955, recommended that they should be used only in the climatic sense and in the type area.

again around 10000 years ago, especially in the south, and even more
different and in other ways, as a consequence of the lowered temperature, around 15000 years ago.

For full recognition of the complexity of the changing ecological
influences that have pressed upon modern species of plants and animals
it must be recalled that, on a multiplicity of evidence, though by far the
greater part from higher latitudes in the northern hemisphere, there
were four main glaciations in the course of the last 1 million years or so,
in each of which about the same minimum temperatures appear to have
been attained (Oakley, 1964), though the extension of the ice-sheets
varied under the influence of other factors. They were separated by
interglacial periods, which were warmer than the present, and each
glaciation consisted of two or more phases of varying severity separated
by milder interludes ("interstadials"). Datings are somewhat uncertain
(see review in de Heinzelin, 1963), even with the new techniques of
potassium-argon and of radiocarbon (which has, however, an extreme
limit of 70000 years and diminished reliability beyond 40000) except
for the latter part of the Last (Würm=Wisconsin) Glaciation, which at
its worst was probably at least as severe as any of its predecessors. The
available data on the sequence of temperature changes in the course of
the Last Glaciation in different parts of the world have been assembled
by Woldstedt (1960), Flint and Brandtner (1961) and Flint (1963a). It
appears that the glaciation began about 70000 years ago and that present world temperatures were not again attained until about 8000 years
ago. The last severe phase of the glaciation was between about 25000
and 15000 years ago, with its culmination around 19000. During these
10000 years the state of central Europe, for example, was fully arctic. For
two other, briefer, periods the climate may have been nearly as bad,
perhaps around 52000 years ago and again around 35000. There were
numerous minor fluctuations, but during the intervals between the "lows"
the temperatures seem to have averaged about half-way between the
worst and the present, with no interval which came close to the present.

From about 16000 years ago temperatures rose fairly steadily; except
for a marked but brief cold relapse* around 10500, temperatures rose
to the so-called "climatic optimum" or "hypsithermal", when for a
period broadly between 8000 and 4000 years ago temperatures generally reached something like 2°C higher than they are now, and the
conditions of an interglacial period were approached. Thereafter temperatures dropped again, so that between A.D. 1400 and 1850 they were
a little below the present (Lamb, 1963). Since similar climatic sequences
recurred again and again in the Pleistocene, it can be inferred that whatever can be learnt of the changing conditions of Africa in the last 18000

---

* The Younger Dryas of European nomenclature. As an example of the potential effects
of a relatively minor climatic change on vegetation one may cite the evidence that during
this cold snap the timber-line in central Europe appears to have retreated south by about 500
miles in as many years (Firbas, 1951).

years or so is applicable to successive earlier stages. There is, however, one caveat. The climate we are experiencing today may be regarded as intermediate, nearer to interglacial than to glacial conditions. Such an intermediate climate will have occurred several times before in the Pleistocene, but the present condition of the world's vegetation is unique owing to its unprecedented impoverishment by human interference.

Because the montane flora and bird fauna of tropical Africa differ so markedly from the lowland (see also Chapter 5) and because so much of the African surface lies above the critical boundary of about 1500 m between montane and lowland, it is of prime importance to assess the effects of the glacial periods on African temperatures. It is now generally admitted that the glaciations were simultaneous at the higher latitudes in both hemispheres (see, for example, Godwin, 1960); and in view of the magnitude of the temperature changes the intermediate and equatorial latitudes cannot fail to have been cooler at the same time. Thanks primarily to pollen studies, a beginning has now been made in the reconstruction of climatic sequences at low latitudes during the latter part of the Pleistocene. First, near Bogotá, within 5° of the Equator in South America (Colombia), van der Hammen and Gonzalez (1960a, b), working at an altitude near the timber-line and hence providing a sensitive indicator, have established a sequence that is in good agreement with that for the North Temperate Zone. Still more recently Coetzee (1964) has derived similar conclusions from a series of pollen data from a lake at 2440 m in the forest on Mt Kenya. Moreover, material from a lake close to the lowest moraine on Ruwenzori suggests from its radiocarbon dating $(14\,700 \pm 290)$ that the glaciers on this mountain began to shrink around 15000 years ago (Livingstone, 1962), which concurs well with the sequence elsewhere. On the whole, then, it can on all grounds now be accepted that temperature changes at the lowest latitudes were in step with those at higher.

It remains to determine the amplitude of the equatorial temperature changes (see references in Moreau, 1963). Data obtained in the tropical Atlantic from isotope measurements of shells and from the fauna of seabottom cores indicate temperatures during the last phase of the Last Glaciation about 5°C below the present, with a subsequent rise concurrent with that worked out for the northern land-masses. (But for some caveats on the interpretation of deep-sea cores, see Davies, 1963.) In tropical Africa inferences from the descent of glaciers on the great mountains (especially Ruwenzori, a block that has been steadily rising) suggest that the temperature reduction on the continent was at least as great as in the neighbouring Atlantic. In their Colombian locality van der Hammen and Gonzalez (1960a) inferred an even more marked cooling, by between 7° and 8°C. In Africa four determinations of the extent to which vegetation belts descended below the present levels are now available: by 1000–1100 m on Mt Kenya about 17000 years ago; by at least 700 m, and probably much more, on the Cherengani Mountains in western Kenya some 16000 years ago; by 800 m on the Tangan-yika–Zambia border around 18000 years ago; and by at least 850 m in north-eastern Angola probably about 14000 years ago, i.e. when the glaciation had already relaxed considerably (Bakker, 1962b, 1964a, b, and *in litt.*). Now, as a world-wide average temperature falls by about 0·5–0·6°C with every increase of 300 m in altitude. Therefore these African determinations point to a reduction in the temperature of the local tropics by at least 5°C during the last phase of the Last Glaciation. This figure is rather less than that arrived at in Colombia and is in accord with that for ocean temperatures at low latitudes. I feel, then, that it can be accepted whole-heartedly, and I may perhaps express a special satisfaction in this demonstration because in my 1952 paper 5°C was the reduction in temperature that I postulated as the least necessary to account for certain features in the distribution of African birds.

It follows that we can accept that during the period 25000–18000 years ago, and also during a number of earlier periods of the Pleistocene, the climate at about 1500 m, which is the present lower limit of the montane biomes in most of tropical Africa, would have been encoun-tered at only about 500 m above present sea-level. Owing to the way in which the relief of the continent is distributed, the effect on the ranges of various biomes would have been revolutionary. As will be seen from Fig. 2, the montane zone, instead of forming a number of ecological islands, some of them very remote from each other, as it does today, would have occupied a continuous block from Abyssinia to Angola and the Cape Province, and moreover there would have been a montane connexion, though tenuous, round the north end of the Congo basin to the mountains of the Cameroons at the head of the Gulf of Guinea. Further west, tropical West Africa would hardly have been affected, owing to its low relief; but elsewhere the lowland biomes of the Ethio-pian Region, instead of being greatly predominant over the whole of it, would have been limited to the middle of the Congo basin, to another patch in the Sudan, and to a rim all round the coast—which reached its greatest width of some 200 miles in the east in Mozambique and in Kenya/Somalia.

Such conditions would have provided a huge montane area in which the characteristic montane communities could develop, with no barrier except that of mere distance to interchange between even the most isolated stations where montane birds appear today. Conversely, those species which are most typically lowland and are incapable of living in a montane climate were, except in West Africa, crowded into a small fraction of the area now open to them and into the limited sections of the continent mentioned above, between which interchange was diffi-cult. These differences from the present were fully effective for the last time from 25000 to 18000 years ago and to much the same extent also during the preceding phases of the Last Glaciation. It is, in fact, true to say that for the greater part of the last 70000 years conditions in

Africa have been nearer to those outlined above than to those of the present. Moreover, as can be seen from Fig. 2, even in the intermediate stage, no more than 12000 years ago, with only half the maximum drop in temperature, say 2·5°C below the present, which connotes a fall in the montane limit from about 1500 m to about 1000 m a.s.l., a great block of montane conditions would still have stretched from Abyssinia through East Africa and the southern edge of the Congo basin to South West Africa, except for a narrow gap between Abyssinia and Kenya in the neighbourhood of Lake Rudolf. However, at this intermediate climatic stage the contemporary montane area of Rhodesia would have been cut off from the East African by the Zambezi Valley, and also from the South African; the latter would have been isolated by some 300 miles from the highlands of South West Africa as well as from those of Rhodesia; while the Cameroon highlands would have been nearly as widely isolated from the East African as they are at present.*

Lowered temperature has the immediate consequence of decreased evaporation, and with a difference of 5°C this is considerable. This alone could, in fact, bring about the effects normally regarded as "pluvial", even without any increase in precipitation. Such damper conditions, of course, favour the expansion of evergreen forest and hence of forest birds, but this forest cannot normally survive in a typical monsoon climate, with a long dry season, however high the total annual rainfall. This factor has not been given weight in the reconstructions of vegetation in Zambia and Rhodesia on the hypothesis of a rainfall half as great again as the present (Clark, 1957, Summers, 1960); for this area is subject to a six-month dry season, the effects of which can only be offset locally where the topography is such as to favour the production of mountain mist and instability rain—nowadays chiefly on the mountainous eastern edge of Rhodesia. However, in the view of H. H. Lamb (personal communication), it is possible that during a glacial period the dry season would be mitigated by occasional rain. This would certainly tend to extend the area of evergreen forest, but although during a glaciation one factor after another would favour the spread of evergreen forest, there is no reason to suppose that it filled the montane block from Abyssinia to the Cape and it is impossible to delineate its actual boundaries. In Rhodesia itself Lawton (1963) has described some relict patches of "dry" evergreen forest.

Apart from such ecological consequences, the temperature fluctuations of the Pleistocene have had another consequence that has had a direct though marginal effect on the geography of Africa. During a

* The causeways to the Cameroon highlands, both south and east, are little above 500 m, so low that, although during the last, most severe, phase of the Last Glaciation the inferred reduction of 5°C in temperature would have sufficed to make them "montane", if, as appears, the earlier "maxima" of the glaciation were less severe than the last, they are unlikely to have established montane connexion over these routes. But it may be noted that 5°C is the minimum drop to be inferred from the geological data, and van der Hammen and Gonzalez have, as noted above, deduced about half as much again for a locality in South America.

glaciation the amount of water locked up in the ice-caps has appreciably lowered the level of the oceans of the world and with the subsequent warming this has risen again. It appears that during the last glacial maximum and up to about 16 000 years ago, the level of the sea was at least 100 m (330 ft) below what it is now. This has considerable effects on the periphery of the continents; for example, Great Britain was joined to Eurasia, Fernando Poo and Zanzibar to Africa. Based on inferences from radiocarbon datings, the rate of rise has been depicted by Godwin *et al.* (1958), Fairbridge (1961, 1962), and Shepard (1964). No two of these expositions are in exact agreement, but their general consensus is very strong. As a working formula it appears that an average rise of nearly 1 m a century from about 16 000 to 6000 years ago can be accepted, the subsequent fluctuation amounting to only about 3 m. Hence, in the absence of evidence of tectonic disturbance, it is possible to calculate the approximate date by which islands, such as those quoted above, were cut off from the mainland by the post-glacial rise of the ocean level.

During an interglacial and to somewhat the same extent during the brief hypsithermal, the climatic effects on tropical Africa would have been in certain respects the reverse of those of the glaciation. The higher temperature would both have increased the evaporation and have had the effect of raising the lower limit of the montane zone in tropical Africa by something like 400 m, from 1500 m to 1900 m. On a mountain much higher than 1900 m the biological result would merely have been to contract somewhat the area available for montane species, but, except where the local climatic conditions were abnormal, on an isolated mountain not much exceeding 1900 m the result would be to eliminate the montane species it had carried over from the Last Glaciation. Though the temperature has now fallen in the 6000 years since the height of the hypsithermal, there has been no opportunity for such a mountain to be recolonized by a montane community from a reservoir on some bigger mountain, except by individual incursions such as populate an oceanic island. This could help to account for the fact that, as will be seen by comparing Figs. 2 and 3, the areas of montane vegetation today are fewer than the areas which exceed the montane altitudinal limit of 1500 m. Moreover, the human factor is an important one here; man, as a fire-raiser and especially as a cattle-keeper in modern times, has tended to push upwards the lower edges of the montane forests and greatly to change the nature of the other montane vegetation.

It remains to mention one other possible effect on tropical Africa that may be inferred from a recent world temperature change, namely, the cool spell A.D. 1400–1850. Its amplitude in the tropics is unknown, but a cooling of as little as 1°C, equivalent to an altitude change of 200 m, could be important in affecting the local ranges of montane versus non-montane organisms in such an area as the Lake Victoria basin, much of which lies at critical altitudes between about 1150 and 1500 m, which is the lower limit of the montane forest biome. Incidentally, during this period the flood levels reached by the Nile were low, from which Lamb (1963) has inferred that the equatorial rains were either weak or displaced south.

While the temperature effects of glacial and interglacial alternations in the tropics can be accepted as clear, the situation as regards the rainfall and humidity is very different. One reason is that several factors are operating, independently and all the time, with fluctuating net results in any given area. The simplest humidity factor is that already mentioned, the direct effect of temperature, in that during cooler periods evaporation is reduced. The other factors affecting humidity influence various parts of the continent differentially. First, under astronomical control, shifts in the location of the caloric equator involve changes of up to 10 degrees of latitude in the position of the equatorial rain belt and of the associated monsoon belts. Second, the world's atmospheric circulation and consequently the rainfall total and pattern, together with the amplitude of the climatic belts, change with the alternation of glacial and interglacial periods. So far as meteorological theory is concerned, it is still a matter of disagreement whether at low latitudes pluvials have been synchronous with glaciations, with interglacials or with transitional periods; and the theoretical palaeometeorology of the Sahara is particularly complicated because that area must always have been, to an extent varying in different parts, under the influence of the northern weather system directly affected by the size of the ice-cap and of the monsoon system advancing from time to time from the south. Leading references in this connexion are Zeuner (1958), Willett (1953), Bernard (1962), Büdel (1963), Flint (1963b) and Fairbridge (1962, 1963). In the present state of the subject it must be regarded as impossible to infer solely on theoretical grounds what the ecological condition of any particular part of Africa has been at successive stages in the Pleistocene, and chief reliance must be placed on data from geological sources. The value of this has of recent years been enormously enhanced by the use of the radiocarbon technique.

The available evidence of climatic changes and the inferences to be drawn from them will be described below, successively for southern Africa, the Sahara, and the rest of northern tropical Africa, but first it may be remarked that two areas of the continent seem to have been climatically more stable than any others through the Pleistocene, the south-west and the Horn of Africa. The first area, including the Namib desert on the coast and Bechuanaland in the interior, "embraces a desert focus which has probably existed since the mid-Tertiary, and which has contracted and expanded as long-term climatic changes took place" (Bond, 1963). For Somaliland, through the Pleistocene if not longer, there is no reason to suppose that the climate was any better than semi-arid (Clark, 1954). Thus in at least a part of each of these dry areas there seems to have existed something of that climatic stability, and hence the postulated opportunity for the subdivision of niches, which is commonly regarded as typical of the wet tropics, but seems to have been so imperfectly realized in most of the remainder of Africa.

The Congo basin is occupied at the present time by the most extensive and impressive forest of the Old World tropics, some 600000 sq. miles in extent. It would seem on present appearances that, as Chapin (1932) has remarked, "the equatorial forest of Africa has been in the past one of the most stable environments". However, soil surveys have now shown that nearly the whole of this forest is rooted in Kalahari sand, redistributed by the wind at a time placed by the geologists towards the end of the Middle Pleistocene (cf. Brain (1962) and the dating 75000–52000 years ago suggested by de Heinzelin (1963) for aridity in the equatorial latitudes of Africa). It is inconceivable that this could have happened under anything like the present forest climate; and we must therefore face the tremendous fact that during the worst of this dry period semi-arid plants and animals, probably coming from the south, for a time replaced those of forest, while at the intermediate stages of deterioration and amelioration of the Congo climate savanna birds would temporarily have been in possession. What happened meanwhile to the plants and animals of the Congo forest can only be conjectured; presumably some found a refuge in relict forests along rivers; others on the Atlantic coast, towards Gabon and the Gulf of Guinea; others on the eastern edge of the basin, where the rise in altitude would have mitigated the aridity; and others also on the northern edge of the basin if the incursion of the Kalahari sand was associated with a northward withdrawal of the equatorial rainbelt rather than its rupture. Although forest was subsequently re-established over a wide area, it has since fluctuated considerably in extent; at any rate in the west from the sea as far inland as Leopoldville, marked semi-arid periods, with steppe and even "steppe subdésertique" in place of forest, have been reported around 40000 and around 10000 years ago (Mortelmans and Monteyne, 1962; de Ploey in Bakker, 1964a).

We do not know how far the influence of this latter arid phase extended north and east of the lower Congo, but southwards it was evidently wide. Sand was blowing in north-eastern Angola 13000–11000 years ago (radiocarbon dating; cf. Clark, 1963), and in Zambia west of the Victoria Falls,* so the whole intervening area was presumably affected. This connotes a great increase in the area occupied by plants and animals now characteristic of arid South West Africa and the Kalahari. At any rate in Angola the arid spell was over by, or perhaps

* Bond (1963) has inferred from physical characters of deposits that rainfall in the Victoria Falls area amounted to less than 15 in. per year (about two-thirds of the present) during the Magosian period of prehistoric culture. Three ages of material associated with this culture have been fixed by the radiocarbon method, namely 9550 ± 210 years (Kalambo on the Tanganyika–Zambia border), 9400 ± 110 and 15800 ± 100 (both in (Southern) Rhodesia) (Bakker, 1962a; J. D. Clark, in litt.).

well before, about 6800 years ago (Clark, *in litt.*; Bakker, 1964a), and
further north, across the lower Congo, the now lush evergreen forest of
the Mayombe is regarded as a product of the ensuing humidity (Mortelmans and Monteyne, 1962; de Ploey in Bakker, 1964a). However, up-
stream of the Victoria Falls, round Sesheke, a land-surface dated as
4078 ± 300 years ago is under 5 ft of sand (Clark, 1962a) which may be
due to purely local influences (Clark and Fagan, 1965) or may mark
yet another temporary extension of the Kalahari.

Working on the distribution of the isohyets as they are today, Cooke
(1962) has mapped the hypothetical vegetation pattern of southern
Africa as it would be if the annual rainfall throughout that area were
reduced to 50–60% of its present total. He has shown (Fig. 23) that this
reduction would cause an enormous extension of the desert and semi-
desert, which are at present confined to a narrow strip on the coast of
South West Africa and up the lower Orange River valley, across the
whole of the present site of the Kalahari (which is acacia steppe) and
north-east just far enough to cross the Zambezi in the neighbourhood
of the Victoria Falls. It therefore seems probable that at the time, referred to above, when sand was blowing in this neighbourhood and
aridity extended also right up through Angola, especially around 12 000
years ago, rainfall over this part of Africa was not much more than half
what it is now. The corollary is that, as shown on the same map, a great
area of what is now brachystegia woodland in Angola, Zambia and
Rhodesia, Mozambique and the south-eastern Congo, and of "Bush-
veld" (=mopane and other dry woodland) in the Transvaal, would
have been replaced by "Kalahari grassland or mixed open Acacia
wooded steppe"—the sort of vegetation that dominates Bechuanaland
today, but hardly extends further. This, under hypothetical conditions
of not much more than half the present rainfall, would have dominated
the continent practically from the coast of Angola to the coast of
Mozambique and would, moreover, have extended far up through the
eastern Congo to within 5° of the Equator. The geological evidence
that that section of this hrpothetical vegetation picture which covers
the Victoria Falls was actually realized not much more than 10 000
years ago makes it necessary to envisage the rest of it as a genuine
possibility; there is, of course, no guarantee that the same proportionate
reduction of rainfall applied to the whole of Africa south of 5° S., but
the contemporary evidence from north-east Angola suggests that the
vegetation there may have been predominantly of an even more arid
type than Cooke's hypothetical map suggests.

In the Union of South Africa the scanty information about Pleistocene climatic changes has been summarized by Bakker (1964c) and
Cooke (1964), see also Bond (1963). On evidence from one locality in
the Orange Free State (Florisbad) and one half-way along the southern
coast of Cape Province (Groenvlei), it is tentatively inferred that the
climate in this part of Africa may have been much wetter than it is now

around 45000 and 20000 years ago, much drier around 31000 and
10000. If this last dry spell was indeed contemporary with that further
north referred to above, then conditions must have been severe over a
large proportion of southern Africa at this time.

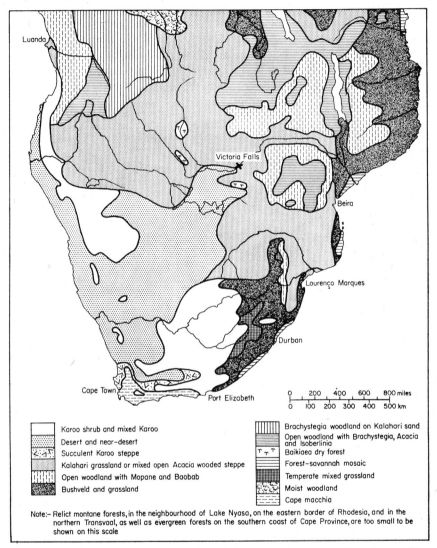

FIG. 23. Hypothetical reconstruction of the vegetation zones of southern Africa
under about half the present rainfall. (After Cooke, 1962.)

North of the Equator changes of the utmost importance have affected
the 3 million sq. miles of the Sahara (Fig. 24), which, as described in
Chapter 2, is today too severe a desert for plants or terrestrial animals
to traverse (see especially the comprehensive review of the Sahara in
the Pleistocene by Monod, 1963). For the past, highly significant data

come from the mammal fossils of the Mahgreb (Morocco, Algeria and
Tunisia), which show that species and genera characteristic of tropical
Africa had repeatedly been able to get across what is now the Sahara.
From the appearance of new species in the north, one such movement
evidently occurred in the Late Middle Pleistocene, and it was not limited
to the corridor along the Atlantic coast, because the implements of
Acheulian man are widespread and such species as the white Rhinoceros
Ceratotherium simum and a *Kobus* antelope reached the Ahaggar massif,
some 400 miles into the Sahara. Moreover, communication was open
to the north, because Mediterranean plants including oak (*Quercus*)
occurred in the same mountains, and they were reached also by the

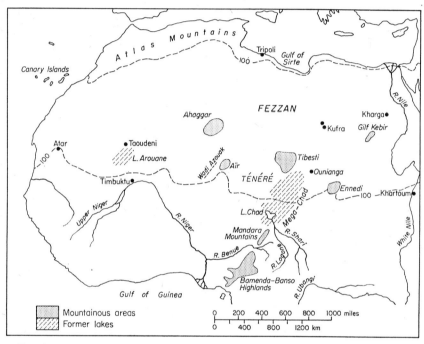

Fig. 24. The Sahara, to show some Pleistocene features. (After Moreau, 1963.)

typically European *Bos primigenius*. At least one thoroughly Palaearctic
type of mammal, a member of the deer family (Cervidae), otherwise
found in Africa only in the Maghreb, seems actually to have penetrated
much further south and to have survived there until around 6000 years
ago, for a rock-painting in Tibesti, probably Neolithic, shows an elk-like
animal resembling *Megaceroides algericus*, which was described from fossils
in the Maghreb (Arkell, 1964). At about the same time, in the Neolithic,
the surface of the neighbouring lake of Wanyanga (Ounianga) Kebir,
fed by local run-off, stood some 40 ft above its present level (Capot-Rey,
1961).

Much of the later part of the Last Glaciation seems to have been cold and dry in parts of the Sahara, but pollen has shown (Quézel, 1960) that in Mousterian-Aterian times,* Ahaggar was clothed with the same species of trees as now dominate the Algerian mountains, with the addition of the European lime (*Tilia*), which no longer occurs in North Africa. So rich a community of Mediterranean vegetation is not likely to have reached Ahaggar by chance colonization, but had presumably been able during a cool humid spell to spread across the relatively low ground of the northern Sahara; and of this there is now pollen evidence (Beucher, 1963).

Around the same period, some 5000 years ago, pollen shows that a more xerophytic Mediterranean woodland occupied also mountains much further south, Aïr (where indeed ancient trees of *Olea laperrinei* still survive), Tibesti and, unless the pollen evidence is fallacious, even the low ground about 200 miles east of the present north end of Lake Chad and only 14° N. of the Equator (Hugot et al., 1962). The extent of the contemporary rainfall rather further north, around 20° E., is vouched for by the Neolithic† fish-harpoons found in the Wadi Azouak, a dry watercourse that winds south-west from the mountains of Aïr. These facts can only mean that for some period extending to within the last 5000 years the western Sahara, most of which is now hardly capable of supporting any vegetation whatever, must have been in sufficiently good condition for Mediterranean scrub and dry woodland to flourish right across to what is today its southern edge, though perhaps discontinuously and partly in dependence on the major drainage lines. Furthermore, artifacts show that Neolithic man and his stock could move around widely in the Sahara (see especially Clark, 1962b, 1964; and also general review by Monod, 1963). It is difficult to avoid the conclusion that at this stage no barrier of real desert existed in the western Sahara, and that typical Mediterranean vegetation was in contact with Ethiopian in tropical West Africa. Probably farther east the situation was different. The relatively indistinct drainage channels in the eastern Sahara, between about the line Tripoli–Ennedi and the Nile, suggest that that area has for a long time been much drier than the western; but the Neolithic period is believed to have been humid in the Sudan and Ethiopia (Butzer, 1961).

Before the end of the Neolithic the Tibesti pollen deposits show that a striking change took place there. The Mediterranean trees were

* Aterian material has been dated at over 30000 years ago and in its last phase at 20000 (Oakley, 1964).

† Radiocarbon datings of material contemporary with Neolithic remains at Adrar Bous, east of Aïr, from three localities in the Ahaggar area and from Khartoum are all near 5000 years ago (Delibrias et al., 1959; Delibrias and Hugot, 1962; Clark, 1962a), and this is consistent with datings around 9000 for the preceding culture of the Capsian Mesolithic (Bakker, 1962a). How long before 5000 years ago the Neolithic began in the southern Sahara does not seem yet to be known. It was presumably later there than in the neighbourhood of the Mediterranean, where the date 6370 ± 103 years ago has been cited for the earliest Neolithic and the first appearance of domestic animals in Cyrenaica by C. M. McBurney in Bakker (1964a), 7000 for the Fayum and 9000 for Jericho (de Heinzelin, 1963).

abruptly replaced by such typically Ethiopian elements as *Acacia* and *Balanites* (Quézel and Martinez, 1958, 1962; Hugot *et al.*, 1962). This process must have been connected with some climatic change, the nature and the cause of which are both unknown. The Ethiopian species must have advanced over several hundred miles of country which later became incapable of supporting such plants, as it is today. The subsequent ecological changes in this part of Africa have been catastrophic. Apparently the whole western Sahara has become desert in the space of the last 5000 years, with virtual elimination of the plant and animal life except in the mountain masses and in the occasional oases (which depend on subterranean water). In both the bird fauna is, however, today very poor (see Chapter 4).

Different opinions have been offered on the meteorological background to this humid Neolithic period in the Sahara; and since the area of the present desert is as large as the whole of Australia or the U.S.A. oversimplification is all too easy. It remains uncertain to what extent this good spell depended on rains penetrating from the north further than they do at present and on the monsoon system temporarily advancing from the south; nor does it seem to be settled how arid the climate was immediately before the Neolithic period in various parts of the Sahara. Presumably it was not as severe as it is now, in which case the survival of the Mediterranean vegetation, at least in the mountains and along drainage channels, could have been assisted by residual effects of that earlier period of good climate which made possible the movement of Mediterranean vegetation southwards across the Sahara. The rather sudden deterioration late in the Neolithic may have been aided by the raised temperature and hence increased evaporation of the hypsithermal, while simultaneously Neolithic man and his domestic stock could have damaged the vegetation at a critical stage. But on any hypothesis of the underlying nature of the Saharan climate during the earlier part of the hypsithermal, there is clearly reason to suppose that no considerable barrier of desert existed between Mediterranean vegetation and Ethiopian (Sahelian) vegetation in at any rate the western part of north tropical Africa for a period around 6000 years ago.

Immediately relevant to all this is the late Pleistocene history of the Chad basin. It has been known for some time that an enormously expanded lake appeared here late in the Pleistocene, and the well-marked shorelines show that the water extended for 400 miles north of its present limits, as well as some distance farther south (Fig. 24). It reached the Bilma area, 19° N., 13° E., in the north and nearly to 20° E. in the east, while in the west an arm extended as far as Bouloum Gana at 15°01′ N., 10°37′ E. Mega-Chad attained, in fact, three times the size of Lake Victoria and was as big as the Caspian. Now that radiocarbon datings have been published by Faure *et al.* (1963) it is known that after a long dry period in this area Mega-Chad was already full 22 000 years ago, which is in the middle of the last glacial maximum, and was still of full size as late as 8500 years ago, nearly 10 000 years after the climate began to warm and just before the hypsithermal reached its peak. But in the next 3000 years the lake must have shrunk very much, for what had been part of its bed was occupied by Neolithic man. This was presumably the route by which Ethiopian vegetation was able to reach Tibesti and so to succeed the Palaearctic vegetation there, as related above, before the increasing desiccation of the area made even the lake bed uninhabitable by plants. Incidentally, as an illustration of the danger of trying to extrapolate results from one part of Africa to another, it may be mentioned that, for whatever reason, the sequence of lake levels in Kenya seems to have been quite different from that of Lake Chad; about 2800 years ago Lake Naivasha stood 60 ft above its present level and Lake Nakuru 125 ft (Leakey, 1964).

The development and persistence of Mega-Chad have interesting implications, for it has been calculated (Grove and Pullan, 1963) that, allowing for reduced evaporation, sixteen times the present intake of water into the basin would have been necessary. Nowadays the inflow is all from the southern quadrants, and mostly from the Shari and Logone Rivers in the south-east. The topography and the absence of tectonic disturbance show that this held good during the existence of Mega-Chad, and evidence for inflow from the south is further provided by the fact that its invertebrate fauna was all Ethiopian (Faure et al., 1963). Meanwhile, as noted above, the flora of the surrounding country was Mediterranean, though probably not that of its actual banks. Clearly, although allowance must be made for some run-off from the northern part of the Chad basin which never occurs now, the rainfall in the southern part must have been much greater than now, and on a scale that postulates the northward extension of the equatorial rainbelt over the ridge dividing the Congo from the Chad basin. Probably an advance by some 300 miles would have sufficed, and with it would have come the equatorial forest and its birds. Such an advance could have been part of a latitudinal shift or of a widening of the rainbelt, with consequent actual extension of the area occupied by the forest. Some soil evidence for the advance and retreat of evergreen forest in West Africa already exists (Schnell, 1950, for example).

To complete the picture of the lushness of much of the country on the southern side of the Sahara during and after the last phase of the Last Glaciation, it may be added that around 8000 years ago a lake existed at Adrar Bous, 20°18′ N., 9°02′ E., extending into what is now the utterly hopeless desert of Ténéré. This lake was presumably maintained by the run-off from the mountains of Aïr, as was that of the Wadi Azouak (on the opposite side of the Aïr mountains) some 3000 years later (p. 55) and it is not yet known how long the good conditions in this area lasted. More important, an immense area west and north-west of Timbuktu was covered with water, the "Lac d'Arouane" (Fig. 24), an inland basin fed by what is now the upper Niger. It finally rose so high that it broke through its sill in the south-east and drained into what had hitherto been the separate system of the lower Niger (Voute, 1962). Thus the present great river with the anomalous bends in its course is the product of the very late Pleistocene, perhaps not much more than 10 000 years ago.

The foregoing gives us an impression, not yet clear, of the mitigation of the Sahara desert at various stages in the Pleistocene. At least the western half seems to have been readily traversible from time to time by animals and plants, for the last time as late as the Neolithic around 5000 years ago, while, whatever the conditions of the main body of what is now the desert, its southern verges were vastly changed by the intrusion of great lakes for periods prior to about 8000 years ago. By contrast, at some stage in the late Pleistocene, desert of the present severe type came much farther south towards the Guinea coast of West Africa than it does now. As shown in Fig. 25, a line of dunes, now long

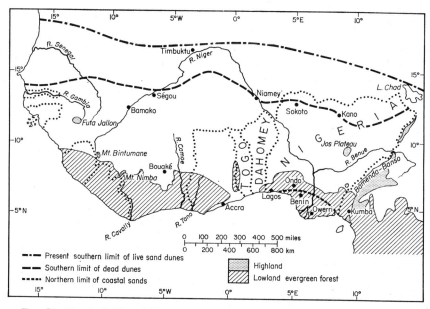

Fig. 25. Tropical West Africa: the evidence for the advance of the desert in the Late Pleistocene. (After Moreau, 1963.)

fixed and overlaid by the primitive local agriculture, has been located as running from Senegal eastwards some 300 miles south of the present limit of moving sand (Grove, 1958). These dunes have not been dated, but they are presumably older than the wet period at low latitudes north of the Equator that nourished Mega-Chad. That is, the dunes must have been left by a southward advance of the desert prior to 22 000 years ago, when Mega-Chad is known to have been already full. It is not clear how they can be fitted into the theoretical climatic sequence and they may be a relic of another shift, this time southwards, in the equatorial rainbelt.

Whatever the date of these dunes, their existence can only mean that at some stage in the late Pleistocene the vegetation belts north of the Gulf of Guinea, the Semi-arid, the Savanna and the (Upper Guinea) Forest, with a total depth between them of 800 miles at the present day, were then restricted to some 500. The southernmost, that including the evergreen forest, compressed against the coast, cannot fail to have suffered more severely than the others. As a result of extrapolation on the map one might well conclude that the whole of the evergreen forest of West Africa was pushed into the sea and temporarily eliminated, but the existence of species endemic to these forests shows that this did not happen. It is, however, certain that the forests, divided today by the "Dahomey Gap", must have been much reduced in area and further cut up, as already postulated by Booth (1958) working on the ranges of the forest primates in West Africa.

CONCLUSION

To summarize the most salient features of the ecological vicissitudes of the Pleistocene, it can be said that during the glaciations a continuous block from Abyssinia to South Africa had a montane climate, so that typically lowland species intolerant of montane conditions can have occupied only a very minor part of the surface except west of the Cameroons. This situation was fully developed for the last time from about 25 000 to 18 000 years ago, but during most of the last 70 000 years the ecological picture presented by Africa has been nearer to that associated with the glacial maximum than to that of the present day; and the balance between montane and lowland that we now see in Africa is the result of changes between about 16 000 and 8000 years ago. Of the actual extent to which evergreen forest occupied the montane block during any stage of the glaciation it is impossible to be sure.

The Sahara desert at one stage extended 300 miles farther south than it does now, but at others, and on the last occasion no more than around 6000 years ago, was to some extent clothed in its western half with vegetation, apparently of Mediterranean, not Ethiopian type; and it could then have interposed no important barrier to north-and-south exchange by many species of animals. Great lakes were formed in the southern Sahara somewhat earlier, in particular a vastly enlarged Chad lasting from at least 22 000 years ago until after 8000. This may have been associated with a northward shift or extension of the equatorial rainbelt and consequently of the forest. Changes in the eastern Sahara were probably much less marked than in the western. The present extreme desert conditions throughout the Sahara belt were developed after about 5000 years ago.

The evergreen forests of West Africa were reduced and fragmented during some period prior to about 22 000 years ago, those of the Congo basin to an enormous extent for a period prior to about 50 000 years and again to a limited extent around 12 000 years ago. At this latter stage Kalahari conditions extended so far north as to cross the mouth of the Congo River and perhaps to reach 5° S. in the east of the basin, while they also spread widely east, nearly to the Indian Ocean. This means that at that comparatively recent date the great block of *Brachystegia*, which today lies across the continent from Angola to Mozambique, was greatly reduced and broken up.

In sum, it is hardly possible to exaggerate the sweeping nature and, on the geological time-scale, the rapid succession and recent date of the ecological changes on the surface of Africa. Those we have been discussing form only the latest section of a succession that has filled the Pleistocene; but a species need be only some 20 000 years old to have witnessed the full range of the continent's ecological vicissitudes.*

* Since the foregoing was written, I have received the paper by Morton (1961) in which from West African botanical data he postulates that a drop of "about 4°C to 6°C" must have occurred in the climate of tropical Africa. In discussion J. P. M. Brennan commented on the lack of any morphological divergence, even at the varietal level, between the species of the Cameroon Mountain flora and those of East Africa–which leads to the further postulate that such a drop in temperature was comparatively recent. It will be seen how closely all this ties in with what has been said about the effects of the glaciation in tropical Africa.

4

The Bird Faunas of Mediterranean and Saharan Africa

In preceding chapters it has already been stressed that the vegetated belt of Mediterranean Africa is separated from tropical Africa by a desert of great severity about 1000 miles wide. The Mediterranean strip is itself discontinuous, for the Maghreb (=Barbary: Morocco-Algeria-Tunisia) and western Tripolitania are almost isolated from Cyrenaica by the desert reaching the southern end of the Gulf of Sirte; and Cyrenaica is connected vegetatively with the Nile Delta only by a narrow belt of low scrub. Nevertheless, it is of interest to discuss the zoogeography of Mediterranean and Saharan Africa as a whole; and this does not seem to have been done hitherto, although Heim de Balsac (1936) considered the mammals and birds of the north-western part.

For critical discussion it is convenient to divide the subject into three sections, and to deal with the Sahara, Egypt and the rest of Mediterranean Africa in that order. For this purpose I take the Sahara, as have other writers, to consist of that great area—nearly 3 million sq. miles—between the 100 mm isohyets, i.e. with an annual rainfall that averages less than 4 in. Actually most of the Saharan rainfall averages much less than this and its irregularity, with whole years devoid of any precipitation whatever, further accentuates the aridity. "Egypt" is here limited to that part of modern Egypt within a couple of hundred miles of the Mediterranean. This includes a zoologically important appendage to the Nile Valley, the Fayum depression (see below), but excludes Upper Egypt. This latter area is all Saharan except for the thread of the Nile Valley, which is virtually devoid of natural vegetation, and the Gebel Elba massif in the extreme south-east corner of the country, nearly 600 miles from the Mediterranean. This massif is biologically Ethiopian, its elements being a reduced form of those occupying the Red Sea Hills of the north-eastern Sudan. The "rest of Mediterranean Africa" I take as the non-Saharan parts of the Maghreb, including Atlantic Morocco, and of the Kingdom of Libya.

THE SAHARA

Perhaps the best general account of the Sahara as a whole is that of Capot-Rey (1953), but I do not think that any author has covered all parts of the vast area equally. A valuable ecological picture of part of the centre, through western Libya and the Fezzan to Tibesti, is given by Guichard (1955a).

Much of the literature that purports to deal with the Sahara is concerned with the area that includes such Algerian oases as Colomb Bechar, Touggourt and Ouargla, which are within about 300 miles of the Mediterranean and not much more than 100 miles short of the 100 mm isohyet; really, in fact, only within the fringe of the true Sahara. Even these oases, however, like all the others, are not dependent on local rainfall but on artesian water, derived from underground seepage from great distances and in part "fossil", a legacy of the less severe conditions of the late Pleistocene. By far the greater part of the oasis vegetation consists of date-palms, trees which have extraordinarily little to offer to birds.

Over most of the 3 million sq. miles of the Sahara conditions are altogether more rigorous than in this relatively favoured northern area. Oases are few and much more widely isolated from each other, and although over parts of the desert ephemeral vegetation flushes when, at intervals of years, some rain falls, in others there is no response because no seeds exist to germinate. A great deal of the whole surface is more or less flat, but three considerable mountain masses diversify the surface, Ahaggar, rising to 7000 ft, Tibesti (10000 ft) and Aïr (7000 ft). All contain some rock pools ("gueltas") in deep clefts, where water accumulates in the rare rains, but the vegetation even of the higher mountains is extremely poor and scanty. Only on their flanks and in the vicinity of some of the northern oases is it possible for any nomad grazing at all to be practised. Outside the mountains and the oases, the surface of the desert varies considerably. There are huge areas of shifting sand, especially the Sand Sea south of Siwa, with dunes rising to 300 ft, where no life is possible except on the occasional tiny outcrop of rock or on strips of solid desert temporarily swept clean by the winds. There are the flat pebbly deserts to the west of the Nile, in the northern part of which Bagnold (1931) saw two scraps of living vegetation in 270 miles, and in the southern part of which Newbold and Shaw (1928) in $6\frac{1}{2}$ days of camel travel "saw no signs of plant-life except for a few dozen age-old tufts of *nussa* grass". North-west of Timbuktu over a vast area of what Monod (1958, 1963) called "l'empty quarter", he found practically no trees, no bushes, no rocks and no birds, though Addax antelopes were not unknown, dependent on the scraps of harsh *Aristida* grass; and in 250000 km² not half a dozen species of Phanerogam. Again, in the 60000 sq. miles of the Ténéré, the low desert east of Aïr, in December, Heu (1961) saw no resident species of bird except rarely a falcon (which implies prey species), a chat *Oenanthe leucopyga* or a raven *Corvus ruficollis*. Sporadically, of course, the better parts of the desert carry a small population of drought-adapted lizards and small rodents, with fennec foxes to prey on them. Monod (1963) has given diagrammatic examples of food relations in deserts, with some birds included incidentally.

One part of what is climatologically the Sahara is untypical, namely the strip parallel to the Atlantic in the Sahara Español, so admirably discussed by Valverde (1957). Here, with a negligible rainfall, the dew is great enough to sustain a considerable bushy vegetation, where a few Mediterranean species, such as the Sardinian Warbler *Sylvia melano-cephala*, breed, into which even the Magpie *Pica pica* penetrates, and in the open places of which enough hares *Lepus* live to support a few pairs of Golden Eagles *Aquila chrysaetos*. This vegetated strip of the Sahara Español maintains a unique projection of Palaearctic species south-wards; none of them appears in the Saharan mountains.

When Heim de Balsac (1936) discussed the breeding birds of the Sahara he listed some seventy species, but by far the greater part of them, which include such birds as Blackbird *Turdus merula*, Reed Warbler *Acrocephalus scirpaceus*, Blue Tit *Parus caeruleus* and Little Ringed Plover *Charadrius dubius*, have nothing to do with the desert as such; they live only in ecological islands within it, and indeed most of them are represented only in the comparatively lush oases of the northern Algerian Sahara. There are also water-birds, especially the Moorhen *Gallinula chloropus* and Little Grebe *Podiceps ruficollis*, in the few spots of open water. By contrast with the abundance in the Algerian oases, the Egyptian oases of Siwa, Bahariya, Kharga and Dakhla (though the last of them is comparatively rich, with gardens of olive and citrus trees in addition to the usual date-palm groves) seem between them to have at most four resident birds: Little Owl *Athene noctua*, Rock Dove *Columba livia*, Palm Dove *Streptopelia senegalensis* and the chat *Oenanthe leucopyga* (Moreau, 1927, 1941; Meinertzhagen, 1930). (The Great Grey Shrike *Lanius excubitor* has been credited to Kharga Oasis by Meinertzhagen (1930), but I saw none there.)

In addition to the foregoing residents three species come to the oases in the summer in considerable numbers to breed: Turtle Dove *Strepto-pelia turtur*, Rufous "Warbler" *Erythropygia galactotes* and Olivaceous Warbler *Hippolais pallida*. One odd thing about this list is the absence of aerial-feeding birds. There seems to be no breeding swift, no fly-catcher, no hirundine and no nightjar. Another surprise is the accumula-tion of breeding doves, two species as a rule, but three in some oases; it would be interesting to work out the food relations of these birds. In any case the ornithology of the Egyptian oases as a whole should be re-examined; recent developments may well have induced changes.

The mountain masses rising out of the Saharan lowlands potentially form ecological islands of far greater size and importance than the oases. Moreover, the dry water-courses that lead off them show how important a part they played in the ecology of this part of Africa in the past. For example, the Wadi Igharghar leads all the way from Ahaggar to the schotts of Tunisia. In resident birds the mountains are, however, surprisingly poor. The worst seems to be Ahaggar: although it reaches 9000 ft and has "a remarkable extent of floral growth at the higher elevations", mostly of Mediterranean affinities (Meinertzhagen,

1934), it seems to support hardly any species of birds beyond those just mentioned and those regarded below as typically Saharan. Tibesti, an even more impressive mountain mass, has a somewhat less impoverished bird fauna (Malbrant, 1954; Guichard, 1955b; Tuck, 1959; Simon, 1965), which include several typical Ethiopian species, for example the bustard *Neotis denhami*, dove *Streptopelia (decaocto) roseogrisea*, the turdine *Cercomela melanura*, and the Golden Sparrow *Passer luteus*. The mountains of Aïr, less completely isolated from the acacia steppe which forms the northernmost vegetation belt of tropical Africa, have a much better bird fauna than the other mountains and are typically Ethiopian (Bruneau de Miré, 1957).

If now discussion is to be concentrated on typically desert birds, I think they should be limited to the twenty-five shown in Table I.* It will be seen that nearly all of them have a wide west-to-east extension, from the western end of the Sahara to Iraq, Iran or India, in fact through the Saharo-Sindian subtropical desert belt. A few reach Somaliland, but by edging down the desertic shores of the Red Sea, not by passing inland through the Sudan. Two of the species, the partridge *Ammoperdix heyi* and the chat *Oenanthe monacha*, stand apart from all the others in not extending west of the Nile, but for this there is a direct ecological explanation, since the rocky hills of great barrenness that they seem to need do not exist in the western desert of Egypt. The apparent gaps, within Egyptian territory, in the range of the three birds marked † in Table I are not easy to explain; I think the birds might be found if carefully looked for in the few spots that might be suitable for them. I should try the Gilf Kebir.

The allocation of this small typically desert bird fauna to a "Region" has bothered zoogeographers, and phytogeographers have been no less bothered by the flora (Monod, 1957). Heim de Balsac (1936) concluded that most of the bird species were "strange to the Palaearctic fauna", while Meinertzhagen (1954, p. 33) held the bird fauna to be "predominantly Palaearctic at the specific level" with by far the majority "purely Palaearctic at the generic level". Actually, for the group of twenty-five species listed in Table I no decided conclusion is justified at all; most of the birds do not provide evidence as to faunal affinities one way or the other and those which do show a fairly even balance. *Ammoperdix*, *Chlamydotis*, *Rhamphocorys* and *Scotocerca* are not represented outside the Saharo-Sindian desert belt and the fact that all are mono-

* *Calandrella rufescens* seems to have some claims to be regarded as a desert bird, but not enough. *Pterocles lichtensteini* is not included, because it seems to be known in the Sahara only from the Ahaggar mountains, while it occupies a wide range south of the desert as far as Kenya. Also *P. exustus (senegalensis)* is typically an Ethiopian bird, to as far south as Malawi and its penetration into Egypt is apparently in close dependence on the Nile. There are several cases of taxonomic difficulty: *Hirundo (Ptyonoprogne) obsoleta*, *Corvus ruficollis* and *Eremophila bilopha* have been accepted, following Vaurie (1959), as specifically distinct from *H. rupestris* (or *rufigula*), *C. corax* and *E. alpestris*, while *Bubo ascalaphus* would appear in Table I if one did not follow Meinertzhagen (1930) and Vaurie (1959) in treating it as conspecific with *B. bubo*.

TABLE I

The Sahara Desert bird fauna

(WS, West Sahara in the area Morocco–Mauretania)

	Western limit	Eastern limit
Group B		
Falco concolor	Tripolitania	? India
Group C		
Ammoperdix heyi	Egypt east of Nile	Muscat
Chlamydotis undulata	WS	Central Asia
Cursorius cursor	WS	India
Pterocles coronatus	WS	India
P. senegallus	WS	India
Group E		
Alaudidae		
Certhilauda alaudipes	WS	India*
Ammomanes cincturus	WS	India
A. deserti	WS	India
A. dunni	WS	S.W. Arabia
Eremophila bilopha	WS	Iraq
Eremopterix nigriceps	WS	India*
Rhamphocorys clot-bey	WS	N. Arabia
Corvidae		
Corvus ruficollis	WS	India*
Emberizidae		
Emberiza striolata	WS†	India
Fringillidae		
Rhodopechys githaginea	WS	India
Hirundinidae		
Hirundo obsoleta	WS	India*
Ploceidae		
Passer simplex	WS†	Transcaspia
Sylviidae		
Scotocerca inquieta	WS	India
Sylvia nana	WS†	Turkestan
Timaliidae		
Turdoides fulvus	WS	Red Sea
Turdidae		
Oenanthe deserti	WS	Gobi
O. leucopyga	WS	Iran
O. lugens	WS	Iran
O. monacha	Egypt east of Nile	India

* Also reaching down the Red Sea coast to Somaliland.
† Apparently absent from Egypt.

typic genera, i.e. with no close relatives, emphasizes their specialization to that area. *Falco, Corvus, Hirundo, Passer* and *Oenanthe* are well represented in both the Palaearctic and the Ethiopian regions. Only *Eremophila, Rhodopechys* and *Sylvia* are otherwise purely, or overwhelmingly, Palaearctic, while *Turdoides* is otherwise Indo-Ethiopian, *Eremopterix* and *Pterocles* predominantly Ethiopian, and *Ammomanes* has, in addition to its Saharan species, two in South West Africa. Clearly, at the generic level the list is as inconclusive as it could be. At the specific level, *Eremophila bilopha* and *Corvus ruficollis* are definitely of Palaearctic affinities, but the affinities of the other species are equivocal. These independent analyses give, in fact, full support to the conclusion of Darlington (1957) that the Saharan fauna as a whole is "very limited, specialized . . . with mixed relationships, part (most?) Ethiopian, and part Eurasian".

Most of the twenty-five desert species show no subspecific change between the Atlantic coast of the Sahara and the far side of the Red Sea, a distance of 3000 miles. This is easily understood if the amelioration of the western Sahara as late as the Neolithic period (Chapter 3) was great enough to exclude the desert species from that area; with the consequence that it has had to be repopulated in the course of the last 5000 years from a persistent reservoir of desert forms in the more consistently arid east of the Sahara (cf. the hypothetical isohyet map for the Neolithic in Butzer, 1958). The exceptional species are interesting. In the bustard and three of the chats different subspecies are recognized east and west of the Nile, and in only one of these, *Oenanthe leucopyga*, have transitional specimens been reported. These differences at first suggest that the Nile Valley has been of importance as a barrier, but this I doubt, because for most of its course south of the Delta the vegetated area is less than 10 miles wide. I am inclined to regard the subspecific differences as having arisen in the conditions of the Neolithic period (and/or the preceding humid periods of the Pleistocene); then, the mountainous desert east of the Nile would have been too well vegetated to be suitable for desert species and so would have interposed a much more formidable barrier than the Nile Valley itself, but the low shores of the Gulf of Suez and the Red Sea could well have provided a refuge for desert creatures.

One of the larks, *Ammomanes deserti*, typically associated with stony areas, shows high local variation in correspondence with the prevailing colour of the ground, a phenomenon which, as Meinertzhagen (1930) noted, is in interesting contrast to the unchanging plumage colour of *Ammomanes cincturus*, which is a bird of nondescript flattish desert. The little scrub warbler *Sylvia nana* is a special case in that its range falls into two hugely sundered halves, one the hinterland of Algeria and Tripolitania, the other 2000 miles away, from the Volga to the Gobi, all the more surprising because in winter these eastern birds move west as far as the Nile Valley.

Not all the species comprising Table I are everywhere confined to the desert. For example, in the western Sahara the bunting *Emberiza striolata* (one of those which seems to miss Egyptian territory) has become locally a village bird. The chat *Oenanthe leucopyga*, at home in the desert wherever there is broken ground, has in the oasis of Siwa become *hajjemoleyn*, the "friend of the house" to the inhabitants, singing from the piled-up dwellings of mud and palm trunks like a starling in England. In Dakhla Oasis, too, I found it frequenting the houses, but in Baharia it seemed to prefer the cemeteries, over 3000 years' accumulation, on the edges of the oasis.

The converse of such opportunism, displayed by typically desert species, is provided by certain birds of prey. *Bubo bubo*, though normally a bird of well-vegetated country, has penetrated the Sahara thoroughly, especially if *ascalaphus* is regarded as conspecific. Again, Lanner Falcons *F. biarmicus* and Peregrines, which are not typically desert birds, not only frequent the neighbourhood of oases but have been found breeding on fragmentary outcrops of rock in full desert, even in the Sand Sea, by the man who later became Major-General Orde Wingate and also by P. A. Clayton (Moreau, 1934a), who was later in the Long Range Desert Group. For a discussion of these falcons' ecology see Niethammer (1959) and Jany (1960). It is surprising that they should have found the food supply sufficient to attract them to such surroundings, unless they depend on the spring migrants and depart after breeding. Presumably they do so and thus alternate seasonally with another falcon, *concolor*, which has a breeding range from the latitude of Cairo to the southern Red Sea, and unlike the Lanner is not an opportunist but a specialist confined to the Sahara and to desert islands (Clapham, 1964). In the Sahara it is capable of breeding in the hottest time of the year and in the utmost extremity of desolation, as shown in a photograph by Booth (1961). In a tiny cairn of stone slabs alongside what was once a camel-track at about 26° N., east of Kufra, he found a pair with eggs in August 1957. The surrounding desert is "one vast featureless gravel sheet . . . It is devoid of all vegetation, either living or dead, and its utter barrenness is complete." Passing that way in the spring of 1961, Jany (1963, and *in litt.*) found Lanners breeding in the same cairn or one very like it. Nothing could suggest more strongly that this falcon and the Lanner alternate in their depredations on the migrants, the former depending on the spring movement and the latter on the autumn.

It appears that *concolor* in its breeding season resembles another falcon, *eleonorae*, of rocky islets in the Mediterranean and off the Atlantic coast of Morocco, hitherto believed to be the only extra-tropical Old World species breeding regularly well after the summer solstice and at no other time of the year. It has been thoroughly established that *eleonorae* lays in the latter half of July and early August, and while the adults eat insects as well as birds, the young are fed "almost entirely on small birds" (Vaughan, 1961). Evidently much the same applies to *concolor*, since this species has been found breeding in the late summer also on the Dakhlac Islands off Eritrea (Clapham, 1964), and there can be little doubt that the extraordinary breeding season of these two falcons is adapted to take advantage of the temporarily abundant food supplied by the passage of the autumn migrants from Europe and Asia to the African tropics. From about November to April both species are present in Madagascar, where at least *concolor* is insectivorous (Stresemann, 1954; Rand, 1936). Although all the birds concerned presumably pass through Africa both coming and going, there is so far no satisfactory evidence that any of these falcons spend the off-season on the African continent. The entire "world" population of both species must be very small, and it would seem that they are entirely dependent for winter quarters on the island of Madagascar (for which see Chapter 17). In this respect also *F. eleonorae* and *F. concolor* have no parallel.

Little seems to be known in detail about the distribution of the sand-grouse in the Sahara. It probably varies to take advantage of the occasional local downpour and consequent temporary standing water. But in much of the Sahara conditions probably never are good enough and certainly in any given year the sandgrouse must be very widely scattered, because they seem to be under the necessity of drinking regularly and their radius from accessible water may not be more than about 50 km (Valverde, 1957). There is one other species whose distribution in the Sahara we shall now never know, the Ostrich *Struthio camelus*. I left it out of Table I because south of the Sahara the species ranges widely over steppe and open savanna in tropical and southern Africa. Probably under early Neolithic conditions much of at least the western Sahara suited it well and in the extreme east as late as the end of the eighteenth century, when Napoleon invaded Egypt, the men of his army hunted ostriches between Cairo and the Bitter Lakes. It would be particularly interesting to know the ecology of the apparently dwarf Ostrich which, if the ancient Egyptian bas-reliefs are to be believed, was so thoroughly domesticated that it could be patted on the rump (bas-relief figured in Meinertzhagen, 1930, p. 71).

Lower Egypt

It is useful here to amplify the incidental references to Egypt in the general sketch of Africa and the sequence of past climates that have been given in the introductory chapters. At the present time the poverty of natural habitats and of marginal habitats in Egypt is great. Except for the narrow coastal belt of scrub between Alexandria and the frontier of Cyrenaica, and the northern Delta, where there are alkaline flats with patches of tamarisk as well as beds of rushes and reeds, Lower Egypt is a country of violent contrasts, utterly clean agriculture, dead flat, against extreme desert. Moreover, nearly all of the 27 million people are crammed into the 13 500 sq. miles reached by the Nile waters (Church *et al.*, 1964). Only in the Fayum depression is there such a thing as a slope of land that is green. This is possible because Lake Moeris, which once filled the depression to about the level of the sea or a little over, has become so shrunken that the surface of the relict, now called Lake Karun, stands today at 45 m below; with the result that most of the old lake-bed, with its gentle slopes and valleys, is now under crops (irrigated like all others in Egypt). Elsewhere south of the Delta the width of the "sown" averages less than 10 miles, between the mountainous desert on the east and the flattish rolling desert on the west. In these wastes, apart from the major oases, which have been referred to incidentally in the discussion of the desert birds, there is a minor one, the Wadi Natrun, which is hardly more detached from the Nile than the Fayum, and which occupies a special place in Egyptian ornithology. It is a shallow depression, which contains some seepage of water, at its eastern end less than 30 miles from the western edge of the Delta. The seepage maintains a chain of red alkaline lakes, thickets of rushes and tamarisk and incidentally a group of Coptic monasteries.

In the Nile Valley and the Delta the habitats are mostly limited to crops, palm groves, which have remarkably little to offer small birds, and watery vegetation. As listed in the flora of Egypt (Täckholm et al., 1956), besides acacias and tamarisks, there are not half a dozen species of native tree; in fact, the only ones likely to be noticed are the sycamore fig and the zizyphus, and there are neither woods nor thickets. This poverty of species is worse than it was under the more humid conditions of the Neolithic, but even then the flora was not in the least comparable with that of the Maghreb. For this period Butzer (1959b) reported only four additional species of tree (three of them of Ethiopian origin), although in one way and another prehistoric plant remains are not scanty. There were, however, at that time only some 7000–5000 years ago, much less exiguous habitats lining the Nile, seasonally inundated flats "with a lush grass and shrub vegetation" and also, especially on the raised banks of the water-courses, groves of acacias, tamarisk and sycamore. Not only are these gone but also the papyrus, now extinct in Egypt.

For compiling a list of the birds breeding in Lower Egypt today the records in Meinertzhagen (1930) can be used,* but for the present purpose we want to include any other birds for which there is evidence of a population in the recent past. Species in this category are the Sacred Ibis *Threskiornis aethiopica*, a favourite subject for mummification in dynastic Egypt and a survivor in the Delta until 1876, the cormorant *Phalacrocorax africanus*, gone since 1875 from Lake Karun in the Fayum, where it was over 1000 miles from the bird's present African range, and the hawk *Micronisus gabar*, of which two mummies have been recorded. There is also the Shoebill (= Whale-headed Stork) *Balaeniceps rex*, which I believe, as I said in Meinertzhagen's book, is identifiable in one of the Fifth dynasty bas-reliefs. Additional evidence has recently been provided by the publication of another relief by von Rosen (1962). These examples suggest that Lower Egypt in the recent past, and certainly some 5000 years ago, may have abounded in Ethiopian species of birds.† Their

---

* The great increase in population and in "development" since 1930 may have induced changes in the birds, but there has been no documentation.

† For the "Egyptian Plover" *Pluvianus aegyptiacus* (actually one of the Glareolidae, not one of the Charadriidae), there is no evidence of occurrence in Lower Egypt, though it was not uncommon in Upper Egypt in the last hundred years; and the Egyptians with whom Herodotus came in contact were familiar with the bird.

occurrence so far north at that time would certainly have been favoured by the temperatures, bcause it seems generally agreed that the world as a whole was then warmer than it is now, the Northern Temperate zone perhaps even by 2–3°C. Moreover, the Nile and the vegetation on its borders, at that epoch hardly modified or devastated by human beings, would have offered a good channel of communication (as today tropical families of fish can follow the river north nearly to its mouth; Darlington, 1957, p. 56).

Another species, not typically Ethiopian, the Hermit Ibis *Geronticus* (=*Comatibis*) *eremita*, merits special mention, for it provides a classic case of relict distribution. It inhabited Austria, Hungary and Switzerland as late as the sixteenth century, but today the only known breeding colonies are in Morocco (recently Algeria also) and on the upper Euphrates, though it is likely that there are others, yet to be located, near the southern end of the Red Sea. The ancient Egyptians were sufficiently familiar with the bird to use it as a hieroglyphic, and I have given (in Meinertzhagen, 1930) reasons for believing that this is the ibis that Herodotus describes as warring against the "flying dragons" (which I hold to be locusts). The eastern desert of Egypt, with its great rock faces, would have provided an ideal habitat for these birds during the Neolithic humid period and for at least a considerable period thereafter. Now, it seems likely that everywhere in the Egyptian deserts except in the extreme south-east, around Gebel Elba, desiccation has gone too far for these birds to survive. Nevertheless, on 8 May 1921, in what was one of the most remarkable ornithological experiences of my life, I came on eight of these Hermit Ibises on the edge of the desert near Giza. They were 600 miles from the nearest known or suspected colony, that on the Euphrates. It is just possible that they were a remnant from some secluded corner among the mountains of Sinai or of the eastern desert, perhaps the Gelalas, south-west of Suez.

The breeding bird fauna of Mediterranean Egypt, thus arrived at, has an odd composition: at the outside about fifty-seven water-birds, thirteen raptors and scavengers, five game-birds, eleven group D non-passerines and only nineteen passerines. The proportions in each category are thus 0·35, 0·17, 0·06, 0·16 and 0·25, which compare with 0·18, 0·16, 0·06, 0·12 and 0·48 in Morocco, for example (see below). The most striking feature is the paucity of passerine species breeding in Egypt; moreover, of the nineteen species two, the lark *Chersophilus duponti* and the chat *Oenanthe moesta*, are intruders from Cyrenaica into the narrow belt of scrub between the Mediterranean and the western desert of Egypt, while three others seem to have only a precarious foothold within Egyptian limits: the Short-toed Lark *Calandrella cinerea* only in the Wadi Natrun, perhaps merely a temporary occupation and certainly in very small numbers; the Goldfinch *C. carduelis* round Cairo; and the Sardinian Warbler *Sylvia melanocephala* as an endemic subspecies relict in the tamarisks on the edge of the lake in the Fayum.

The differences between the Egyptian and Moroccan bird faunas will be further discussed in a later section. To ascertain the affinities of the Egyptian bird fauna it may be divided between (1) those species which are typically Palaearctic, with little or no representation in the Ethiopian Region, south of the Sahara; (2) those typically Ethiopian, though they may reach the north shore of the Mediterranean in places; (3) those which are widespread in both regions. The results of this analysis are given in Table II, which shows how much the five groups of birds differ in their allocations. In particular the water-birds are equally divided between the three categories, while only one-quarter of the passerines are Ethiopian and three-quarters are Palaearctic.

TABLE II

The affinities of the Egyptian bird fauna

| | No. of species | | |
Group	Palaearctic	Ethiopian	Not allocable to either
A	9	9	9
B	4	3	6
C	3	1	1
D	5	4	2
E	14	5	—
	—	—	—
	35	22	18

THE REST OF MEDITERRANEAN AFRICA

Comprehensive lists of the birds of Mediterranean Africa have recently become available in the works of Heim de Balsac and Mayaud (1962) and of Etchécopar and Hüe (1964). Unpublished records kindly communicated by Major B. D. MacD. Booth add a Swift *Apus pallida* and the Hoopoe *Upupa epops* to the list of Cyrenaican breeding birds.

All the Maghreb States have a wide range of habitats, from desert through cultivation and macchia of various types to mountain forests and, in Morocco, open country above the timber-line. Among the relevant literature, which is fully listed by Heim de Balsac and Mayaud, attention should be drawn to the especially ecological studies of Snow (1952), who worked the mountain forests that form an extensive disconnected series from Tunisia to Morocco, and of Brosset (1961), who was resident in the north-east corner of Morocco. Through the greater part of the Maghreb States a belt of vegetated country 200 miles or so wide intervenes between the sea and the desert, and in all of them opportunities for the breeding of water-birds have been considerable. East of Tunisia the whole ecological aspect changes. There are no mountains, the desert everywhere approaches to within a few miles of the sea. Natural tree vegetation almost ceases to exist and birds not adapted to desert or very dry steppe find their only habitat in a coastal strip, which eastward becomes a line of disconnected patches of palms, with some olive, citrus and grain cultivation. In eastern Tripolitania desert reaches the sea on a wide front, but on the east side of the Gulf of Sirte, in western Cyrenaica, the terrain becomes more varied, with the plateau of the Gebel Akhdar rising to some 3000 ft and carrying relict juniper woods. Eastwards again to the Egyptian frontier the country becomes once more low and dry, with the vegetation reduced to a narrow coastal strip of sparse low scrub. Freshwater habitats between Tunisia and the Egyptian Delta, a distance of 1300 miles, do not exist except for a few hundred acres in Tripolitania and a few score in Cyrenaica.

The great difference in range of habitats as one travels from west to east through Mediterranean Africa is naturally reflected in the number of bird species. Excluding species in Table I, Morocco has about eighty passerine breeding species, Tunisia sixty-nine, Tripolitania twenty-five and Cyrenaica twenty-six. All the species in the last three territories also breed in Morocco, so that, in fact, the Tripolitanian and Cyrenaican bird faunas consist merely of fragments of that of the Maghreb. It is interesting that three woodland species, which are on the list for Cyrenaica but not for Tripoli, Blue Tit *Parus caeruleus*, Chaffinch *Fringilla coelebs* and Wren *T. troglodytes* all survive there in the relict juniper woods, of which there is no equivalent in Tripolitania.

If the bird fauna of the Maghreb as a whole is analysed in the same way as that of Egypt on a preceding page, the result is as in Table III.

TABLE III

The affinities of the Maghreb bird fauna

Group	No. of species Palaearctic	Ethiopian	Not allocable to either
A	17	2	15
B	14	4	12
C	7	4	1
D	15	3	5
E	79	5	2
	132	18	35

It will be seen that, taking all the species together, only one-fifth are not allocable to one "region" or the other, and that in every group of birds the typically Palaearctic species greatly outnumber the Ethiopian. Indeed, excluding desert species, the bird fauna of the Maghreb is remarkably like that of Spain. It has only one endemic species, the turdine *Diplootocus moussieri*, which has resemblances to both *Saxicola* (the Stonechat genus) and *Phoenicurus* (the Redstart genus). Its total of 185 breeding species lacks only eight that breed in the southern half of Spain and includes only thirteen that do not. Most of these thirteen are typically Ethiopian, with a big range south of the Sahara, and these are further considered in the discussion section below. These facts indicate the small significance of the water-gap in the bird geography of this part of the world. However, its influence is not entirely negligible, because, of the eighty-six non-desert passerines breeding in the Maghreb, twenty-six have been accepted as subspecifically distinct from the Spanish populations (Vaurie, 1959), although often slightly, compared with only nine that show subspecific differentiation within the range Morocco–Tunisia.

Half the Ethiopian species that reach the Maghreb exist only in Morocco (Table IV), but this may be in part due to local extinctions in recent historical times, since at least one of the species, the owl *Asio capensis (helvola)*, was known in Algeria a hundred years ago. All these Barbary populations of Ethiopian species are isolated by some hundreds of miles from their relatives south of the desert. The four passerines, the Rufous "Warbler" *Erythropygia galactotes*, bulbul *Pycnonotus barbatus*, shrike *Tchagra senegalus* and martin *Hirundo paludicola*, together with the guinea-fowl *Numida meleagris* and the francolin *F. bicalcaratus*, have all been distinguished subspecifically. They may well have been isolated only since the post-Neolithic desiccation of the western Sahara, i.e. for some 5000 years.

Besides providing the extreme north-western outpost of a number of Ethiopian species, the Maghreb, and in particular Morocco, serves as a refuge for a number of other species. The Hermit Ibis has already been referred to, with its Moroccan population apparently sundered from that on the Euphrates by over 2500 miles. The *Rhodopechys sanguinea* of the tops of the Grand Atlas have their nearest neighbours on the Lebanon. They surely represent a relict of a more extensive range during the last glaciation. However, even then the whole 1500 miles from the Tunisian mountains to Sinai would have been unsuitable for the bird, so that the most likely communication between these populations would have been round the north of the Mediterranean and through the plateau of Asia Minor. The relict population of the Demoiselle Crane *Anthropoides virgo* in the Maghreb is separated by 1600 miles from the nearest survivors in recent historical times, close to the mouth of the Danube.

General Discussion

The analysis above of the depauperate Egyptian bird fauna gives indeterminate results, except that the affinities in the water-birds are equally Palaearctic and Ethiopian, but analysis of the Maghreb bird fauna indicates a preponderance of Palaearctic over Ethiopian in every group, and on the whole bird fauna a ratio of 7 : 1. This situation is strikingly at variance with that for the mammals. For the Maghreb, Heim de Balsac (1936) listed 103 species of mammals, which included some typical of the desert. Many cannot be allocated to either the Palaearctic Region or the Ethiopian, but of the remainder he regarded only eighteen as "European", i.e. Palaearctic, against forty-two African.

TABLE IV

Ethiopian species reaching Mediterranean Africa

(Records of locally extinct species are given in parentheses)

	Morocco	Algeria	Tunisia	Egypt*
Group A				
Alopochen aegyptiaca†				x
Ardeola ibis	x	x	x	x
Balaeniceps rex				(x)
Charadrius pecuarius				x
Fulica cristata	x	x	x	
Hoplopterus spinosus				x
Phalacrocorax africanus				(x)
Porphyrio porphyrio‡	x	x	x	x
Rostratula benghalensis				x
Threskiornis aethiopica				(x)
Group B				
Asio capensis	x			
Elanus caeruleus	x	x	x	x
Falco biarmicus	x	x	x	x
Melierax metabates	x			
Micronisus gabar				(x)
Group C				
Burhinus senegalensis				x
Choriotis arabs	x			
Francolinus bicalcaratus	x			
Numida meleagris	x			
Turnix sylvatica	x	x	x	
Group D				
Apus affinis	x	x	x	
Centropus senegalensis				x
Ceryle rudis				x
Clamator glandarius	x	x	x	x
Merops orientalis				x
Streptopelia senegalensis	x	x	x	x
Group E				
Erythropygia galactotes	x	x	x	x
Cisticola juncidis	x	x	x	x
Nectarinia metallica				x
Prinia gracilis				x
Pycnonotus barbatus	x	x	x	x
Hirundo paludicola	x			
Tchagra senegalus	x	x	x	

* It is possible that the Skimmer *Rynchops flavirostris* should be included in the Egyptian list, as it may have been resident on the lakes of the Delta in the nineteenth century (cf. Allen, 1864).

† There are records of this species in the Maghreb, but only odd ones in winter, which are impossible to account for.

‡ Includes *P. madagascariensis* as conspecific.

If bats are excluded these figures become ten and forty, increasing the Ethiopian proportion to four-fifths. The fossil record (summarized by Cooke, 1962) indicates that a preponderance of Ethiopian mammals has persisted throughout the Pleistocene. Meanwhile, some time prior to about 20 000 years ago—presumably during a glacial maximum—there was an invasion of a few Palaearctic species (e.g. deer (Cervidae), a species of which still survives in the Maghreb, and a bear *Ursus*, now unrepresented anywhere in Africa). Heim de Balsac argued for their coming along the Mediterranean coast from the east (cf. Arambourg, 1962) rather than directly from the north; there is in any case no evidence for a land-bridge, by way of either Gibraltar or Sicily, in the latter part of the Pleistocene (cf. Fairbridge, 1962).

For Egypt, Flower (1932) listed the existing mammals, but no analysis of them has been published. R. W. Hayman has, however, kindly made a rough assessment for me, and again the result shows predominance of Ethiopian affinities. Moreover, in neither the Maghreb nor Egypt do the modern faunal lists include some of the most characteristic Ethiopian mammals that reached Mediterranean Africa and survived there until well after the Neolithic period. According to Hopwood (1954), buffalo, elephant, giraffe, hippopotamus and rhinoceros disappeared from the Maghreb, where localities providing evidence for their occurrence have been plotted by Mauny (1956), only at dates varying from 2000 to 300 B.C. Most of the large Ethiopian mammals seem to have disappeared from Egypt rather earlier than from the Maghreb (Butzer, 1959a), though the hippopotamus lingered in the Delta until 300 years ago. In general, however, it is certain that in the recent past the proportion of typically Ethiopian mammals on the southern shore of the Mediterranean was even higher than it is now.

The question that now arises, and in the absence of fossils cannot be answered, is whether in Mediterranean Africa Ethiopian species have ever preponderated in the birds as they have in the mammals. Presumably at those stages when conditions were suitable for mammals to get across the Sahara, as they evidently were more than once during the Pleistocene and as late as the Neolithic, they were no less suitable for birds. The surviving guinea-fowl and francolin in Morocco show that the crossing was accomplished even by the most sedentary species, which are unlikely to have done it by any process other than gradual expansion of range. If there were indeed massive northward movements of Ethiopian birds, as there were of mammals, why have the birds gone again?

The discussion could be carried further if we knew more of the vegetational history of the Maghreb, but this is most poorly documented, with only two relevant fossil floras described (Quézel, 1957). Of one in Tunisia ascribed to the Villafranchian (at the beginning of the Pleistocene), 26% consists of Ethiopian elements not now found in the Maghreb, 52% Mediterranean (still in Tunisia) and 21% more northern and no longer in the Maghreb. The other fossil flora, ascribed to the second glaciation, suggests a climate more like that of Provence today, with no Ethiopian clcmcnts at all. These findings, of course, leave open the possibility of long periods when Ethiopian plants could have predominated in the Maghreb, especially during interglacials. At present the plants of the Maghreb are purely Palaearctic, apart from an acacia and a couple of other species of Ethiopian affinities, as I am informed by R. D. Meikle and Dr J. P. M. Brenan of the Kew Herbarium, Surrey, England. Moreover, we know, as mentioned in Chapter 3, that late in the Pleistocene the present temperate forest of the Algerian mountains had crossed the northern Sahara at least to Ahaggar and some less moisture-loving Mediterranean elements as far as Tibesti and even the latitude of Lake Chad. It is reasonable to suppose that Palaearctic birds, at any rate of woodland, went with these Mediterranean plants; and it seems certain that for some time in the late Pleistocene there was in north-western Africa a pleasing conjunction of such birds as blue tits with elephants and magpies with white rhinos. They may indeed have coexisted all across what is now the western Sahara and have disappeared from the lower latitudes only when first Ethiopian vegetation and then desert supervened around the end of the Neolithic, some 4000 years ago, as described in Chapter 3.

But the central puzzle of the unconformity between the bird fauna and the mammals in the Maghreb in recent times, and perhaps all through the Pleistocene, is why the Ethiopian mammals have been able to hold their ground there through the climatic vicissitudes of the glaciations, while the Ethiopian birds have not—if indeed they ever established themselves there. At first sight a possible explanation might be found in the persistence of the water-barrier on the north, narrow as it is. As Arambourg (1962) has pointed out, no typically African species of mammal appears to have reached Spain during the latter part of the Pleistocene except the hippopotamus; but on the other hand, as noted above, the Palaearctic bats, the mammals most able to cross a water-gap, have entered the Maghreb so freely that, like the Palaearctic birds, they predominate there. But then, according to Darlington (1957), the amphibians of the Maghreb are also predominantly Palaearctic, while its "very limited" freshwater fish fauna is "closely related to that of Europe"; and neither of these two classes crosses salt water easily. The puzzle is likely to remain unless the discovery of more fossils can solve it.

It is interesting to consider the distribution of typically Ethiopian birds breeding in Mediterranean Africa. In Table IV occurrences are given in parentheses where a species is locally extinct or its breeding is not certain. Two other Ethiopian species are virtually certain to have lived in Egypt in the not distant past: the swift *Apus affinis*, which lives in Palestine, and the darter *Anhinga rufa*, which has survived in the Jordan Valley (Lake Huleh) and is most unlikely to have got there except by way of the Nile. It may be also that *Aquila verreauxii* should be added to the Egyptian list. Meinertzhagen (1930) had a single record of a specimen captured on the Suez Canal in 1916 and mentioned also a Palestine record. These are about 1000 miles from the nearest part of the bird's tropical African range (Sudan and Abyssinia) and they have been difficult to evaluate. However, recently Undeland (1964) observed one of these unmistakable eagles in Sinai and two nests have been found in Israel (A. Zahavi, personal communication). This eagle is everywhere characteristic of rocky hills, where hyraxes (*Procavia, Heterohyrax*) are a staple food. These animals are obtainable in the mountains of Sinai and Palestine and probably also sporadically in the mountains between the Nile and the Red Sea. Therefore I suspect that a very scanty population of *A. verreauxii* may have been maintaining itself there. This makes an interesting counterpart to the Golden Eagles *Aquila chrysaetos* (a typically Holarctic species), living on hares in the open steppe-desert of the Sahara Español, as mentioned on p. 63, *verreauxii* being perhaps a relict of the warm humid period of the Neolithic and *chrysaetos* of the glaciation.

It will be seen from Table IV that the difference in Ethiopian representation at the two ends of Mediterranean Africa is remarkably strong, for of the thirty-three species concerned only ten are common to both Morocco and Egypt. There is no instance of an African species in either area being replaced by a closely related Palaearctic one in the other, except that on the classification current until recently this would have applied to the *Porphyrio* populations. The magnitude of this difference is probably due to two factors. One has no doubt been differential extinction, in correlation with changes in climate and habitat destruction by human means. The other is the different means of access from the Ethiopian reservoir of species, for the western Sahara has several times been sufficiently ameliorated for animals to get across, while in Egypt the Nile has provided a permanent though tenuous line of communication, especially for water-birds. The river connexion explains the occurrence in Egypt of nine water-bird species out of a total of twenty-three Ethiopian, compared with three out of nineteen in the Maghreb. With these Egyptian water-birds should be bracketed the coucal *Centropus senegalensis*, which in the tropics typically inhabits swamp vegetation, though it does without it in its modern Egyptian refuges. On the other hand, it remains a mystery why the Crested Coot has not been found in Egypt; and why such birds as the Senegal Stone Curlew, the Spurwing Plover and *Charadrius pecuarius*, inhabiting Egypt, should not be on the list for the Maghreb; they should have had no difficulty in reaching there, at any rate up the Atlantic coast during one of the periods of ameliorated conditions.

It will be seen that only one Ethiopian species of group C, the Senegal Stone Curlew, is on the Egyptian list, compared with four in Morocco. This may be due to habitat destruction, for it is inconceivable that the guinea-fowl or francolin could survive in modern Egypt now, and it is possible that even in Old Kingdom times, around 4000 years ago, the sort of scrub they require was not adequately provided. In the birds of groups D and E the differences in the lists of the two areas seem unaccountable except that *Ceryle rudis* evidently owes its northward extension in Egypt to the Nile.

Reviewing the whole picture of the North African bird faunas in the light of the Pleistocene ecological changes, it seems that when vegetation extended over much of the Sahara there would have been refuges for the desert species at least in the eastern half, from which they could spread once more west to the Atlantic as soon as aridity returned. This is what has presumably happened in the last 5000 years. During the "good" spells north and south expansion of range across the western Sahara would have been easy for many species; further, Mediterranean vegetation appears to have extended south to the latitude of Lake Chad and thus to have been in contact with Ethiopian vegetation on a broad front across what is now sub-Saharan West Africa. These premises leave us with two puzzles. One already considered is the unconformity between the bird fauna and the mammal fauna of Mediterranean Africa, especially of the Maghreb, with its overwhelmingly Palaearctic birds and predominantly Ethiopian mammals. The other is why the bird faunas on the northern and southern edges of the Sahara should differ so profoundly, only thirty-three of the 1481 Ethiopian species (Table V in Chapter 5) appearing in Mediterranean Africa. If indeed a Palaearctic (Maghreb) bird fauna and an Ethiopian bird fauna were in contact along a broad front, then there are two possibilities. Either more interchange took place than is now evident and there has been subsequent elimination, or the interchange was never much more than we see now (with, for example, only five Ethiopian species among the passerines of the Maghreb). In the first case it is conceivable that a community composed of a mixture of species indeed existed temporarily in a belt across what is now the Sahara; that the site of it was overtaken by the post-Neolithic aridity presumably spreading from the east; and that the survivors were increasingly pushed apart, northwards and southwards. The southern birds pushed to the north of the arid wedge would then have been isolated among northern species, in an environment of Palaearctic vegetation and would shortly have been faced with the drop in temperature of some 2°C from the hypsithermal. These factors are all likely to have been unfavourable to them and in case of competition between northern and southern species in Mediterranean Africa the southern are the more likely to have suffered. South of the Sahara the opposite applies, except that the temperature factor would not be operable. On the other hand, if little or no interchange of species took place when Palaearctic and Ethiopian bird faunas were in contact, this could have been because each community was too closely integrated, saturated, and hence unreceptive. No species can establish itself in a community unless a niche becomes available for it, either because it is unoccupied or because the new-comer can compete successfully with a species already there. Neither condition was likely to be satisfied when the communities in question, those of the Mediterranean (Palaearctic) scrub and woodland and those of the Ethiopian thorn steppe or savanna, were continental and had presumably been developing for a long time. It seems to me therefore that on present evidence either hypothesis to account for the present striking difference between the bird faunas north and south of the Sahara is tenable; while, since the two hypotheses are not mutually exclusive, they could both have contributed to the present avifaunal situation.

5

The Composition of the Ethiopian Bird Fauna

The families represented by breeding birds south of the Sahara and the number of species in each are given in Table V, with their distribution between certain broad ecological categories.* In case some of the families in groups D and E are not familiar, it seems worth while to give a summary indication of their food habits, as in Table VI. For each family that type of food which is judged to be at least as important as any other is shown by a solid line; that which is a minor source of some importance for the family generally and/or the staple source for a section of it is shown by a broken line. Food sources that are believed to be incidental or utilized by only an inconsiderable proportion of a family—as ground invertebrates by *Geocolaptes* alone among twenty-nine species of woodpecker—are not specified. Table VI is, of course, extremely crude and overgeneralized and is not to be taken as quantitative in any way, except that it brings out that fruit is of prime, or considerable, importance to a wider variety of birds than seeds. It might be possible to draw up a more elaborate chart of this kind on the basis of genera or even species and a more detailed subdivision of food sources, but this will serve for the present purpose.

In pursuance of the scheme outlined in Chapter 1, in Table V a primary division is made between the birds of evergreen forest and the others, and a secondary division between lowland and montane. I have accepted as forest species those whose typical habitat is described in the standard works as primary or secondary forest (or both), whether or not they extend in some degree into "gallery" or "fringing" forest. But I have not counted as forest birds those species which typically inhabit this last vegetation type (which is normally dependent on ground-water rather than rainfall) and do not inhabit rain-forest (whether primary or secondary); here I am actuated partly by my personal experience that typical rain-forest birds are lacking in ground-water forest; for example, in northern Tanganyika, from the Rau and the Kahe at the foot of Kilimanjaro, and the Gonja at the foot of the South Pare Mountains. A similar reservation applies to some of the records in the current Northern Rhodesian list (Benson and White, 1957), in which

---

* Table V was compiled when no up-to-date comprehensive list of African birds existed and I drew on several sources. The revised check list recently completed by White (1961–65) arrives at slightly different totals of the number of species in some families, owing mainly to adjustments of species limits; but the totals for the different categories of birds distinguished in Table V are practically unchanged.

some species are ascribed to "evergreen forest" that are not typically birds of rain-forest elsewhere. The explanation is that, as can be seen from the description on page xiii of the Northern Rhodesian list, these "evergreen forests" (locally called *mushitu*) are no more than patches in damp depressions, "often forming . . . belts as much as 100 yards across". They are, in fact, of ground-water type and extremely small. In Zambia there are, moreover, other evergreen formations, *Cryptosepalum* and *Marquesia* woodlands, which are anomalous, since, although dominated by evergreen trees, "they lack the physiognomic features of tropical rain forest" (Cole, 1963).

Species typical of secondary rain-forest raise a special evolutionary problem, to be discussed later; and much the same problem is raised by a few that seem to be confined to forest edges. I have therefore grouped secondary and edge species in a separate minor category of Table V. A few species, described as typically found in "clearings", although by inference in proximity to rain-forest, I do not classify as forest birds, since they are not dependent on forest vegetation but on the rank growth induced by the humidity prevailing in those localities where forest is dominant.

As mentioned in preceding chapters, the altitudinal change in biota occurs at somewhat different levels in different mountainous areas of the African tropics. Chapin (1932), thinking particularly of birds in the Congo, placed it around 1500 m and, apart from certain abnormalities, this level appears to be critical at least as far south as Malawi at 16° S. At higher latitudes than this the few species typical of montane areas in the tropics that still occur so far south come lower and finally reach nearly to sea-level in the Knysna forest of Cape Province.

Two tropical localities that show abnormal relations between temperature and the montane/lowland boundary are the East Usambara Mountains, about 50 miles inland in the north-east corner of Tanganyika, and Mt Cameroon close to the Gulf of Guinea. Both these massifs have an exceptionally high rainfall and humidity. Figure 26 gives relations between altitude and temperature for the East Usambara Mountains in comparison with those for Kenya generally. It will be seen that while the hot season maxima are high in the East Usambaras, the cool season minima as low as 900 m above the sea are the same as they are in Kenya at nearly 1500 m. Conformably, in the East Usambara Mountains montane bird species predominate (as residents) as low as 900 m (Moreau, 1935a), though some lowland species also persist there and not all the montane species from higher altitudes in the neighbouring West Usambara Mountains occur. The East Usambara forest is floristically rich and peculiar. Few typically lowland trees exist and the most characteristic montane species are also absent, but there are numerous endemics. On the south-east face of Mt Cameroon, where at 900 m the temperature is slightly lower than even that in the East Usambaras, Eisentraut (1963) and Serle (1964) have found that montane species come even lower than in the East Usambaras. Serle has shown that eleven of these occur regularly, and not merely seasonally, below 2000 ft (a little over 600 m), though hardly any below 1700 ft. He is inclined to connect this abnormality in altitude range with the great prevalence of heavy mist in a belt down to about 1700 ft, but neither he nor Eisentraut discusses the floristic composition of this forest and it would be interesting to know whether it changes in the same way as the birds.

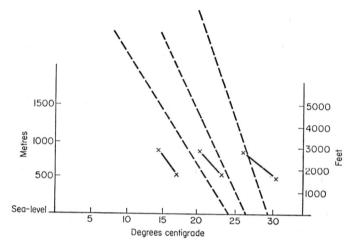

Fig. 26. The relation of temperature to altitude in certain African areas. Solid lines, East Usambara Mountains: left to right, mean minima of three coolest months, annual mean temperature, mean maxima of three hottest months. Broken lines, means of a number of Kenya localities: left to right, annual mean minima, annual mean temperatures, annual mean maxima. (After Moreau, 1934b, 1935a.)

Making due allowance for these abnormal areas, I have for the present purpose allocated as "montane" in Table V all species that do not normally occur below about 1500 m in the African tropics, and all those not normally surpassing that altitude as "lowland". But I have not been rigid about this; it is not unusual for a normally lowland species to live locally as high as 1800 m and I have still classified it as "lowland". (Since making this decision I find that Weber (1943) in discussing the mountain ants of the world has allowed the same margin; while Carcasson (1964), in dealing with African butterflies and citing 5000 ft as the approximate "dividing line" between his Lowland Forest and Highland Forest faunas in the tropics, notes that "highland species do not normally descend below 3000 ft and lowland species do not rise above 6000 ft".)

I have accepted as "montane" a few birds that nowhere go very high, but nowhere go very low; an example is the estrildine weaver *Nesocharis ansorgei*, 3000–6000 ft (say 900–1900 m). There are, however, some largely lowland species of the inner tropics which in an appreciable part of their range regularly transcend 1500 m to a considerable extent, and these are classified in Table V as "both" lowland and montane. They are of two types: either there is a continuous population from sea-level (as in *Estrilda astrild* to 8000 ft, some 2400 m) or there is an isolated highland subspecies (as of the whydah *Euplectes progne* above about 7000 ft (2100 m) in Kenya, while the Rhodesian population is below 1600 m). It is difficult to know how many species to admit to the "both" columns and there is no doubt that, very locally, more species show altitude adaptability than I have allowed for in Table V.*

In making allocations to highland and lowland respectively I have been especially indebted to the precise altitude data given for individual species by Chapin (1932–54), with particular reference to the Congo, and by Benson (1953) for Malawi. In accepting such altitude records one has, of course, to assume that they refer to the breeding range unless there is reason to suppose otherwise. Individual birds wander un-accountably to high altitudes; on a single afternoon at 10 500 ft on the upper edge of the Mt Kenya forest I saw a Hamerkop *Scopus umbretta* and a small hornbill *Tockus alboterminatus* (=*Lophoceros melanoleucos*), both some 4000 or 5000 ft above their normal range.

It will be seen that in Table V I do not attempt to divide the montane forest species into those of primary growth and others. This is because the distinction tends to break down; montane forests are as a whole less dense than lowland forests and are broken by the precipitous nature of much of the ground on mountain slopes cut deeply by ravines. In many places the discontinuous nature of the forest has been accentuated by fires, so that a mixture of species results. Thus, in the list for the montane forest on Mt Moco in Angola, Hall (1960a, Table 2) includes a number of species that are characteristically non-forest, for example the "grass-warbler" *Cisticola erythrops*; this particular forest is, however, described as "limited to patches lining precipitous gullies", the trees "nowhere very tall or very dense". Hence, in allocating species to forest or non-forest I have discounted records from such an area.

The last column of Table V contains species that I cannot to my own satisfaction allocate as either "forest" or "non-forest". The forty-two species are of two different types which I combine in Table V, mainly because each is too small in total number to be worth separating: (*a*) species whose typical habitat I cannot define, such as *Bias musicus*; (*b*) species that have an abnormally wide habitat range. The half-dozen extreme examples of this last category, two (parasitic) honey-guides, two doves and two white-eyes *Zosterops*, are worth listing in detail. It should be added that the inclusion of the two *Zosterops* species depends on a highly polytypic interpretation of a most complicated taxonomic situation (Moreau, 1957). On the green pigeon (*Treron*) there has been

---

* For example, several species that are normally regarded as typically lowland appear to be resident in a corner of Basutoland (Jacot-Guillemard, 1963), a country which all lies at over 1500 m.

TABLE V

The composition of the Ethiopian land bird fauna, by families, and the ecological distribution of the species

Family	Total no. of species	Forest					Non-forest				Uncertain or unspecialized
		Lowland Primary	Lowland Secondary and edges	Montane	Both	Total	Lowland	Montane	Both	Total	
Group A											
Anatidae	18	1	—	—	—	1	15	1	1	17	—
Anhingidae	1	—	—	—	—	—	1	—	—	1	—
Ardeidae	20	1	—	—	—	1	19	—	—	19	—
Balaenicipitidae	1	—	—	—	—	—	1	—	—	1	—
Balearicidae	3	—	—	—	—	—	3	—	—	3	—
Charadriidae	16	—	—	—	—	—	14	1	1	16	—
Ciconiidae	7	—	—	—	—	—	7	—	—	7	—
Heliornithidae	1	—	—	—	—	—	1	—	—	1	—
Jacanidae	2	—	—	—	—	—	2	—	—	2	—
Laridae	3	—	—	—	—	—	3	—	—	3	—
Pandionidae	1	—	—	—	—	—	1	—	—	1	—
Pelecanidae	2	—	—	—	—	—	2	—	—	2	—
Phalacrocoracidae	2	—	—	—	—	—	2	—	—	2	—
Phoenicopteridae	2	—	—	—	—	—	2	—	—	2	—
Podicipitidae	3	—	—	—	—	—	—	2	1	3	—
Rallidae	20	2	—	—	—	2	15	2	1	18	—
Recurvirostridae	2	—	—	—	—	—	2	—	—	2	—
Scolopacidae	2	—	—	—	—	—	1	—	1	2	—
Scopidae	1	—	—	—	—	—	1	—	—	1	—
Threskiornithidae	8	1	—	—	1	2	4	2	—	6	—
Total	115	5	—	—	1	6	96	8	5	109	—

Group B

Aegypiidae	8	—	—	—	—	—	7	1	—	8	—
Aquilidae	47	6	—	1	1	8	31	2	2	35	4
Falconidae	12	—	—	—	—	—	11	—	1	12	—
Sagittariidae	1	—	—	—	—	—	1	—	—	1	—
Strigidae and Tytonidae	26	9	—	2	1	12	11	3	—	14	—
Total	94	15	—	3	2	20	61	6	3	70	4

Group C

Burhinidae	3	—	—	—	—	—	3	—	—	3	—
Glareolidae	10	—	—	—	—	—	10	—	—	10	—
Otididae	16	6	1	6	—	—	16	5	—	16	2
Phasianidae	47	—	—	—	—	13	25	—	2	32	—
Pteroclidae	10	—	—	—	—	—	10	—	—	10	—
Struthionidae	1	—	—	—	—	—	1	—	—	1	—
Turnicidae	3	—	—	—	—	—	3	—	—	3	—
Total	90	6	1	6	—	13	68	5	2	75	2

(continued overleaf)

TABLE V (cont.)

Family	Total no. of species	Forest					Non-forest				Uncertain or unspecialized
		Lowland Primary	Lowland Secondary and edges	Montane	Both	Total	Lowland	Montane	Both	Total	
Group D											
Alcedinidae	17	3	1	—	—	4	12	—	—	12	1
Apodidae	13	3	—	—	—	3	8	1	—	9	1
Bucerotidae	26	8	3	1	1	13	12	—	—	12	1
Capitonidae	42	10	4	2	1	17	23	1	1	25	—
Caprimulgidae	17	2	—	—	—	2	14	1	—	15	—
Coliidae	6	—	—	—	—	—	6	—	—	6	—
Columbidae	28	4	1	4	1	10	15	1	—	16	2
Coraciidae	7	—	—	—	—	—	6	—	—	6	1
Cuculidae	22	4	2	—	2	8	13	—	1	14	—
Indicatoridae	11	4	—	—	3	7	2	—	—	2	2
Meropidae	16	—	3	—	—	3	11	1	1	13	—
Musophagidae	18	1	1	5	1	8	9	—	—	9	1
Phoeniculidae	8	—	1	—	1	2	6	—	—	6	—
Picidae	29	4	5	2	2	13	13	1	2	16	—
Psittacidae	19	2	—	—	1	3	13	2	—	15	1
Trogonidae	2	1	—	1	—	2	—	—	—	—	—
Upupidae	1	—	—	—	—	—	1	—	—	1	—
Total	282	46	21	15	13	95	164	8	5	177	10

Group E										
Alaudidae	47	—	—	—	—	—	44	—	3	47	—
Campephagidae	7	2	—	2	1	5	2	—	—	2	—
Corvidae	9	—	—	—	—	—	6	3	—	9	—
Dicruridae	3	2	—	—	—	2	—	—	—	—	1
Emberizidae	8	—	—	—	—	—	8	—	—	8	—
Eurylaemidae	4	2	—	1	1	4	—	—	—	—	—
Fringillidae	26	1	—	2	—	3	14	8	1	23	—
Hirundinidae	29	—	—	—	—	—	19	1	6	26	3
Laniidae	55	4	3	7	4	18	30	—	5	35	2
Motacillidae	24	—	—	—	—	—	18	2	4	24	—
Muscicapidae	71	21	5	6	3	35	27	2	3	32	4
Nectariniidae	66	9	6	8	—	23	29	9	2	40	3
Oriolidae	6	2	—	2	—	4	1	—	—	1	1
Paridae	15	2	—	1	—	3	9	1	2	12	—
Pittidae	2	1	—	—	—	1	1	—	—	1	—
Ploceidae											
Estrildinae	66	6	3	4	2	15	43	3	3	49	2
Others	101	16	2	4	—	22	69	1	7	77	2
Promeropidae	2	—	—	—	—	—	2	—	—	2	—
Pycnonotidae	48	17	6	7	5	35	12	—	1	13	—
Salpornithidae	1	—	—	—	—	—	1	—	—	1	—
Sturnidae	41	3	1	4	—	8	29	2	1	32	1
Sylviidae											
Cisticolae	36	—	—	—	—	—	25	2	9	36	—
Others	100	7	10	21	5	43	49	5	1	55	2
Timaliidae	33	8	2	6	—	16	17	—	—	17	—
Turdidae	95	13	2	21	2	38	41	7	6	54	3
Zosteropidae	5	—	—	—	—	—	2	1	—	3	2
Total	900	116	40	96	23	275	498	47	54	599	26
Grand total	1481	188	62	120	39	409	887	74	69	1030	42

some taxonomic controversy, but the other three species are straight-forward and show the minimum of subspecific variation.

> *Indicator variegatus*. Sea-level to 11000 ft. "Scrubby woods in savanna" (Congo); "rain-forest, rich *Brachystegia*, thick acacia" (Malawi); rain-forest (Usambara Mts); "thick bush and ravine forest" (South Africa).
>
> *Indicator minor*. Sea-level to 8000 ft. "From forest to sparsely wooded areas [i.e. thorn] in Damaraland" (South West Africa); "more or less open bush" (West Africa); "rain-forest" and "any woodland" (Malawi); "patches of large trees and scrubby woods in savanna" (Congo).
>
> *Streptopelia semitorquata*. Sea-level to nearly 8000 ft. "Open thornbush" (Somaliland); "savanna country and cleared forest" (West Africa) to "secondary forest" (Congo).
>
> *Treron australis* (=*Vinago calva*). Near sea-level to about 8000 ft. "Forest and savanna" (Congo); *Brachystegia* (Malawi); "throughout" (Zambia); also mangroves (West Africa).
>
> *Zosterops senegalensis*. Sea-level to 10000 ft. Acacia steppe to montane forest.
>
> *Zosterops virens*. Sea-level to over 7000 ft. Rain-forest to acacia country (South Africa).

It is worth mentioning two peculiar cases of birds which I have not definitely allocated to either forest or non-forest. One is the montane starling *Onychognathus tenuirostris*, which has been reported as nesting on rocks either behind waterfalls or on moorlands above the timber-line. It feeds partly on the ground at these high altitudes, but also depends a good deal on the fruit of montane-forest trees. The other bird is the weaver *Ploceus olivaceiceps*. The nominate form has a restricted range in *Brachystegia* in the vicinity of Lake Nyasa and probably only between about 3000 and 5000 ft. In 1931 a weaver that superficially resembled nothing else was collected on a few occasions, always on the edge of evergreen forests at 3000–5500 ft, in the Usambara Mountains of northern Tanganyika, 500 miles north of the range of *Ploceus olivaceiceps*, and was named as a new species, after my old friend Michael Nicoll. Later, when revising the Ploceidae (Moreau, 1960a), I was impressed by the fact that *nicolli* was mainly distinguished from *olivaceiceps* by its heavy infusion of melanin, which was what might have been expected under Gloger's rule in the humid Usambara climate. I therefore decided to treat the two populations as conspecific; but the extreme rarity of cases of this kind, apart from any other consideration, makes me think my taxonomic treatment may be wrong.

It is extremely unusual for any typically evergreen-forest bird to adapt itself to a man-made habitat. Where I lived at Amani, in a clearing in the East Usambara forest, the only forest species which were to be seen in exotic vegetation were all species of the forest edge: the fly-catcher *Muscicapa adusta* among coffee-bushes (with an overhead shade of introduced trees), the sunbirds *Anthreptes collaris* and *A. rectirostris* in bushes and trees in gardens that were close to forest. Similarly Benson

(1953) noted that in Malawi very few of the forest birds "are able to survive destruction of forest", though at Dedza (5000 ft), where cultivation and plantations of exotic pines have replaced forest except for a few relics, three forest species have shown some adaptability: the arboreal warbler *Apalis thoracica* to the trees, the warbler *Bradypterus mariae* and the estrildine *Mandingoa nitidula* to the undergrowth of the pines. The adaptation on the part of the *Apalis* is the most interesting of all the six cited and one would like to know what food the pines provide. The two undergrowth species would no doubt find much the same dull light among the tangled scrub composed of native species under the pines at certain stages of growth as in the native forest. Finally, in Nigeria, at Ibadan, which has over 200 terrestrial species in a mosaic of habitats, Elgood and Sibley (1964) have provided a most interesting study of the extent to which the bird species of the local forest, which is secondary, and those of the local savanna respectively make use of intermediate types of habitat, especially farms and gardens. They show (Appendix 2) that of thirty-nine species whose "most preferred habitat" is forest only five also occur "regularly" in "farms or savanna". Moreover, of these five at most only one is elsewhere a bird of primary forest, and two others, although preferring secondary forest at Ibadan, would not on broad African experience be classed as typically birds of forest of any kind. Moreover, out of thirty-one species listed as having "farms or savanna" as their preferred habitat, only a single one is shown as occurring regularly also in "secondary forest" (Appendix 3). A somewhat higher proportion of both "forest" and "savanna" species occur regularly in "gardens" at Ibadan, which have "a mixture of hedges, shrubs and open lawns with various shade and fruit trees". Nevertheless, the data given by Elgood and Sibley emphasize to a striking degree the distinctiveness of the savanna bird fauna from that of forest, even secondary forest, and their extremely limited adaptability from one habitat to the other.

The dichotomy between forest and non-forest species indeed goes very deep in Africa. As will be seen from Table V, there are only forty-two species out of a total of 1481 which I have not allocated to one ecological category or the other, though I am aware that others who analysed the bird fauna in the same way might arrive at a rather bigger proportion. For a large part of the difference between the bird faunas there are obvious ecological reasons: forest provides no habitat for species adapted to life in a number of environments; for example, open stony ground or tall grass. Moreover, in the lower part of evergreen forest the eco-climate is special, damp and dark to a degree found nowhere else. In a Tanganyika forest the light-intensity near the ground is less than 1% of that outside (Moreau, 1935b), which especially on dull days could hinder birds' foraging significantly, unless they were adapted to those conditions. Such difficulties cannot, however, operate with birds dependent on insects and fruit taken in trees. It would seem

TABLE VI

Main foods of Ethiopian bird families
(For explanation, see p. 80)

	Vertebrates	Invertebrates			Fruit	Seeds*	Nectar and misc.
		On and near ground	On trunks and foliage	In air			
Group C							
Burhinidae: Stone curlews		- -					
Glareolidae: Coursers and Pratincoles				—			
Otididae: Bustards		—					
Phasianidae: Game birds						—	
Pteroclidae: Sandgrouse						—	
Struthionidae: Ostrich						—	
Turnicidae: Button quails		—				—	
Group D							
Alcedinidae: Kingfishers	—						
Apodidae: Swifts				—			
Bucerotidae: Hornbills	- -	- -			—		
Capitonidae: Barbets			- -		—		
Caprimulgidae: Nightjars				—			
Coliidae: Colies					—		
Columbidae: Pigeons					- -	—	
Coraciidae: Rollers		- -		—			
Cuculidae: Cuckoos		- -					
Indicatoridae: Honey-guides			—				
Meropidae: Bee-eaters				—			
Musophagidae: Turacos					—		
Phoeniculidae: Wood-hoopoes			—				
Picidae: Woodpeckers			—				
Psittacidae: Parrots					—	—	
Trogonidae: Trogons				- -	—		
Upupidae: Hoopoes		—					

Group E

						Misc.
Alaudidae: Larks						
Campephagidae: Cuckoo shrikes						
Corvidae: Crows						
Dicruridae: Drongos						
Emberizidae: Buntings						
Eurylaemidae: Broadbills						
Fringillidae: Finches						
Hirundinidae: Swallows						
Laniidae: Shrikes						
Motacillidae: Wagtails						
Muscicapidae: Flycatchers						
Nectariniidae: Sunbirds						Nectar
Oriolidae: Orioles						
Paridae: Tits						
Pittidae: Pittas						
Ploceidae: Weavers†						
Estrildine						
Others						
Promeropidae: Sugar-birds						Nectar
Pycnonotidae: Bulbuls						
Salpornithidae†						
Sturnidae: Starlings						
Sylviidae: Warblers						
Timaliidae: Babblers						
Turdidae: Thrushes						
Zosteropidae: White-eyes						Nectar

* For group C birds "seeds" must be taken as amplified by other vegetable matter, especially bulbous roots.

† Near tree creepers.

that the basic requirements of a bird living in this way in deciduous trees could equally be found in evergreen forest. The extreme rarity of species that do occupy both forest and non-forest show that factors are operating against this. The immediate answer is probably to be found in interspecific competition, each bird community being so closely knit, so "saturated", that entrance by an outside species is almost impossible.

In this connexion it is interesting that, of the six species mentioned in detail above as showing abnormal ecological adaptability and living in both forest and other habitats, four are less likely to suffer from inter-specific competition than are most birds. The honey-guides are unique in relying for much of their food on wax, which no other birds can utilize, because it cannot be digested without the aid of such symbiotic bacteria as the honey-guides have acquired (Friedmann, 1955). The Zosterops have an abnormally wide range of diet; they take insects, swallow very small fruits and, aided by their brush tongues, eat the pulp and juice of large soft fruits, lick honeydew and take nectar. It may be noted also that none of these four species, nor the two uncommonly adaptable species of doves cited above, frequent the ground stratum of the forest, so that they do not encounter those extreme eco-climatic con-ditions which might specially deter a species attempting to establish itself in this sub-habitat.

Special interest attaches also to cases (mostly listed by Moreau, 1948) where some particular species has spread out of its normal non-forest habitat into a niche in forest which for reasons unknown is not occupied by its normal tenant. Examples may here be quoted from the forests on Cholo Mt in Malawi (Benson, 1953), Hanang Mt (Fuggles-Couchman, 1953) and the Mbulu highlands (Moreau and Sclater, 1937), the last two being in northern Tanganyika. None of these forests is remote geographically from other, comparable, forests in which the missing species occur. Hence it is more likely that the absences are due to local extinction, of the type to which isolated populations are prone (cf. Mayr, 1942, on instances in the Pacific islands), than that the missing species never reached the forests where they are lacking today.

On Cholo Mt the paradise flycatcher *Terpsiphone viridis*, normally a bird occupying "woodland", but "most usually riparian forest or thickly leafed trees, e.g. figs or mangoes", at low altitudes (Benson, 1953), replaces another flycatcher *Trochocercus albonotatus*, which is typical of all the other montane forests. On Hanang Mt, where the forest tends to be rather ragged, the situation may be somewhat analog-ous to that on Mt Moco in Angola, referred to above (p. 83), for Fuggles-Couchman found several non-forest species inside the forest at 7000 ft, but it is noteworthy that the sylviine *Bradypterus mariae*, typical of East African montane forests from Kenya to Malawi, was missing and *B. cinnamomeus*, also a montane species, but typical of heath and bracken, had permeated the forest in its stead. Again, in the Oldeani and Ufiome forests of Mbulu I found two non-forest species taking the places of forest birds, the turdine *Cossypha heuglini* in the stead of *C. semirufa*, and the weaver *Ploceus ocularis* where either *P. insignis* or *P. bicolor* would have been expected. More remarkably, in the Mbulu mountain forests generally I failed to find either of the two woodpeckers *Campethera taeniolaema* and *Mesopicos griseocephalus* which, without anywhere meeting, between them occupy all the other mountain forests of East Africa. In their absence the woodpecker was *Dendropicus* (= *Yungipicus*) *obsoletus*, typically a species of open deciduous woodland, which in the Mbulu forests has developed a peculiar dark subspecies—a clear example of Gloger's rule. *D. obsoletus* is the only bird among those just mentioned in which subspecific differentiation has taken place in correlation with change in habitat. This may or may not mean that the woodpeckers have occupied their peculiar forest niche in Mbulu longer than the other birds in question; it certainly shows that a case of opportunist ecological substitution of this kind can persist long enough for subspecific change to develop, though this might be measured only in a few score generations (Chapter 1).

The evergreen forests south of the Zambezi, especially those in extratropical South Africa, furnish several examples of what appears to be specific replacement of this kind. In particular, from these forests all the bulbuls (Pycnonotidae) typically associated with evergreen forest in the tropics, whether lowland or montane, are absent, and in their places appear two other bulbuls which are widespread in other vegetation types of the tropical lowlands: *Andropadus importunus*, there a bird of bushes growing without overhead shade, and *Phyllastrephus terrestris*, an inhabitant of dense scrub along streams. Again, in the South African forests the oriole is *O. larvatus*, the drongo is *Dicrurus adsimilis*, in the absence of those congeneric species which occupy the forests further north. It is far from clear how this South African situation can have come about.

By contrast with the foregoing, certain species that at low latitudes are exclusively associated with forest, though with the edges rather than the depths, are more adaptable in South Africa. The flycatcher *Batis capensis* appears also in acacia bush well away from forest and the warbler *Apalis thoracica* is even more enterprising. In the western Cape Province it is found in low macchia, even in isolated gorges in dry mountains 150 miles or more from forest or any area that might have carried forest in recent times (M. K. Rowan, personal communication). It appears that, unlikely as it might seem, the non-forest bird fauna in this area had not occupied fully the niches available to insectivorous passerines.

Turning now to the statistics in Table V, it will be seen that in Africa south of the Sahara a negligible proportion of water-birds are confined to forest and only about one-fifth to one-sixth of the raptor and ground-bird groups. These results are what might have been expected, since watery habitats are merely incidental in forest, raptors and scavengers have life made difficult for them by the density of the foliage screen and the abundance of obstacles to rapid movement, and the ground-bird (group C) families typically belong to open and grassy habitats, except for some of the Phasianidae. The same applies to some of the families in group E (passerines), which are absent from forest but well represented elsewhere, the most important being the Alaudidae (larks), Corvidae (crows), Emberizidae (buntings). There are other families which, though not absent from forest, contribute a much smaller proportion to the total of species comprising the forest bird fauna in their particular group than they do to the non-forest, namely, the Caprimulgidae (nightjars), only 2% compared with 8% in non-forest, Meropidae (bee-eaters) 3:7%, Estrildidae 5·5:8%, Fringillidae (finches) 1:4%, Ploceidae 8:13% and Sturnidae (starlings) 3:5·5%. The first two families are insectivorous, but they consist of birds that need plenty of room for manoeuvre, which is not easy in forest; the next three families are predominantly eaters of seed, a form of food not nearly so well provided in forest as outside (see also Chapter 15). But it is not easy to suggest the ultimate reason why the starlings, which are so largely fruit-eaters, should bulk so much smaller in the forest than in the non-forest bird fauna.

With the foregoing, several families, that depend on insects or fruit or both, provide a contrast. The Bucerotidae (hornbills) contribute 14% of the group D forest total and only 7% of the non-forest; the Muscicapidae *sensu stricto* (flycatchers) 17:5·5%; Pycnonotidae (bulbuls) 12:2%; Sylviidae (warblers), omitting the cisticoline grass-warblers, 16:9%; and Timaliidae (babblers) 6:2·5%. Birds having the habits of these families are all well catered for in forest.

It will be seen from Table V that the non-forest bird fauna of the Ethiopian Region comprises 109 species in group A, 70 in group B, 75 in group C, 177 in group D and 599 in group E, the passerines. The considerable differences between these totals are due partly to the fact that, omitting watery habitats (group A), there are more species of passerines than of any other group in any particular spot or any particular community (see examples in Chapter 15), and partly to the fact that individual passerine species have on the average much smaller geographical ranges than birds of other groups. The extent to which this holds good in Africa is shown by Tables VII and VIII. The first of these analyses shows the number of species in each group that are common to certain geographically remote parts of Africa, (a) Cape Province and West Africa, (b) Cape Province and (late British) Somaliland. Group A is omitted from the latter comparison because Somaliland lacks freshwater habitats. The proportions must, of course, be calculated on the number of species in the smaller bird fauna of each pair; the alternative method encounters the "artificial" limitation imposed by the size of the smaller bird fauna. Details of the bird faunas concerned and of the statistics of shared species are given, by families, in Table XIV of

Chapter 7. Cape Province totals are smaller than West African in all groups except A and Somaliland smaller than Cape Province throughout.

TABLE VII

The proportions of non-forest species common to certain geographically remote parts of Africa

| Group | Cape Province | | | Somaliland | | |
| | Total no. of species | Shared with West Africa | | Total no. of species | Shared with Cape Province | |
		No.	Proportion		No.	Proportion
A	74	44	[0·59]*			
B	41	30	0·73	28	21	0·75
C	28	14	0·50	21	6	0·33
D	51	25	0·49	47	14	0·30
E	180	56	0·31	120	22	0·18

* Calculated on the West African total of sixty-three species.

It will be seen that in both pairs of territories the proportions of raptors shared is higher than in any other group and passerines much the lowest, with ground-birds and group D birds ranking about the same. Another aspect of this is brought out by Table VIII (top half). This shows that no less than two-fifths of all the non-forest water-bird and raptor species breeding south of the Sahara range from West Africa through to the Cape, compared with one-sixth of the ground-birds, one-seventh of the group D non-passerines and less than one-tenth of the passerines.

TABLE VIII

The comparative pervasiveness of different groups of Ethiopian non-forest birds

| Group | No. of species in Ethiopian Region | Species common to West Africa and Cape Province | |
		No.	Proportion of total Ethiopian
A	109	43	0·40
B	71	31	0·44
C	75	13	0·17
D	177	25	0·14
E	599	56	0·09
Ardeidae	19	14	0·74
Strigidae and Tytonidae	14	7	0·50
Cuculidae	14	7	0·50
Estrildinae	49	8	0·16
Hirundinidae	26	4	0·15

As further illustrations of these differences it is interesting to quote (Table VIII, lower half) the figures for families which in each group have a particularly high proportion of species common to West Africa and the Cape. (Some very small families, of course, have a higher score; for example, the Scopidae, the single species of which ranges throughout Africa south of the Sahara, and the Turnicidae, two of whose three species occur in both West Africa and Cape Province.) It will be seen that three-quarters of the heron species have this wide range, half the owls and cuckoos, while none of the passerines have more than about one-sixth. The low proportion among the swallows is the more surprising because of the high individual mobility of these birds and the prevalence of long-distance migration among the African species (see Chapter 13).

Several factors could have contributed to the fact that group A species, the water-birds, and group B, the raptors, etc., have on the average so much wider ranges within the continent than the other groups. First, there is probably much individual movement among them (and their paucity of subspeciation is consonant with this), since many of them have notable powers of flight. Second, the large average size of group A and group B birds would tend to give them greater tolerance of the temperature differences encountered in ranges that comprehend many degrees of latitude; group C birds are, of course, nearly all large-bodied also, but they are for the most part exceptionally sedentary. Third, the passerines, as apparently the youngest and most rapidly evolving group of birds, would be the quickest to speciate in response to the vicissitudes of African ecology. Indeed, the proportion of passerine species which belong to superspecies, nearly two-thirds (Hall and Moreau, in preparation), probably greatly exceeds the proportion in other groups.

The extent to which species are limited to an altitude zone, being either typically lowland or typically montane, is shown in Table IX. The figures are extracted from Table V, with the omission of water-birds, since their habitat is so deficient in the montane zone. It will be seen that in both forest and non-forest about nine-tenths of the species fail to extend upwards into the montane zone; the montane-forest bird fauna is nearly as specialized, since three-quarters of the species there do not occur in lowland; while by contrast the montane non-forest bird fauna is only half peculiar to it, an equal number of species being shared with the lowlands. Evidently the montane- and the lowland-forest communities are each well integrated, making it relatively difficult for a species from one zone to penetrate the other. It is perhaps true that the opportunity for the specialization of the montane-forest bird community would have been especially good during a glaciation, when montane forest would have been far more extensive and continuous than it is now, but it is not clear how the same argument could be applied to lowland forest. Certainly both types of forest have a history of such dis-

TABLE IX

The altitudinal specialization of Ethiopian birds of groups B–E

Habitat	No. of breeding species			Proportion of breeding species limited to altitude zone
	Limited to altitude zone	Others	Total	
Lowland forest	245	38	283	0·86
Montane forest	120	38	158	0·76
Lowland non-forest	791	64	855	0·92
Montane non-forest	66	64	130	0·51

ruption, as explained in Chapter 3, that they have not known the long periods of stability commonly accepted as necessary for the elaboration of communities rich in species. It is noteworthy that, as Chapin remarked, no typically montane species of bird is particularly close to any typically lowland one. Hence, although in that minority of species which occupy both lowland and montane forest some subspecific differentiation has taken place with altitude, cases in which this has proceeded to the specific level are not obvious. From this it may be inferred that, although instances can be cited in which today lowland and highland populations of the same species are isolated by some hundreds of miles, such cases have not in the past commonly led to speciation.

Another feature that emerges from Table IX is that among the forest birds there are half as many typically montane species as typically lowland (120:245), whereas among non-forest birds (groups B–E) the proportion of typically montane species is only one-twelfth (66:791). The great difference between one-half and one-twelfth is no doubt mainly due to the fact that while lowland and montane forest both provide much the same range of (sub-)habitats, lowland non-forest is far more varied than montane non-forest. It offers everything from subdesert steppe to humid savanna, with a variety of herbaceous cover from the scantiest and most ephemeral to grass several feet high, and a range of woody growth from open thorny scrub to semi-evergreen thicket and tall trees along water-courses. By contrast, the non-forest (sub-)habitats of the montane zones range only from short turf and occasional coarse grass (but no tall vertical grasses) to woody and ericaceous scrub, with or without small scattered trees (see also Chapter 12).

It is interesting to consider the statistics of the different categories of species in relation to the area available to them today. Africa south of the Sahara consists of about 8 million sq. miles. Of this total, lowland evergreen forest occurs over some 700 000 sq. miles and montane evergreen forest, taking all its isolated areas together, over probably some 20 000 sq. miles. Thus, the 120 typically montane-forest species (plus thirty-eight others) are accommodated in 20 000 sq. miles, compared with 245 lowland forest species (plus thirty-eight others) in an area at least thirty-five times as large. This great difference is not due in any way to the local montane communities being richer in species than the lowland; in fact, observational data show that they are much poorer (Chapter 15). But owing to the sporadic and relict distribution of the montane forests the areas occupied by individual species of their birds are small and geographical replacement is prevalent among them. Indeed, thirty-three of the montane-forest species are each restricted to areas not exceeding 500 sq. miles and fifteen of these to between 10 and 100 sq. miles (Hall and Moreau, 1962). There is nothing comparable to this in the temperate zones and the African situation is due to a combination of its peculiar topography, with plenty of isolated highlands, of its recent climatic vicissitudes and of the very prolonged persistence of such montane refuges as there are under present conditions, as explained in Chapter 3.

If, again omitting water-birds, we compare the numbers of lowland forest and lowland non-forest species we get respectively $245 + 38 = 283$ species in some 700 000 sq. miles and $791 + 64 = 855$ in over 7 million sq. miles—one-third as many forest species as non-forest in one-tenth of the area. To this result two factors contribute. The major factor must be a smaller geographical range (on average) on the part of the forest species. Secondly, as shown in Chapter 15, the lowland forest is the richest of the main climax communities in bird species, having about 1·4 times as many as brachystegia, 1·5 times as acacia and 2·4 times as many as mopane or montane forest.

The lowland non-forest figures may be further considered on the background of the condition of Africa during a glacial maximum, as described in Chapter 3. Whereas now the montane areas are isolated and individually circumscribed, then they formed one great block, much of it forested, extending from Abyssinia to South Africa with an arm reaching across to the Cameroons, as shown by the 500 m contour in Fig. 2. At the same time, apart from West Africa, birds intolerant of montane conditions would have been confined to the southern Sudan, to part of the Congo basin and to a strip round the coast nowhere more than 200 miles wide. The contrast between then and now is brought out by Fig. 27, which is based on Fig. 2. ("Intolerant" is the operative word: there are certain species, such as the dove *Streptopelia capicola*, typically lowland in the tropics, that in inland South Africa in winter survive conditions as rigorous as those high in the montane zone near the Equator.) Each of the lowland areas in Fig. 27 would have been to some extent under evergreen forest. The coastal strip, essentially linear, would have consisted of sections of wetter and drier country, with and without evergreen forest, so that bird populations would have been isolated in segments all round the rim of the continent. Moreover, the three other lowland areas, in the Sudan, the Congo and West Africa were each isolated. All in all, the conditions for the speciation of birds intolerant of montane conditions were peculiarly favourable.

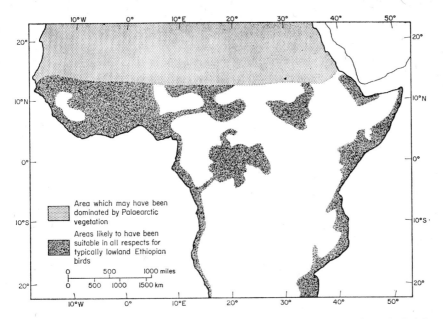

FIG. 27. Tropical African areas available to exclusively lowland birds at the height of the Last Glaciation.

These conditions began to give way to those of the present only about 18 000 years ago. There is no reason to suppose that there has been speciation on a considerable scale since that date and consequently it can be accepted that at least as many species of birds were present in Africa then as there are now. At the present time only 159 forest and 143 other species out of a total of 1481 are either specialized to, or tolerant of, montane conditions, which shows that adaptation from one zone to the other has not been common. Perhaps a larger proportion of species were tolerant of both lowland and montane conditions during the glaciation than now; and perhaps montane non-forest offered a bigger range of habitats than it does now, after so many years of organized human interference. Neither of these possibilities seems very likely; and if indeed they were not realized, then as many as the 1075 strictly lowland species, including 887 non-forest, now encountered south of the Sahara, would have been compressed into West Africa, part of the Sudan, the central Congo basin and the coastal strip.

For so many lowland species to be accommodated in so much smaller an area than now—and the comparison is much more crucial for the non-forest species which are now so much more widely spread than those of forest—it is necessary to postulate either that the individual communities were richer in species, i.e. the niches were more subdivided, or that the geographical ranges of the birds concerned averaged much smaller than they do now. The second alternative is likely to have been the more important, if not the only one operating. Individual sections of the low-lands were probably rich in endemics—a state to which the isolation or near-isolation of each one would be conducive—as indeed Somalia and South West Africa still are today (see Chapter 10), after thousands of years of opportunity for their species to spread over the post-glacially expanded lowlands. A total of 921 non-forest species is not so great when it is recalled that a single sample bird community in Acacia steppe can total eighty-six species and one in brachystegia woodland more still, as shown in Chapter 15.

West Africa, which offered during the glaciation much the biggest lowland area and may have had as complete a series of habitats, from rich forest to desert, as it has today, naturally suggests itself as the most important refuge for non-forest species. Now we come up against the paradox that today West Africa has a non-forest bird fauna of only 418 species that is poor in endemics and is poorer in species than most other territories of Africa (see Chapter 7, Table XIV), including the Rhode-sias, which presumably were nearly all montane during the glaciation and have received their lowland population to a great extent since then. It is not clear how this paradox is to be resolved; moreover, it does not apply only to birds, for West Africa is less rich in plants than are corre-sponding vegetation formations south of the Equator (see quotation in Moreau, 1952a, p. 900, and further discussion in Chapter 7).

All in all, if the lowland non-forest bird fauna that we see today is essentially what survived the Last Glaciation and has not been much supplemented by species evolved since that came to an end, then its richness is surprising. But when we turn to the montane bird faunas we are faced with the reverse situation. The montane area was many times as large as it is now, practically continuous from Abyssinia to the Cape. Moreover, if the precipitation were no less than it is now the reduced evaporation would have favoured the growth of evergreen forest. We must be on our guard against extending evergreen forest over areas where, however great the total annual rainfall, the dry season is long; but the strength of the affinities between montane-forest bird faunas so isolated today as those of Mt Cameroon, Ruwenzori and Kenya (see Chapter 11) provides the strongest possible circumstantial evidence for the former continuous extension of montane evergreen forest right across Africa at about the Equator, while similar considerations postulate its extension from Kenya to Malawi and Angola. To what extent the subsequent shrinkage of montane forest to its present remnants has been accompanied by the extinction of bird species we shall never know. Some extinction seems inevitable, but the very same process that brings this about also sets the stage for new speciation. As indicated in Chapter 3, those islands of montane forest which we see now have been very dependable refuges, for they have probably never been smaller and more scattered than they are today, when for the first time in the history of the world vegetation is exposed not only to both the effects of the heightened post-glacial temperature and to large-scale destruction by human beings. In each period of minimum forest, such as the present, the isolation of each surviving bird population is greatest, to the furtherance of its differentiation and speciation. By this process it has been possible that, when the time comes for each resurgence of montane forest, with a pluvial or perhaps with the falling temperatures at the end of an inter-glacial, new species, especially within the superspecies, here and there may become available (alongside old ones) to repopulate the increasing new areas. Perhaps the best examples are today members of the *Francolinus* superspecies, with *F. camerunensis* on Mt Cameroon, *F. nobilis* on the mountains of the east Congo, *F. jacksoni* in Kenya, *F. castaneicollis*, *F. erckelii* and *F. ochropectus* on different mountain masses in north-east Africa, and *F. swierstrai* in Angola (Hall and Moreau, 1962; Hall, 1963a). As such new species expand and meet, they come into potential competition and some may achieve such ecological adjustment as enables them to live alongside one another—to the enrichment of the local bird community—rather than to be mutually exclusive. Examples of such mutual adjustment are provided by the thrushes *Geokichla gurneyi* and *G. piaggiae*, which coexist in the forests of Mt Kenya (but the details of their ecological relationship are unknown). Such indeed has been the pattern of evolution and community building among the montane-forest birds in the past. Now, with the explosion of the human population the pattern has presumably recurred for the last time. If the climatic conditions associated with the glaciation were to return to Africa tomorrow, such is the present demand for agricultural land, which implies the destruction of trees, that the spread of montane forest from its strictly defined reserves would be impossible.

A special evolutionary problem is raised by those species which are typical denizens of secondary forest and forest edges. Table IX shows that they account for one-quarter of all lowland forest species (67 out of 255). Thirty-seven genera contribute species, and most of these genera typically inhabit forest or have representatives in both forest and savanna, but nine of them, each consisting of no more than one or two species or superspecies, are known only from forest edges or secondary forest. The problem is, of course, general to animals and plants in this context, but it is most acute for birds because of their individual mobility. The prevalence of bird species, that are characteristic of secondary forest but not of primary (though those of primary often occupy secondary also), has been discussed especially by Chapin (1932). Birds of both habitats may occupy gallery forest to varying degrees. From the nature of the case secondary forest in every stage of transition can be met with. The advanced stage described by Chapin (Fig. 3) is typically not more than about 60 ft high, compared with the 150 ft of full forest, is without a dense and continuous canopy and consequently possesses more abundant and often matted undergrowth. The characteristic tree species of the first stages of regenerating forest are soft and quick-growing, comparatively short-lived and superseded in the course of years by the long-lived hardwoods of true forest. Such secondary forests are of wide extent today. As Chapin said, in the Congo "the virgin forest has been devastated over wide areas" and in Nigeria it is stated that no virgin forest remains. Both territories, however, contain great areas of secondary forest, in the climatic belt humid enough for tree-growth to regenerate directly a clearing is abandoned.

If such great areas of secondary forest were produced by natural causes the evolutionary problem would be comparatively simple, but they are, in fact, the result of the activities of a few generations of agricultural man. Before him, the only opportunities afforded for the growth of secondary and the production of edge conditions were in spots and in belts no more than a few yards wide, in openings caused by the fall of a big tree, by the passage of a stream, and still more by the changing of its course, by the shifting of the border, climatically determined, between forest and savanna and by the incursion of a fire (*ex hypothesi* of natural origin) into the edge of the forest where it abuts on grassland. In sum, the area of all these adventitious snippets of secondary conditions might be appreciable, but it is impossible to see them as providing that effectiveness of geographical isolation and hence of reproductive isolation which most of us accept as necessary to enable speciation to proceed in ordinary sexually reproducing species. It must be emphasized also that for speciation to be fully achieved isolation would have to be maintained over a long period of years, yet secondary growth is of its nature transitory.

Now, the parent stock of secondary and edge birds must have belonged to either the forest or the neighbouring savanna, and from either or both of these habitats individuals living in the edges of the forest or in the spots or exiguous strips of secondary growth must under natural conditions have been separated by only a few yards. Given the individual mobility of birds, it seems that such conditions would altogether fail to cut them off from their parent population. The only tenable hypothesis seems to be that the species now typically inhabiting forest edges or secondary forest were originally evolved in geographical isolation as either typical savanna or typical forest species. Thereafter three possibilities present themselves.

(1) Some species, e.g. the sunbirds, the *Meropogon* bee-eaters or the *Macrosphenus* spp. (? Sylviidae), found themselves particularly well suited with edge situations, providing, compared with the forest interior, more room for manoeuvre and, at intermediate heights above the ground, denser vegetation and stronger light, between them conducive to more insect-food, flowers, fruit and nesting-sites.

(2) Some species, e.g. hornbills and barbets, found the fruit carried by the dominant species of the secondary particularly acceptable.

(3) The five Picidae of secondary, two *Campethera* spp., two *Dendropicos* spp. and the tiny *Verreauxia*, may conceivably be adapted to secondary

because in that habitat alone the dominant trees are of soft wood, easy to excavate. (But this is purely speculative, since we know nothing of the comparative capabilities of woodpecker species in this respect.)

(4) In contrast to what may be called the three preceding positive hypotheses, this last is negative, for it is that when formerly isolated species met one was forced into a marginal habitat. This possibility is suggested by the case of the two *Baeopogon* bulbuls, which overlap throughout the great area from the Cameroons to the Ituri in the north-east corner of the Congo. According to Chapin, *B. indicator* is a bird of secondary forest and is replaced in "heavy forest" by *B. clamans*.

It remains to recall that nine small genera consist entirely of species that are typical inhabitants of secondary forest: in the Capitonidae, *Trachylaema* (1), in the Meropidae, *Meropogon* (3), in the Muscicapidae, *Artomyias* (2), *Megabyas* (1) and *Pedilorhynchus* (2), in the Picidae, *Verreauxia* (1), in the Pycnonotidae, *Calyptocichla* (1), in the Sylviidae (perhaps), *Macrosphenus* (4), in the Timaliidae, *Phyllanthus* (1). It is noteworthy that although in recent years so many genera have been merged, all these small genera except *Pedilorhynchus* are still accepted in the latest critical lists. This means that this element in the bird fauna is composed of conspicuously distinct species, which accordingly may be comparatively old. If their adaptation to secondary forest owes anything to the fourth hypothesis above, either their competitors have ceased to exist or their competitors were extra-generic.

6

The Affinities of the Ethiopian Bird Fauna

Even on Wetmore's or Storer's classification, neither of which, as noted previously, uses subfamilies, only a very small proportion of the eighty or so families can be credited to the Ethiopian Region as endemic to it; while on the classification of Mayr and Amadon only the following groups at any level above the genus can be cited as endemic (number of species in parentheses):

One order: Colii (Coliidae, 6).

Three families: Musophagidae, turacos (18); Sagittariidae, Secretary Bird (1); Scopidae, Hamerkop (1).

Eight subfamilies: Bubalornithinae (in Ploceidae), buffalo weavers (2); Buphaginae (in Sturnidae), oxpeckers (2); Malaconotinae (in Laniidae), bush-shrikes (39); Numidinae (in Phasianidae), guinea-fowls (7); Phoeniculinae (in Upupidae), wood-hoopoes (6); Prionopinae (in Laniidae), helmet-shrikes (9); Promeropinae (in Meliphagidae), sugar-birds (2); Viduinae (parasitic weavers, in Estrildidae (9).

One tribe of a subfamily: Picathartini (in Timaliinae, Muscicapidae), picathartes (2).

To the foregoing, however, the Shoebill *Balaeniceps rex* has a good claim to be added, since its systematic position is uncertain and, although Mayr and Amadon reduce it (by inference) to a genus of the Ciconiidae, Wetmore rates it as high as a suborder and Storer as a family. Also the Struthionidae (ostriches) formerly widespread in Asia have, since the recent extinction of *Struthio camelus* between the Red Sea and Iraq, become confined to Africa. Examples of all the fifteen groups that have been mentioned are given in Figs. 28–32.

It will be noted that eight out of the fifteen groups mentioned above are very small indeed, consisting of only one or two species. Prima facie one might suppose that they were relicts, but three of them are so specialized ecologically that it seems doubtful whether the species comprising the group can ever have been much more numerous. The two oxpeckers (with particularly sharp claws and laterally flattened beaks) were until the advent of domestic stock completely and directly dependent for food on what they could find on the hides of big game living in open country. The sugar-birds, with their curved beak and brush tongue (but accepted as having no close affinity with the sunbirds), feed almost exclusively on the flowers of *Protea* spp. The picathartes, of doubtful familial relations and represented by two isolated species in different

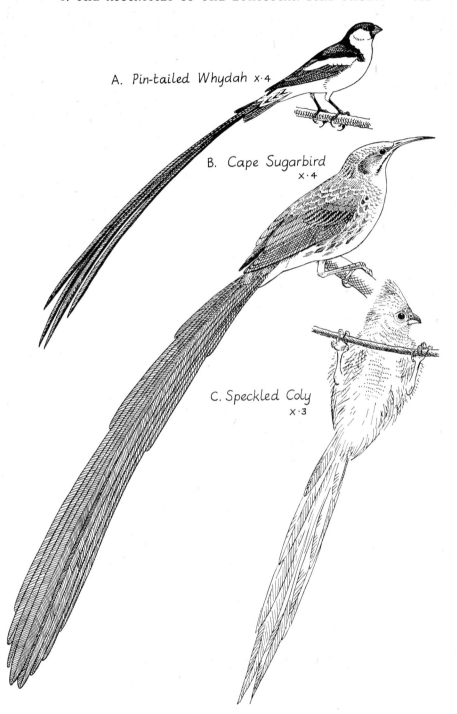

A. Pin-tailed Whydah x·4

B. Cape Sugarbird x·4

C. Speckled Coly x·3

FIG. 28. Characteristic birds 1. A. *Vidua macroura*, VIDUINAE. B. *Promerops cafer*, PROMEROPINAE. C. *Colius striatus*, COLIIDAE. Robert Gillmore, *del.*

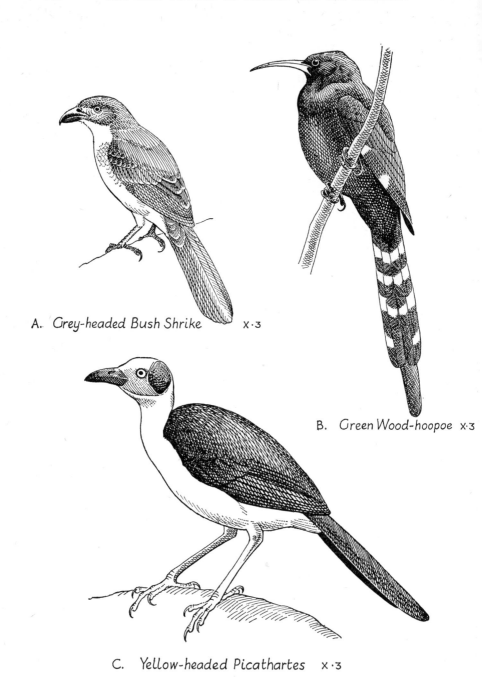

A. Grey-headed Bush Shrike   x·3

B. Green Wood-hoopoe x·3

C. Yellow-headed Picathartes   x·3

FIG. 29. Characteristic birds 2. A. *Malaconotus blanchoti*, MALACONOTINAE. B. *Phoeniculus purpureus*, PHOENICULIDAE. C. *Picatharthes gymmocephalus*, PICATHARTINI. Robert Gillmor, *del*.

A. Red-billed Oxpecker x·5

B. White Helmet Shrike   x·5

C. Buffalo Weaver   x·4

FIG. 30. Characteristic birds 3. A. *Buphagus erythrorhynchus*, BUPHAGINAE. B. *Prionops plumata*, PRIONOPINAE. C. *Bubalornis albirostris*, BUBALORNITHINAE. Robert Gillmor, *del*.

A. Hartlaub's Turaco x·2

B. Hamerkop x·2

C. Tufted Guinea Fowl x·2

Fig. 31. Characteristic birds 4. A. *Tauraco hartlaubi*, MUSOPHAGIDAE. B. *Scopus umbretta*, SCOPIDAE. C. *Numida meleagris*, NUMIDINAE. Robert Gillmore, *del.*

A. Ostrich x·06

B. Shoebill x·07

C. Secretary Bird x·06

FIG. 32. Characteristic birds 5. A. *Struthio camelus*, STRUTHIONIDAE. B. *Balaeniceps rex*, BALAENICIPITIDAE. C. *Sagittarius serpentarius*, SAGITTARIIDAE. Robert Gillmor, *del.*

parts of the very wet Guinean forests, build a mud nest under a rocky overhang, so that the acceptable sites are decidedly limited and localized. It may be added that the guinea-fowls, the wood-hoopoes, the helmet-shrikes and the turacos all have representatives in both forest and drier vegetation, but each of these groups is represented by different species and usually different genera in the two types of habitat. The colies, though eating exclusively fruit and vegetation, do not enter forest at all. In one respect the turacos are the most remarkable of African birds, since some of their members, mainly forest dwellers, elaborate a brilliant red pigment which is a copper uroporphyrin unique in the whole animal kingdom.

As regards the sugar-birds *Promerops*, there seems to be no objection, in either their morphology or their habits as described by Broekhuysen (1959), to putting them with the Meliphagidae (honey-eaters), as Mayr and Amadon have done. But some reluctance to accept this classification must be felt, because the geographical difficulties are great. The Meliphagidae are typically an Australo-Papuan family, with no other extension westward except that a single species reaches the island of Bali, just east of Java. The admission of *Promerops* into the Meliphagidae gives this family a range that is unique. Clearly, to account for the present situation three possibilities suggest themselves: (1) movement direct across the Indian Ocean; (2) expansion of the ancestral stock round the Indian Ocean, with subsequent enormous extinction in intermediate areas; (3) convergence. I think the first of these possibilities most unlikely and I favour the last. Elsewhere in the animal kingdom the most wonderful cases of unquestionable convergence have occurred, for example between placental mammals and Australian marsupials, such as *Notoryctes* with all the adaptations of a mole *Talpa*, or *Thylacinus* with "dog-like form, digitigrade stance, and the canine appearance of its skull and dentition" (Troughton, 1959). In birds the great external similarity of the sub-oscine *Neodrepanis* of Madagascar with the oscine sunbirds *Nectarinia* (Amadon, 1951) might be cited. Compared with such convergence from more distant starting-points, such a one as between *Promerops* and the Meliphagidae, both in the same suborder, the Oscines, would be minor. Provisionally I keep the Promeropidae as a family.

While, as shown above, the Ethiopian Region can claim as endemic few groups of status higher than the genus, it may be borne in mind that several families are much more richly represented in Africa than elsewhere. Of the twenty-two species of Otididae (bustards) in the world, sixteen live in the Ethiopian Region; ten out of sixteen Glareolidae (pratincoles, coursers and the "Egyptian Plover"); ten (plus two Saharan) out of fifteen Pterocletidae (sandgrouse); eleven out of thirteen Indicatoridae (honey-guides); forty-seven out of sixty-nine Alaudidae (larks); fifty-five out of seventy-four Laniidae (shrikes); 101 out of 112 Ploceine weavers and sixty-six out of about 100 Estrildines.

Several of these groups, the bustards, coursers, sandgrouse and larks, are typically birds of dry open country, of which Africa possesses, and probably within relevant time always has possessed, large areas. The other families are well represented in a wide range of habitats containing trees, except that the estrildines are few in forest. In honey-guides the predominance of the Ethiopian Region is remarkable, especially compared with the Oriental Region, which possesses only two out of the thirteen species in the world, yet is rich in both Capitonidae (barbets) and Picidae (woodpeckers), the two families on which, as nest parasites, the honey-guides mainly depend. Moreover, it is difficult to believe that the two groups of insects, the bees and the scale-insects (Coccidae), on which the honey-guides in Africa rely for food, are significantly deficient in, for example, the Oriental Region.

Another remarkable feature of the Ethiopian bird fauna is the status of the Scolopacidae. Their sole breeding representative is the Ethiopian Snipe, which in the inner tropics is a high-altitude bird, breeding above the timber-line, even, apparently, at 12000 and 13000 ft. This shortage of breeding Scolopacidae (as distinct from Charadriidae) is, of course, a pan-tropical, not merely an African, feature—of the seventy-nine species listed in Peters's "Check List of Birds of the World" under the Scolopacidae, only eleven breed in the tropics, and of those five breed only at high altitudes, and two only on tiny Pacific islands. But it is posed here with especial point, since over a score of Palaearctic Scolopacidae winter in the Ethiopian Region, many of them in enormous numbers, most of them partly or wholly on inland water, and with an appreciable number of non-breeding individuals lingering throughout the northern summer. As discussed later in this chapter, the only other species for which breeding south of the Sahara has ever been claimed is the sandpiper *Tringa hypoleucos* and that as an extreme rarity. I am at a loss to suggest any explanation for the fact that no scolopacid species except the one snipe has been evolved—or has survived—to take advantage of tropical conditions in Africa. To the human observer, it looks as if immense food resources go unutilized during the four or five months when the great bulk of the Palaearctic Scolopacidae are absent.

At first sight the level of endemicity in the suprageneric categories, as described above, is remarkably low for a great land-surface that must have been a centre of evolution throughout the existence of the class Aves. By comparison, South America, which with sixty-seven families has about the same total as Africa, scores one-third of them as endemic, and Australia scores one-quarter (10/44). But the Oriental Region in this respect ranks even lower than Africa, and the reason is doubtless the mutual accessibility of these two regions and the accessibility of each of them to the Palaearctic, so that there has been much interchange of birds between them in the past. The affinities merit separate discussion in detail. It is convenient to dispose of South America first; then to discuss the affinities between the Ethiopian Region and India, as the most relevant part of the Oriental Region; and finally the affinities between the Ethiopian Region and Europe, as the most relevant part of the Palae-arctic. The affinities with Madagascar, a neighbouring island so large and peculiar as to be ranked by many as a separate zoological region, will be dealt with in Chapter 17.

AFFINITIES WITH SOUTH AMERICA

Even allowing for the claims made for continental drift, there is no reason to suppose that any direct land connexion has existed between Africa and South America (now over 1500 miles apart by the shortest route) in the last 130 million years, i.e. since long before the evolution of any modern birds (Mayr, 1952). For working out the ornithological affinities between the two continents we can use the catalogue of Hell-mayr and Conover (1932–49), as amended above the generic level by the revision of Mayr and Amadon, already quoted. Among the Ethio-pian families, all those of water-birds also appear in South America, and half the other non-passerine families. Several factors have presum-ably contributed to this, especially the ages of these groups of birds and the potential mobility of water-birds over seas.

With the passerines, the situation is very different. For one thing, in South America most of them are sub-oscines, and of these Africa has only two families, the Pittidae and the Eurylaemidae, both shared with the Oriental Region but not with South America. In the result, the only passerine families common to South America are the Corvidae, Frin-gillidae, Hirundinidae, Motacillidae, Sylviidae and Turdidae. Most of these are represented in South America only thinly and marginally, but strongly in the Nearctic and Palaearctic, so that the means by which they came to enter both the southern continents seems clear.

Of the genera on both sides of the South Atlantic only a small frac-tion are, on the existing classifications, common to Africa and South America; of the total of thirty-eight in common, twenty-one are of water-birds, fifteen of other non-passerines (made up of five bird-of-prey genera, six owl, two swift, *Columba* and *Burhinus*), and only two of passerines, *Anthus* and *Turdus*. Most of the genera concerned are widely distributed also in the Holarctic, and for these distribution direct across the South Atlantic need not be invoked.

The Ethiopian and South American Regions have ten breeding species of land-bird in common (apart from the Cattle Egret *Ardeola ibis*, which has so recently irrupted into South America from the Old World). Four, namely the Osprey *Pandion haliaetus*, *Falco peregrinus*, the Barn Owl *Tyto alba* and the Moorhen *Gallinula chloropus* are also Hol-arctic; the Stilt *H. himantopus* might perhaps have got round by the north during the warm interglacial; the egret *Casmerodius albus* and the ducks *Dendrocygna bicolor* (=*fulva*), *D. viduata*, *Aythya erythrophthalma* and *Sar-kidiornis melanotos* there is no reason to suppose have made other than a direct crossing, though since they are none of them long-distance migrants it is difficult to imagine what might lead them to take off across the Atlantic. The first three of the ducks are not even differentiated subspecifically on the two sides of the ocean. This is the more surprising in the case of *D. bicolor*, because the birds from California to Mexico differ from the South American, and in Africa the species is confined to the east of about 25° E., so that the gap between the African and the South American populations of *D. b. bicolor* amounts to over 4000 miles.

It is worth mentioning that striking cases of convergence occur between certain birds on the two sides of the Atlantic. The African sylviid *Melocichla mentalis* strongly resembles members of the furnariid genus *Phacellodromus* (Diesselhorst, 1959). This example gives us a lead in evaluating another case, this time within the genus *Turdus*. As pointed out to me by Derek Goodwin, the South African *T. o. olivaceus* looks so like the South American *T. r. rufiventris*, differing only slightly in shade and in wing-length, that if the two birds inhabited the same continent they would undoubtedly be regarded as conspecific. It might be suggested that here we have evidence of recent colonization across the south Atlantic, but especially in view of the somewhat limited repertory of colour and pattern within the genus *Turdus* it seems better to regard the resemblance between these two thrushes as due to convergence.

AFFINITIES WITH INDIA

For a discussion of the relationship between the Ethiopian and Oriental bird faunas the check list of Indian birds recently published by Ripley (1961) provides a valuable basis. For the present purpose I exclude from "India" "the Palaearctic zone in the north and the higher montane areas", i.e. the Himalayas, as was done explicitly by Ripley (1953) when he discussed the origin of the Indian bird fauna. Those species which I have in Chapter 4 regarded as forming the Saharan bird fauna are, of course, also excluded from consideration, both because they do not form part of the Ethiopian bird fauna and because for them the desert forms a highway rather than a barrier.

For other birds the barriers between Africa and India consist of the arid condition of Arabia and the water gaps of the Red Sea and the Persian Gulf. Both these gaps are, however, narrow, the Red Sea at its mouth being hardly wider than the Straits of Gibraltar. The width of the Red Sea would have been little affected by the lowering of the ocean level during the glaciation, but the Gulf would have been almost eliminated. The minor importance of the Red Sea in this connexion is shown by the fact that about forty species of birds otherwise known only in Africa breed also in the south-west corner of Arabia, which has caused some authors to include the area in the Ethiopian Region. Three of the Ethiopian species to appear in south-western Arabia are actually montane birds usually associated with forest, so that they have jumped the ecological barrier of the low arid coastal strips as well as the water-gap:

a dove *Streptopelia lugens*, a warbler *Phylloscopus* (=*Seicercus*) *umbrovirens* and a thrush *Turdus abyssinicus.**

By far the most important obstacle to east-and-west communication of non-desert birds is undoubtedly the condition of Arabia. The only parts of the peninsula with any good vegetation are in the mountains of the south-west corner and those comprising the Gebel Akhdar in the south-east. Most of the rest of the country is too dry to carry even acacia steppe, let alone savanna. Probably during each interglacial period conditions have been as bad as they are now. How far they were ameliorated during the glaciations is uncertain, but it is difficult to envisage anything like a belt of evergreen forest across the peninsula at any stage of the Pleistocene; at its best the climate of the southern, tropical, part of the peninsula is likely to have resembled that of the Horn of Africa, which Clark (1954) thought from geological evidence was never better than semi-arid at any time during the Pleistocene. (See also the discussion of Arabian biogeography by Ripley, 1954.) This does not, of course, apply with full force to the mountain masses of the Yemen and Gebel Akhdar in the south-west and south-east corners of Arabia, and the vegetation on the somewhat elevated southern rim of the peninsula was probably sufficient during the glaciations and also during the Neolithic humid period to provide a corridor for the birds of wooded steppe and savanna, perhaps even with intermittent patches of dry evergreen forest, as on the highest mountains of Somaliland.

Africa and India (as limited above) have about the same number of bird families, most of which appear in both areas. The exceptions are that India lacks the few African endemic families discussed previously in this chapter and on the other hand India alone of the two areas possesses Podargidae (frogmouths), Artamidae (Wood Swallows), Irenidae and Dicaeidae (flower-peckers). For the purpose of investigating the relationships at the generic level I have used the classification of Ripley's (1961) Indian list, excluding those genera which are limited to the Himalayan montane area, without any of their members reaching the plains except as winter visitors. As shown in Table X, three-quarters to four-fifths of the Indian genera in groups A and B are shared with Africa, compared with only two-fifths in the remaining groups. Presumably several factors contribute to this result. For one thing, the taxa of groups A and B are probably older than those of group E, the passerines, if not of the other groups. Next, most of the raptors, etc., have great individual mobility and, at the other extreme, some of the ground-

---

* Although the mountains of the Yemen rise to over 3000 m, they are understood to be thoroughly deforested today, but "luxuriant juniper forests" have been described as flourishing north of the northern Yemen border, i.e. in Asir (Scott, 1942, p. 94). There seems to be no information about the habitat of *Streptopelia lugens* in Arabia today, but both the other two birds have evidently shown adaptability. Bury (1915) found the *Phylloscopus* "among the leafless walnut groves below Menakha", i.e. at about 2000 m, and the thrush "along the coffee gardens [which would be evergreen] and scarped terraces" in the same area.

birds have particularly little. Finally, distribution between Africa and India is probably easier for water-birds to achieve than for others, because many of them can survive in transit on sea coasts, irrespective of the condition of the land.

TABLE X

Genera of breeding birds shared by India and the Ethiopian Region

| Group | Total | Indian genera shared with Ethiopian region | |
		No.	Approx. proportion
A. Ducks and waders	10	8	0·8
Other water-birds	40	31	0·8
B. Raptors, etc.	34	25	0·75
C. Ground-birds	19	7	0·4
D. Other non-passerines	57	21	0·4
E. Passerines	107	43	0·4
	267	135	

There are numerous genera in which one species inhabits Africa, the other India, but the two are sufficiently distinct to be universally treated as species. Examples are *Butastur* (Grasshopper Buzzard), *Aviceda* (Cuckoo Falcon), *Leptoptilos* (Marabou Stork), *Threskiornis* (Sacred Ibis), *Nettapus* (Pigmy Goose). Leaving aside these superspecies, there are about seventy cases in which bird populations in India and the Ethiopian Region are generally agreed to be conspecific, some being identical in the two areas, others showing varying degrees of differentiation. They are worth listing in detail.

Group A. Water-birds, with twenty-five species, including ten Ardeidae, account for more than one-third of the total. (There may be at least one more, for *Ardea goliath* has been reported to breed in Bengal and Ceylon. If indeed it is only a vagrant in these and other Indian localities, as classed by Ripley, it is difficult to imagine where the birds concerned can come from; the nearest admitted breeding-grounds of the species, in Iraq, are far away.)

Group B. Birds of prey and vultures (together eleven species) and owls (three) make up another one-fifth of the total number of species common to India and the Ethiopian Region. Moreover, one more species, the bat-eating *Machaeramphus alcinus*, widespread in Africa but unknown in India, reappears in lower Burma.

Group C. Only three species of non-passerine ground-birds are common to the two areas. These are button quail *Turnix sylvatica*, quail *C. coturnix* and pratincole *Glareola pratincola*. Both the latter are long-distance migrants with a wide Palaearctic range.

Group D. The eleven other shared non-passerines are a mixed lot, the Hoopoe *Upupa epops*, three swifts, two bee-eaters, two doves (the Palaearctic *Columba livia* and the tropical *Streptopelia senegalensis*), a parrot *Psittacula krameri*, a parasitic cuckoo *Clamator jacobinus* and a coucal (non-parasitic cuckoo) *Centropus toulou*. Of all the fourteen species in groups C and D only six have a surviving population in Arabia today, namely the Hoopoe, Little Green Bee-eater *Merops orientalis*, the doves and two swifts. It is not obvious why most of the other eight species should not have been able to retain a foothold in the peninsula, at any rate in the relatively good south-western and south-eastern corners. The cuckoo, unique in that part of the Indian breeding population, apparently performs a regular migration to and from eastern Africa to pass the off-season there (cf. Friedmann, 1964).

Group E. There are at least sixteen passerine species in common, as shown below, nearly all represented by different subspecies in the two continents. Those marked * occur also widely in the Palaearctic Region and some of them elsewhere also.

(a) A swallow *Hirundo daurica**, Crested Lark *Galerida cristata**, Great Grey Shrike *Lanius excubitor**, the longtailed warbler *Prinia gracilis* and the finch lark *Eremopterix nigriceps*, all of which are widely distributed in the Arabian peninsula.

(b) Paradise Flycatcher *Terpsiphone paradisi*, two pipits *Anthus similis* and *novaeseelandiae**, a lark *Mirafra javanica*, Fantail Warbler *Cisticola juncidis*, Silverbill *Lonchura malabarica*, Stonechat *Saxicola torquata**, all of which in Arabia breed only in the comparatively well-vegetated south-western and/or south-eastern corners.

(c) The "tree-creeper" *Salpornis spilonotos*, Wire-tailed Swallow *Hirundo* (*Riparia*) *paludicola* and warblers *Prinia subflava* and *Schoenicola platyure*, none of which now appear in Arabia at all.

Some of these species shared between tropical Africa and India have such remarkable total ranges that they merit comment. The pipit *Anthus novaeseelandiae* occupies the eastern Palaearctic (where it is a long-distance migrant), Australia and also New Zealand, in addition to much of the Oriental and Ethiopian Regions. Stonechats occupy the greater part of the Palaearctic Region whence some of them migrate to winter as far south as Abyssinia, while some eastern birds reach the Philippines; other, sedentary, populations live in parts of India, the Ethiopian Region (usually in montane areas), Madagascar, the Comoro Islands and Réunion. The Paradise Flycatcher actually surpasses this, for besides being distributed over the Ethiopian and Oriental Regions, including Malaysia, it has, at any rate as a superspecies, established itself widely over the islands of the Indian Ocean, reaching as far as Mauritius, Réunion and even the remote Seychelles. *Cisticola juncidis*, however, in certain ways exceeds all these and qualifies as the most remarkable small bird in the world. Though tiny, short-winged and to all appearances everywhere sedentary, it has occupied nearly all the Ethiopian Region, the coasts of the Mediterranean and a succession of stations right across the Oriental Region to Japan and north-east Australia. This last, most easterly, population is so like the most westerly, in West Africa, that in the words of the monographer of the genus "it is hard to find any difference" between these two extreme populations of *C. juncidis* (Lynes, 1930). Whatever the inherent stability of the species and whatever the power of convergence, it seems necessary to accept that most of this bird's immense expansion has taken place in very recent evolutionary time. Such examples of range-extension, which seem unaccountable, are useful to bear in mind as cautionary, when attempting any zoogeographical reconstructions.

In sum, it can be said that at the generic level the affinities between the Ethiopian Region and India are close, even among the passerines, while the number of species in common is appreciable. A good many of the genera, and a few of the species, are found also in the Palaearctic Region, but for the remainder it is necessary to envisage a former continuous range through Arabia. In fact, nothing more lush than steppe or savanna with scattered trees is required by any of the species listed above as common to the Ethiopian Region and India, though it is perhaps surprising that a bird so apparently sedentary and so closely dependent on trees as *Salpornis* should be among them. Thus in all these cases it is possible that their range from Africa to India was continuous across Arabia no longer ago than the Neolithic humid period around 6000 years before the present.

Since none of the birds cited above as common to tropical Africa and India is a forest species, it is difficult to know how Chapin (1932) arrived at the conclusion that while "the savanna forms [of the two areas] show much in common, the resemblance is just as marked in the forest faunas". However, there are at least two forest birds in the Ethiopian Region that raise very difficult distributional problems. An owl in the Congo, *Phodilus prigoginei*, of which the type and sole specimen was discovered only in 1951, is very like the Indian *P. badius*; the warbler known as *Artisornis metopias*, sporadic in Tanganyika and Portuguese East Africa, is now believed (Hall and Moreau, 1962) to be congeneric with the Indian tailor-birds, *Orthotomus*, for it resembles them closely in plumage and also in stitching leaf-edges together for its nest. Both the *Phodilus* and "*Artisornis*" are known in Africa only in montane evergreen forest, while their Indian relatives are not typical of this habitat there (Ripley, 1961). The situation is thus most peculiar. Convergence cannot be ruled out, but I do not favour it, because in *Artisornis/Orthotomus* it involves the assumption that it affected the nest-building technique as well as the plumage. On the other hand, if the relationship between the Indian and African *Phodilus* owls and the Indian and African warblers is indeed as close as it looks, then the explanation is difficult. I regard accidental colonization as impossible, because of the immense distance involved and the sedentary nature of the birds. A former continuous

range of the ancestral stock across Arabia is left as the only possibility.
Since, as discussed above, this seems never to have provided a forest
connexion, a change in the birds' habitat preferences must be envisaged.
If the stocks came from the east, then they must have "taken to" evergreen forest, and secondarily to montane evergreen forest, after arrival
in Africa; if the ancestral stocks were African there are two possibilities.
Either they were typically non-forest birds that somehow became
adapted to montane forest and became extinct elsewhere on the continent after sending colonists across Arabia, or they were typically
forest birds that managed to spread by way of non-forest vegetation.

Affinities with Europe

Since Europe is that part of the Palaearctic Region which is directly
north of Africa, its bird fauna is of the most interest for comparative
purposes, and Voous's (1960) "Atlas of European Birds" is available as
a basis from both the geographical and the taxonomic points of view.
It may be noted that the figures I arrive at below differ from those in
Voous's (1959) Palaearctic/Ethiopian comparison because the areas
taken differ; for one thing, for his purpose the Palaearctic includes the
Sahara and for mine it does not. The list of families of land and freshwater birds is shorter for Europe than for the Ethiopian Region, for,
while the latter lacks divers, grouse, waxwings, accentors, dippers, and
wrens, Europe has no ostriches, darters, hamerkops, secretary birds,
finfoots, jacanas, hornbills, colies, trogons, barbets, honey-guides, pittas,
broadbills, white-eyes or sunbirds; but the proportion of families shared
is very high. Among the genera, the proportion shared varies remarkably
between the different types of birds, as shown in Table XI.

It will be seen that much the smallest proportion of shared genera
is amongst the ducks and waders, both of which are richly represented
at high latitudes and, in the case of the scolopacine waders, hardly at
all in the Ethiopian Region. In the other water-bird genera, of which the
herons and the rails are the most numerous, the great majority are common to the two areas; this applies also to the raptor group and also
to the "other non-passerines" in group D. In this last the genera are
represented in Europe by single species, such as "the" roller, bee-eater,
cuckoo. But the proportion is lower in the few ground-bird genera
(group C) and lowest of all, only two-fifths, in the passerine.

This result is to some extent reflected at the species level. On the basis
of the most generally accepted taxonomic rating, there are established
in at least some part of both Europe and the Ethiopian Region twenty-
nine species of water-birds (but none of duck or gull), twelve of the raptor
group (B), three ground-birds (C), ten other non-passerines (D), and
only nine of the 169 passerines on the European list. These are Stonechat
Saxicola torquata, the crested larks *Galerida cristata* and *G. theklae,* a swallow
Hirundo daurica, the Chough *P. pyrrhocorax,* the Olivaceous Warbler
Hippolais pallida, the so-called Rufous "Warbler" *Erythropygia galactotes,*

the Great Grey Shrike *Lanius excubitor* and a "grass warbler" *Cisticola juncidis*. Only one of these species, the Stonechat, is really widely (though sporadically) distributed in both Europe and Africa south of the Sahara.* Most of the other species are only marginal in the two areas; for example, the Olivaceous Warbler breeds in Spain and the Balkans and also in the acacia steppe that forms the southern edge of the Sahara. All the species are subspecifically differentiated north and south of the Sahara, but the Chough and the Olivaceous Warbler only slightly. It seems probable that during the Last Glaciation and perhaps again as late as the Neolithic humid period each of the species in question had a

TABLE XI

Genera of breeding birds shared by Europe and the Ethiopian Region

Group	Total	European genera shared with Ethiopian Region	
		No.	Proportion
A. Ducks and waders	29	6	0·2
Other water-birds	35	26	0·7
B. Raptors, etc.	25	17	0·7
C. Ground-birds	11	5	0·5
D. Other non-passerines	15	12	0·8
E. Passerines	66	25	0·4
	181	91	

continuous breeding-range across some part of the Sahara. The Chough is peculiar, since it is found in the Ethiopian Region only high in the mountains of Abyssinia; it presumably spread south by way of the hills between the Red Sea and the Nile, during the cool damp of a glaciation.† The Neolithic humid period cannot be ruled out as the occasion for its spread in range, but at that time the temperature along the route would have been higher than that in the stations occupied by the Chough nowadays.

In this connexion it may be noted that there are two species with special Palaearctic affinities in Somaliland. *Columba oliviae* is nearer to the Stock Dove *C. oenas* and the Asiatic *C. eversmanni* than to anything in Africa (Goodwin, 1959); but it must be noted that it inhabits the arid foothills, not the montane heights, of Somaliland. In one corner of this territory appears also the rare fringillid described as *Warsanglia johannis* which is now accepted as a *Carduelis*, apparently the only one in tropical

* Note that two Palaearctic species, the Great Tit *Parus major* and the Blackbird *Turdus merula*, which have penetrated so deeply into the Oriental Region, to Ceylon and Malaysia, have made no impact on tropical Africa at all.

† The same route may be supposed to have served for the parent stock of the Abyssinian Ibex *Capra(ibex) waalie* and certain Palaearctic genera of plants, e.g. *Rosa* and *Primula*, which likewise occur in the Ethiopian Region only in the Abyssinian mountains (Hedberg, 1961).

Africa. The presence of another member of the superspecies on the mountains of the Yemen (Hall and Moreau, 1962) suggests that the Somali stock may have arrived by way of the mountains of western Arabia during a glaciation and then jumped the Gulf of Aden. But equally it could have come along the mountains on the east side of the Red Sea, with subsequent extirpation of the intermediate populations.

In addition to the birds named above, a few more on the European list (a total of less than twenty in all groups) have been suggested by one taxonomist or another as conspecific with what are generally accepted as different species in the Ethiopian Region. Examples are *Hirundo rustica/lucida* and *Motacilla cinerea/clara*, which, as Voous remarks, "can neither be proved nor disproved". Such pairs form, in fact, superspecies which straddle both Africa and Eurasia. From my own general knowledge of the two bird faunas under discussion I should say that when the general taxonomic revision from the point of view of the superspecies comes to be made, the number of these will not be greatly increased, certainly not among the passerines. Hence the number of species breeding widely in both Europe and the Ethiopian Region, and even the superspecies shared between the two areas, is not a large proportion of either bird fauna, especially in the passerines. This is the more remarkable for two reasons. One is the fact that more than once in the Pleistocene, and for the last time only some 5000 years ago, the western Sahara has been nothing like the barrier to interchange of species that it is now—a topic already discussed in Chapters 3 and 4. The other reason is that even now, with the Sahara in its present condition, many species and enormous numbers of individual birds annually cross it to winter in the Ethiopian Region after breeding in the Palaearctic (Chapter 14). But since so few birds of the Ethiopian Region and of the Palaearctic (including the Maghreb) are sufficiently alike for anyone to have suggested that they are conspecific, it is clear that it has been rare in recent evolutionary time for any species typical of one area to establish itself in the other. Hence, as concluded when discussing the bird fauna of Mediterranean Africa in Chapter 4, in view of the opportunities that have recurred in the Late Pleistocene and for many species recur annually now, the extent of the difference between the European and Ethiopian bird faunas must be due to physiological and ecological factors rather than to geographical.

It is remarkable that, as shown in Table XII, the number of species in each group that are established as breeding birds in both Europe and the Ethiopian Region bears no consistent relation to the number of species whose annual visits to areas south of the Sahara gives them repeated opportunities for establishing themselves there. While one-fifth of all the 160 species regularly crossing the Sahara have breeding representatives in the Ethiopian Region, the proportion in the passerines is only two out of fifty-five and in the ducks and scolopacine waders only one out of thirty-two. Moreover, of those migrant species which do have a resident population also south of the Sahara, all but eight (and those water-birds) are subspecifically divided. This means that in twenty-four of the thirty-two species concerned the two populations have been genetically isolated long enough to diverge and that any gene-inter-change there may be between individual migrants from Europe and the conspecific birds breeding in the Ethiopian Region is too limited to be of significance. This limitation of gene-flow will be due in part to the fact that in most species the seasonal rhythms of the European and Ethiopian populations are different, for the case of the Black Stork, discussed below, is exceptional. In a number of species there is also ecological or geo-graphical segregation between the migrants and the residents. For example, in the bitterns *Botaurus* there is no evidence that the Palae-arctic birds get farther south than about the Equator, while the resident birds are all south of the Zambezi. Again, in the buzzards, the residents *Buteo buteo oreophilus* are confined to forested areas, in the tropics only at high altitudes, while the visiting *B. b. vulpinus* range widely over open country.

TABLE XII

Establishment of European species in the Ethiopian Region in relation to opportunity given by migration

Group	No. of European species regularly visiting the Ethiopian Region	No. of species in col. 2 breeding in Ethiopian Region	Total no. of species breeding in both Europe and Ethiopian Region
A. Ducks and scolopacines	32	1	1
Other water-birds	38	16	28
B. Raptors, etc.	21	7	12
C. Ground-birds	3	2	3
D. Other non-passerines	11	4	7
E. Passerines	55	2	8
Total	160	32	59

Another aspect of this subject is that, while altogether thirty-two species are represented by both Palaearctic migrant and Ethiopian populations, nearly as many others (twenty-seven) have breeding popu-lations in both regions, but the Palaearctic individuals do not cross the Sahara regularly. The present situation must, then, even more clearly than in the case of the thirty-two discussed above, be a consequence of conditions different from the present, presumably those of the glaciation or even of the Neolithic humid spell.

It is interesting that several of the species which breed regularly in both Europe and the Ethiopian Region (not merely in its northernmost part) are typically birds of high altitudes and of subtropical southern

Africa; for example, the grebes *Podiceps cristatus* and *nigricollis* (but not *ruficollis*), Alpine Swift *Apus melba*, Quail *C. coturnix*, Black Stork *Ciconia nigra*, Buzzard *B. buteo*, Lammergeier *Gypaetus barbatus*. This applies also to a proportion of those birds which look as if they were closely related, but are not generally accepted as conspecific: the wagtails *Motacilla cinerea/clara*, the rock-thrushes *Monticola saxatilis/explorator* or *rupestris*, the vultures *Gyps fulvus/coprotheres*.

Among the birds mentioned, two especially call for separate remark, the Black Stork and the Rock Thrush. The Black Stork is the only one of those listed above in which the Palaearctic and Ethiopian populations are indistinguishable. In Europe it breeds as far south as the Balkans and southern Spain. All these birds leave for beyond the Sahara and winter south at least to Malawi, where ringed birds have been recovered. Meanwhile other Black Storks nest, where precipices occur, over a vast area of southern Africa from Malawi to the Cape. The breeding-ranges of the populations of this species are thus separated by some 45 degrees of latitude, a distance of over 3000 miles, which is spanned by the annual migration of the Palaearctic birds. The fact that the two populations are morphologically indistinguishable could mean (1) that the two breeding-ranges now so widely sundered formed part of a continuous range not long ago; or (2) that one of them represents a colony that has, on the scale of evolutionary time, only recently been established; or (3) that the Black Stork is not a plastic species; or (4) that stability is maintained by an appreciable gene-flow between the two breeding populations. This last possibility would be facilitated by the fact that the breeding season in Malawi, June and July, overlaps that in Europe, so that modification rather than revolution in its breeding rhythm would be necessitated on the part of a European Black Stork that did not leave its wintering grounds in southern Africa. (For a discussion of the proximate factors influencing breeding seasons see Moreau (1964b) and references therein.)

The Palaearctic Rock Thrush *Monticola saxatilis* breeds in a great area from Spain and Barbary to Mongolia, from the whole of which the birds go to winter in Africa, nearly all north of the Equator, but some ranging to about 8° S. Resident in Africa there are one species of *Monticola* in Abyssinia and neighbouring areas (*rufocinerea*), and four others from Angola and southern Tanganyika southwards. Two of these, *M. rupestris* and *M. explorator*, which are sympatric from the Transvaal to the Cape, are at once the most geographically remote from the European bird and the more like it in appearance than the Abyssinian species. Prima facie this suggests that the Palaearctic and some of the South African rock thrushes have a common origin later than that of the five African species. Since the winter range of the European bird does not overlap that of *M. rupestris* and *M. explorator*, the resemblance between the three is unlikely to be due to continuing gene-exchange. If the African birds represent colonists from Europe, then their ancestral stock arrived long enough ago for its descendants to diverge to specific level between themselves. If the European bird is derived from African stock, it has retained a special resemblance to those African rock thrushes which are geographically the most remote from it.

In this connexion it may be noted that there are a few Palaearctic migrant species that have been reported to breed sporadically in the Ethiopian Region. With one exception, to be discussed below, all the records come from 20 degrees or more south of the Equator. It may be significant that this is the part of the Ethiopian Region where the summer day-length, as well as the temperatures, in extremes and daily range, approximate most to those of at least the southern edges of the Palaearctic ranges of the species concerned, though the seasons are, of course, six months out of step with the Palaearctic. The records in question are the following, as given by McLachlan and Liversidge (1958) except where otherwise specified. White Storks *C. ciconia* nested at Oudtshoorn in the eastern Cape Province for seven successive years around 1940 and another pair in the Western Cape at Bredasdorp for some years before 1961 (Martin *et al.*, 1962). One of the brood raised here in 1961, and ringed in the nest in December, was recovered near the southern border of Tanganyika, nearly 2000 miles to the north, in March 1962 (*Ostrich* (1963) **34**, 48). This looks like an attempt to regain the Palaearctic homeland; perhaps it was stimulated by the movements of other White Storks which had wintered in the neighbourhood and were returning to Eurasia in the normal way; but it is understood that this young bird's own parents did not go.

House Martins *Delichon urbica* have also nested in southern Africa, one pair at Cape Town in 1892, a small colony at 32°40′ S. near King William's Town for some years and another far to the north of these at about 20° S. in South West Africa, also lasting for some years. Finally, and more remarkable in its South African record than either of these species, is the European Bee-eater *Merops apiaster*. It is known to have been breeding sporadically in South Africa, and perhaps consistently in some localities, for at least the last one hundred years. Like some other early Palaearctic migrants, *M. apiaster*, which are passing south through Morocco before the end of July (Heim de Balsac and Mayaud, 1962), appear in South Africa usually in September. The neighbourhoods of Cape Town and Port Elizabeth are perhaps the most favoured at the present time, but R. Liversidge tells me that, among others, Irene in the Transvaal and the banks of the lower Orange River have been breeding localities in recent years. Nesting begins very shortly after the birds appear (September–December in the Cape, October–January at Port Elizabeth) and they are gone again by the end of February. The timing thus makes it possible for the individuals concerned to have a second breeding season in Europe, but whether they do so is unknown. Certainly, though the South African birds have been stated to spend their off-season in tropical Africa (Clancey, 1964), there is as yet no direct evidence that any bee-eaters of this species spend the northern summer anywhere in the Ethiopian Region. A double breeding in the year in localities so distant from one another as sub-tropical South Africa and the Palaearctic, some 70 degrees of latitude, or nearly 5000 miles, would be unique; but there is a possibility that something similar on a smaller scale may be performed by the quail *Coturnix coturnix*. Quail nest in the Maghreb in early spring; in June parents and young cross the Mediterranean northwards and the breeding season in Europe is so extended that the birds from Barbary might well breed there again, but there is no proof (see discussion in Moreau, 1951).

The remaining Palaearctic migrant reported to breed in the Ethiopian Region, and the only one claimed to do so anywhere in the tropics is the Common Sandpiper *Tringa hypoleucos*. According to Mackworth-Praed and Grant (1957–60) this has been recorded breeding in various months and localities in Uganda, Kenya and Tanganyika. Consequently Voous (1960) has shown a large breeding area in East Africa for the species. There seems, however, to have been some confusion. The only first-hand data seem to be (1) a record by the van Somerens (1911) in Uganda, and (2) the statement by Meinertzhagen in Jackson and Sclater (1938) to the effect that he had seen a pair of Common Sandpipers in June on the Kajiado River "with young chicks just fledged". I am indebted to C. W. Mackworth-Praed and others for the information that they know of no other claims that this sandpiper breeds in East Africa and I conclude that at most it does so very rarely, though individuals have not infrequently been seen there in the months of the northern summer.*

It is always possible that other Palaearctic migrants will be found breeding somewhere in the Ethiopian Region. Of those mentioned above, the White Stork, House Martin and Bee-eater are all specially liable to be noticed because of their conspicuousness and/or their association with man. Furthermore, the area in which they have been detected is a part of Africa better supplied with resident ornithologists than nearly all others. By contrast, a bird like the Nightingale *Luscinia megarhynchos*, which sings so vigorously and persistently in winter quarters at low altitudes within a few degrees of the Equator, would, because of the impenetrable nature of the vegetation it inhabits there, be extremely likely to escape detection if it nested. Making some allowance for sporadic breeding by Palaearctic species having gone undetected, the rarity of the occurrence is still very remarkable, and, in conjunction with the low proportion of species which are established permanently in both the Palaearctic and the Ethiopian Region, it emphasizes the difficulty that a migrant must find in establishing itself as a regular breeder south of the Sahara. One hindrance is likely to be the necessity

---

* Another Palaearctic wader, *Charadrius leschenaultii*, has once been reported with small young in Somaliland (Archer and Godman (1937), Vol. 2, p. 384), which may be a southern extension of breeding that is believed to take place along the coasts of the Red Sea.

for adjusting the individual breeding rhythm to enable breeding to take place in an African locality at a favourable time of the year; for example, between September and April for birds of the Swallow family in South Africa. However, we know from the breeding records it can in this case be accomplished, so that this factor would presumably be a minor one compared with that of competing in communities as rich and well established as those of the Ethiopian Region.

Summing Up

Relationships with the other Regions are much closer in water-birds (group A) and raptors, etc. (group B), than they are in the other groups, especially the passerines. This is attributed partly to the greater age of groups A and B, partly to the greater individual mobility of birds of these groups and, in the case of many of the water-birds, to the fact that seas form less of a barrier to them than they do to other birds.

Comparatively few taxa above the level of the genus are confined to the Ethiopian Region because a very large proportion of them are shared with Europe or with India or with both. Among the genera few are shared with South America and only ten species (out of 1481), most of them water-birds. By contrast, half the Indian and half the European genera are represented south of the Sahara. This identity of proportions in the genera as a whole conceals, however, important constituent differences. The first is that in the European ducks and waders the proportion shared with the Ethiopian Region is much lower than that in India or in any other group on either list. The second important difference is that the group D non-passerines, which form a much smaller part (0·06) of the European list than of the Indian (0·25), are nearly all of Ethiopian genera, whereas in India less than half of them are. This suggests that the group D birds of Europe have been derived comparatively recently from Africa, while the Indian have developed more independently. But in the other groups of birds, especially the passerines, the proportion of genera shared with the Ethiopian Region by Europe and by India is the same.

Of the 1481 species on the Ethiopian list, a few less than one-twentieth (69) appear also in India and in Europe (63). Moreover, as shown in Table XIII, in each group the number shared is about the same, except that the shared passerines are fewer in Europe than in India. This result is somewhat surprising. Affinities between Europe and sub-Saharan Africa might have been expected to surpass those of India: firstly, in view of the history of the Sahara, which as recently as 5000 years ago presented no such barrier as it does now; secondly, because each year the migrations of so many Palaearctic species take them into the Ethiopian Region for several months after they have bred each year and thus give them repeated opportunities for settling there. Only one-fifth of the present-day migrant species have, in fact, breeding populations both in Europe and south of the Sahara, and most of these are water-birds and raptors, while hardly any are passerines. There are a few records of European species breeding sporadically in South Africa, and the reason why more migrants do not establish themselves is presumably that the Ethiopian bird communities are too closely knit for additional species to gain admission to breed, even though they find it possible to spend months in Africa during their off-season (see further discussion in Chapter 14).

TABLE XIII

Ethiopian species shared with India and with Europe

Group	No. of Ethiopian species shared with	
	India	Europe
A	25	29
B	14	12
C	3	3
D	11	10
E	16	9
	—	—
	69	63

Of the species shared between India and the Ethiopian Region, some are common to the Palaearctic also, but those which are purely tropical either still occur in Arabia or presumably did so recently, during the Last Glaciation or the Neolithic, when its climate was not so desertic as it is now.

7

Comparisons of Ethiopian Territorial Non-forest Bird Faunas

As will be seen, some interesting results emerge if the numbers of species breeding in different territories are compared group by group and family by family, and if certain limitations are observed. Most important, species of evergreen forest must be reserved for separate treatment (Chapters 9 and 11); the geographical distribution of this habitat is so uneven that territorial statistics that include forest species are not comparable and have little significance. This chapter is therefore concerned only with birds already classified in Table V as non-forest. Marine species and sporadically breeding Palaearctic migrants are also omitted. A further point is that in tropical Africa some territories include montane areas that have an appreciable number of characteristic non-forest species, while others have no such areas; therefore, to make the territorial lists more fully comparable, in Table XIV I have added montane species *en bloc* at the end. The last two columns in this table show for each family how many species are common to certain geographically extreme areas of the Ethiopian Region: (*a*) West Africa and Cape Province, and (*b*) Somaliland and Cape Province; the results have already been discussed in Chapter 5.

The compilation of Table XIV would have been almost impossible without the information that has been provided by the local and sub-regional works of recent years. I have used Bannerman (1930–51) for West Africa; the same supplemented by E. M. Cawkell (unpublished) for Gambia; Cave and Macdonald (1955) for the Sudan; Mackworth-Praed and Grant (1957–60) for Kenya and Uganda; Archer and Godman (1937–61) for Somaliland; Benson and White (1957) for Zambia; Benson (1953) for Malawi; Smithers *et al.* (1957) for Rhodesia; McLachlan and Liversidge (1957) for South West Africa and Cape Province. In the case of the last two (in connexion with which I am indebted to Mrs M. K. Rowan for some interpretations of data), I have read the ranges off the small-scale maps. For convenience the Caprivi Strip has been omitted from South West Africa for this purpose and the northern boundary of Cape Province has been taken as the Orange River (omitting the Gemsbok Park enclave to the north of it). To some of the territorial lists dealt with here a few species have been added by local workers since the publication of the standard works cited above, but for the sake of uniformity I have not included any of them. They would make no significant differences to the territorial totals and in any case these cannot be definitive; in some cases breeding is assumed, not proved, in a particular territory, for a species that seems to be sedentary there. Moreover there are many cases in which a species is recorded only just inside a territorial boundary or is recorded as rare. Since it is impossible to discriminate in such cases, credit has been allowed for all of them. This has probably slightly inflated the totals of breeding species throughout, but the relation between the totals is certainly valid as they stand.

It might be supposed that, especially in the larger territories, the number of species credited to them might be swollen by closely related species, members of superspecies, replacing one another geographically. The numbers involved cannot be estimated precisely, pending a taxonomic revision of the Ethiopian bird fauna from the point of view of the superspecies, but a preliminary sample check I have made with the assistance of C. M. N. White makes it clear that the totals would not, even in the largest territories, be reduced by more than 5% at the outside, if Table XIV were compiled on superspecies rather than species. Hence the statistics in it can be accepted as a basis for discussion that is valid from the taxonomic point of view.

The territories treated in Table XIV have been selected with several considerations in mind, in part the prospect that comparisons between them would be particularly apt and informative, in part that each territory should have been satisfactorily explored and documented. As a preliminary to discussion it is useful to recall some main geographical features of each territory concerned and to refer the reader to Chapter 2 for an indication of its vegetation. For the present purpose it is an advantage, rather than the reverse, that most of them are not enclosed by natural boundaries.

West Africa. This area, west of the Cameroon highlands and south of the Sahara, but excluding that occupied by the Upper Guinea forests, amounts to a little over 1 million sq. miles. It has a wide climatic range, with parallel belts of subdesert steppe, acacia steppe, isoberlinia and other, moister, savanna; but it lacks internal barriers to non-forest species and it lacks any montane areas that have characteristic species.

Sudan. 970000 sq. miles with a similar extensive range of conditions (and also a very small montane area on the southern border).

Gambia. 4000 sq. miles, with an altogether smaller range of conditions than the foregoing, for it is wholly occupied by savanna of various kinds. Because of its very small size Gambia is not fairly comparable with the other territories; it is inserted in Table XIV in order to bring out one particular point of principle.

Somaliland. This is the former British territory, of 68000 sq. miles, which consists of nothing more lush than subdesert steppe and acacia steppe, except for a limited montane area of dry type.

Uganda. 90000 sq. miles, with a wide range of habitats, from acacia steppe to wet savanna and including important montane areas.

Kenya. 220000 sq. miles, with an even more complete range of habitats, from desert alongside Lake Rudolf and much acacia steppe in north and east, to savanna and highland grassland with a montane bird fauna.

Zambia. 290000 sq. miles, with a range of altitude, precipitation and habitats much more restricted than any of the territories mentioned above except Somaliland and Gambia. Practically the whole country is under more or less wooded savanna of different types (acacia, brachystegia, mopane), and also in the extreme west the anomalous dry evergreen *Cryptosepalum* forest.

Rhodesia. 150000 sq. miles, again nearly all rolling wooded savanna of various types, lower and drier in the west, towards the Kalahari, higher and damper on the eastern border.

Malawi. 37000 sq. miles, dotted with spectacular isolated mountains and devoid of anything drier than wooded savanna.

South West Africa. 318000 sq. miles, the greater part of which is very dry, varying only from coastal desert to acacia steppe, though mopane woodland appears in the north.

Cape Province. 272000 sq. miles, with high relief and with a great range of climate from desert (along the lower Orange River in the north-west) to the humid Indian Ocean coast carrying a mosaic of savanna and forest.

It will be seen that of the eleven territories under consideration, four (West Africa, Sudan, Kenya and Cape Province) have practically the widest possible range of lowland climatic and vegetation types, while Uganda lacks only subdesert steppe. All these territories except West Africa have in addition some montane areas with a few characteristic species. By contrast, each of the other six territories is much less varied ecologically. It would have been interesting to give the differences in habitat range numerical expression, but I have been unable to devise a satisfactory formula. It may be added that watery habitats vary greatly in prevalence in the various territories, but it is probably true that only in Somaliland are they so deficient as significantly to limit the number of species in the associated bird fauna (group A).

In considering Table XIV, a comparison must first be made between the figures for West Africa and those for Gambia. It will be seen that, with less than 0·04% of the area of West Africa, Gambia counts nearly three-fifths as many breeding species. Especially since Gambia has a much smaller range of habitats, this is a powerful illustration of the well-known fact that, other things being equal, reduction in area is not accompanied by proportionate reduction in number of species accommodated. The Gambia case shows that in the following discussion the differences between the areas of the other territories can be discounted. With this factor disposed of, it would be expected that to a great extent the avifaunal wealth of a territory would be positively correlated with the range of habitats it provides.

TABLE XIV

Territorial totals (by families) of breeding species not dependent on evergreen forest

Area (in thousands of sq. miles):	West Africa over 1000	Gambia 4	Sudan 970	Kenya 220	Uganda 90	Somali- land 68	Zambia 290	Malawi 37	Rhodesia 150	S.W. Africa 318	Cape Prov. 272	C.P.+ W.A.	Som.+ C.P.
Group A													
Anatidae	7	5	7	12	11	1	13	12	13	12	12	5	
Anhingidae	1	1	1	1	1	—	1	1	1	1	1	1	
Ardeidae	15	12	14	15	16	4	17	15	15	12	16	14	
Balaenicipitidae	—	—	1	—	1	—	1	—	—	—	—	—	
Balaericidae	1	1	1	1	1	—	2	2	2	2	3	1	
Charadriidae	10	3	10	12	12	3	10	7	8	7	7	3	
Ciconiidae	4	3	6	4	6	—	6	6	3	4	1	—	
Heliornithidae	1	1	—	1	1	—	1	1	1	—	1	1	
Jacanidae	2	1	2	2	2	—	2	2	2	2	2	2	
Laridae	3	—	1	3	2	1	4	3	3	3	2	1	
Pandionidae	1	1	1	1	1	—	1	1	—	—	1	1	
Pelecanidae	1	1	1	2	1	—	2	1	—	1	1	1	
Phalacrocoracidae	1	1	2	2	2	—	2	2	2	2	2	1	
Phoenicopteridae	—	—	—	1	1	—	1	—	—	—	—	—	
Podicipitidae	1	1	1	1	1	1	1	1	1	2	3	1	
Rallidae	9	3	5	12	10	—	13	13	12	10	12	6	
Recurvirostridae	1	1	2	2	2	1	2	1	2	2	2	1	
Rostratulidae	1	—	1	1	1	—	1	1	1	1	1	1	
Scolopacidae	—	—	1	1	1	—	1	1	1	1	1	—	
Scopidae	1	1	1	1	1	1	1	1	1	1	1	1	
Threskiornithidae	4	2	4	4	3	1	4	4	2	2	5	3	
	64	38	62	80	77	13	86	75	69	65	74	44	

Group B

Aegypiidae	6	4	7	6	5	5	4	3	3	7	3	2	2
Aquilidae	23	19	23	24	23	12	23	24	23	20	21	16	11
Falconidae	7	3	8	9	7	5	7	5	5	7	6	4	4
Sagittariidae	1	1	1	1	1	1	1	1	1	1	1	1	1
Tytonidae and Strigidae	8	6	9	10	8	5	10	11	10	8	10	7	3
	45	33	48	50	44	28	45	44	42	43	41	30	21

Group C

Burhinidae	3	1	2	3	3	1	2	2	2	2	2	2	1
Glareolidae	6	2	4	7	5	3	5	3	5	6	4	3	1
Otididae	6	2	9	7	4	7	4	2	3	6	8	3	1
Phasianidae	10	5	12	16	17	6	14	13	11	9	9	3	1
Pteroclidae	2	2	5	4	3	2	2	1	3	4	2	—	—
Struthionidae	1	—	1	1	—	1	—	—	1	1	1	1	1
Turnicidae	3	1	2	3	3	1	2	2	2	1	2	2	1
	31	13	35	41	35	21	29	23	27	29	28	14	6

(continued overleaf)

TABLE XIV (cont.)

Area (in thousands of sq. miles):	West Africa over 1000	Gambia 4	Sudan 970	Kenya 220	Uganda 90	Somaliland 68	Zambia 290	Malawi 37	Rhodesia 150	S.W. Africa 318	Cape Prov. 272	C.P.+ W.A.	Som.+ C.P.
Group D													
Alcedinidae	9	8	8	9	8	2	9	9	9	6	7	3	—
Apodidae	5	3	5	8	5	3	6	7	7	5	5	2	2
Bucerotidae	3	4	7	9	6	3	7	6	6	7	2	1	—
Capitonidae	7	4	14	11	12	4	8	7	5	2	3	1	1
Caprimulgidae	9	2	10	8	7	4	7	4	5	4	3	2	—
Coliidae	2	—	2	3	2	1	2	2	2	2	3	1	—
Columbidae	11	8	14	10	12	10	8	8	10	6	7	5	6
Coraciidae	4	4	4	4	4	2	4	4	4	2	—	—	—
Cuculidae	12	5	10	12	13	4	13	11	12	7	8	7	2
Indicatoridae	2	1	3	5	5	2	5	5	5	1	4	2	2
Meropidae	7	3	7	5	5	4	6	5	4	4	1	—	—
Musophagidae	3	3	3	3	5	1	3	3	2	1	1	—	—
Phoeniculidae	2	2	3	4	4	2	2	2	2	2	1	1	—
Picidae	6	4	10	9	8	3	4	7	4	4	5	1	1
Psittacidae	4	3	4	4	3	1	3	4	4	4	1	1	—
Upupidae	1	1	1	1	1	1	1	1	1	1	1	—	—
	87	55	105	105	100	47	88	85	82	58	52	25	14

Group E

	1	2	3	4	5	6	7	8	9	10	11	12	13
Alaudidae	11	3	15	18	8	18	10	5	9	17	19	2	3
Campephagidae	2	2	2	2	2	—	2	2	2	2	1	1	—
Corvidae	3	2	4	4	3	4	3	2	3	3	3	1	2
Dicruridae	1	1	1	1	1	1	1	1	1	1	1	1	1
Emberizidae	4	—	5	4	4	3	4	4	4	4	4	2	1
Fringillidae	3	2	5	8	6	5	6	5	6	5	10	2	—
Hirundinidae	12	5	10	12	12	2	16	14	12	9	9	4	—
Laniidae	14	7	18	28	20	11	15	15	14	13	9	5	2
Motacillidae	5	2	5	13	7	4	12	10	11	5	10	4	3
Muscicapidae	10	8	13	15	12	5	14	14	13	10	11	1	1
Nectariniidae	11	9	14	15	14	5	13	11	11	6	7	1	—
Oriolidae	1	1	1	2	2	—	2	2	2	1	1	—	—
Paridae	3	1	5	6	6	2	5	5	5	5	3	1	—
Pittidae	—	—	—	—	—	—	1	1	—	—	—	—	—
Ploceidae													
Estrildinae	25	15	29	27	28	8	28	20	24	11	11	8	—
Others	26	17	42	49	45	14	33	25	23	14	20	8	3
Promeropidae	—	—	—	—	—	—	—	—	1	—	2	—	—
Pycnonotidae	4	3	4	6	2	1	5	5	4	2	5	1	1
Salpornithidae	1	—	1	1	1	—	1	1	1	—	—	—	—
Sturnidae	11	6	14	17	13	11	12	10	9	7	6	2	2
Sylviidae													
Cisticolae	13	7	14	15	14	2	20	13	13	6	11	3	1
Others	14	8	23	23	22	13	22	15	17	14	17	5	2
Timaliidae	3	4	5	8	4	2	3	1	3	4	1	—	—
Turdidae	13	5	15	14	12	6	13	15	17	14	18	3	—
Zosteropidae	1	1	2	1	1	1	1	1	1	1	1	1	—
Grand total	191	109	247	289	239	118	242	197	206	154	180	56	22
	418	248	497	565	495	227	490	424	426	349	375		
Additional montane species	—	—	14	38	27	3	15	20	6	—	—		

Table XIV shows that this expectation is not realized in general, though it is in some instances. Certainly Kenya, with an unsurpassed range of habitats, has the greatest number of species, 565 (+ 38 montane); but then, with a drop of about 16%, follow Sudan, Uganda and Zambia (497, 495 and 490 respectively), although the last has by far the smallest range of habitats. Moreover, Malawi has much the same limited range, and yet its total of 424 species (+ 20 montane) exceeds the 418 of the vast West African area and still more the 374 of Cape Province, although both these last territories have the full habitat range.

The superiority of Kenya is based very broadly, for it has the largest number of species in every group except the water-birds, of which it has a few less than Zambia. In group D its total predominance is only slight and it is actually surpassed by one territory or another in half the families; but in passerines its predominance is much more consistent, for it is surpassed only in finches, buntings, swallows and estrildines, by a different territory in each case and, except in swallows, by only the narrowest of margins. In general Kenya's wealth of species owes much to the fact that the dry eastern half of the territory is occupied by a typical arid-country bird fauna which extends from Somaliland to central Tanganyika, while its humid west, in the Lake Victoria basin, receives an infusion of Congo species. Moreover, it seems most probable that if eventually the statistics can be based on superspecies rather than species the predominance of Kenya will still not be upset, and a thorough examination of the ecological relationship between the components of each family would be a fascinating study.

The most remarkable feature disclosed by Table XIV is the relative paucity of non-forest species in the vast territory of West Africa, actually fewer than Malawi, even without that little territory's few montane species. Notwithstanding West Africa's comprehensive range of habitats, its weakness is apparent in all the five groups of birds. Probably it is in part only one aspect of a general biological poverty, since the northern savannas of Africa are regarded by botanists as floristically much poorer than those south of the Equator. In this connexion see the citation of unpublished botanical opinion by Moreau (1952a, p. 900), since when Hoyle (1955) has stated even more explicitly that "in most categories the species of plants are something like five times as numerous in the savanna region in East Africa as in the savanna region of northern tropical Africa", i.e. that belt which includes most of both West Africa and the Sudan. No explanation of this seems to be forthcoming; it has been suggested that the southern savannas are the product of a longer period of evolution than the northern, but it is very difficult to postulate the basic meteorological and ecological conditions that could ultimately be responsible for such differences between the two sides of the Equator. In any case, since West Africa possesses a wider range of habitats and fewer species of birds than Zambia, Rhodesia or Malawi, it would seem that on average the birds in these southern territories are specialized to narrower niches than are the birds in West Africa.

So far as birds, and presumably also other organisms, are concerned, there are three factors that may have contributed to the poverty of West Africa in non-forest species. First, it has a remarkably simple topography and vegetation pattern. West of the Cameroon highlands it has no ranges or masses of high mountains, no deep inlets of the sea, no extensive areas of forest that divide the savanna or the steppe into separate blocks. Hence there are few opportunities for populations, especially of non-forest species, to become isolated, which is usually regarded a condition essential to the speciation process. There is no reason to suppose that there has been much change in this respect in the course of the Pleistocene, except in the neighbourhood of the Niger Delta, and that would have affected forest birds, as discussed in Chapter 9, rather than others. Hence the chances for West Africa to evolve species within its own borders have been limited. A minor factor is that, as described in Chapter 3, the location of the southern edge of the Sahara has varied considerably, reaching at one stage as much as 300 miles south of its present position. Such changes are usually regarded as contributing to the extinction of species, but in this case the effect may not have been great. The successive vegetation belts, desert steppe, acacia steppe and savanna, probably only moved north and south, without mutual disturbance and without change in relative width except for the wettest part of the savanna, close to the Guinea coast. At the same time, because of the simple topography, it does not seem likely that the different location of the vegetation belts would have improved the opportunities for the isolation of populations, and hence for speciation.

While circumstances have been against speciation within West Africa, this area may have been sufficiently isolated from the rest of the continent, by both ecological and geographical factors, for the bird fauna to be impoverished by this means. The pronouncement by Hesse *et al.* (1937) that "increased difficulties in the way of immigration of new forms on islands are factors that do not apply to parts of a mainland" was surely ill considered. At the present time West Africa is surrounded by the lifeless Sahara and by the ocean except for its eastern frontage, some 650 miles, between Lake Chad and the Gulf of Guinea. This frontage is to some extent barred today to the passage of lowland non-forest species by forest and by the comparatively narrow line of the Cameroon highlands, except in the north near Lake Chad. Moreover, from what we now know of the Pleistocene ecology (Chapter 3) this barrier may never have been much less. The forest of the Cameroons has probably been as permanent a feature as the mountains and the effectiveness of the mountain barrier would have been accentuated during the glaciation. It is, of course, true that, during a long period late in the Pleistocene, tropical West Africa was not bounded on the north by desert, but its isolation from other Ethiopian populations was probably not diminished thereby, for the vegetation then clothing the western Sahara was Mediterranean macchia, rather than Ethiopian vegetation. The macchia extended south to the neighbourhood of Lake Chad, so that the latitudes comprehending what is now the southern edge of the Sahara may not have been occupied by Ethiopian birds at all, and if it was not, the isolation of the West African bird fauna from that of the rest of the Ethiopian Region would not have decreased through the mitigation of the desert. Furthermore, for a period that lasted at least 15 000 years prior to 8000 years ago, Lake Chad expanded into a formidable water-barrier to east-and-west movement, for it was about 250 miles wide and extended from south of its present limits to the foothills of the Tibesti Mountains, which we know carried Mediterranean vegetation. Hence, with forest and mountain continuing as before to block movement by way of the Cameroons, for some time interchange of Ethiopian non-forest species between West Africa and the rest of the continent must have been very limited (see also Chapter 10).

The hypotheses suggested above to account for the poverty of the West African non-forest bird fauna—and hence for that of the biota in general—have some plausibility, but I do not find them entirely satisfying. For one thing, the Sudan, which is as big as West Africa, straddles the same latitudes and has an equally wide range of habitats, musters 497 species of non-forest birds against 418 in West Africa. This is an increase of about one-fifth, which might be regarded as a measure of the effect of West Africa's isolation; but the Sudan itself has fewer than Kenya, practically the same number of non-forest species as the far smaller territories of Uganda and Zambia, and only sixty-seven more than tiny Malawi, although these last two territories have a much smaller habitat range. This shows the extent to which not only West Africa but the northern tropics of Africa generally are poorer in bird species than the southern. There is nothing to suggest why this should be so, unless it is in some way a result of the parallel fact that the north tropical flora of Africa is much poorer than the south, as already noted.

South of the Equator the progressive reduction in size of bird fauna is remarkable: Zambia, 506; Rhodesia, 431; Cape Province, 374. This is the more noteworthy because the last territory has by far the broadest range of habitats and moreover receives in the west accessions of arid-country species from South West Africa and in the east tropical savanna species by way of the "tropical corridor" along the shore of the Indian Ocean. Cape Province is the only one of the territories under discussion that is outside the tropics, between 28° and 34° S., and notwithstanding its generally genial climate it looks as if we have here an example of the well-known latitude effect, discussed in Chapter 8, whereby in most classes of organism numbers of species diminish with increasing latitude. It will be seen that proportionately the greatest deficiency in the Cape

Province bird fauna lies in the group D non-passerines, especially those families which are dependent on fruit and large insects.

Compared with even Rhodesia, let alone the richer territories farther north, the group D families in which the deficiencies are most important are the barbets (3:5), nightjars (3:5), hornbills (2:6), rollers (0:4), cuckoos (8:12), parrots (1:4), turacos (1:2) and bee-eaters (1:4). Indeed, the Cape Province bee-eater should hardly be counted, because it consists of a sporadic breeding population of European Bee-eaters *Merops apiaster*, undifferentiated and hence probably only very recently derived from Palaearctic migrants (see also Chapter 6). But it is most astonishing that the bee-eater niche in this part of Africa should have been virtually unoccupied.

Relative deficiency in families of birds dependent on fruit or large insects is equally a characteristic of the sample Nearctic and Palaearctic bird faunas when compared with African ones, as shown in Chapter 8, so that in this respect the Cape Province bird fauna is typical of the temperate zones, even though its interchange with the tropics is without impediment and on a broad front. So far as the fruit-eaters are concerned, this is difficult to understand. As Mrs M. K. Rowan has commented to me, in the Cape "fruiting plants do not behave in a strongly seasonal way. For one thing different species . . . give . . . a good succession throughout the year. For another, several species have a remarkably extended season." Moreover, "there does not seem to be a shortage of suitable fruiting plants". When Mrs Rowan was studying colies (a thoroughly frugivorous non-forest group) at a locality in the western Cape, she compiled a list of nearly forty indigenous species on which the colies fed, about one-quarter of which were trees. Moreover, in the Cape "it is rare to see fruiting bushes wholly stripped in the way often described for English environments".

As regards the Cape Province, deficiency in the group D families that depend on large insects, the nightjars, hornbills (*Tockus* spp.), rollers, cuckoos and bee-eaters, it might be suggested that the operative factor is temperature, either directly or indirectly, making their food more strongly seasonal than in even the outer tropics; but although much of the high interior of South Africa indeed has many frosty nights in winter, as mentioned in Chapter 2, this does not apply to much of Cape Province. In fact, a sample of twenty-five meteorological stations between sea-level and 800 m there shows mean minima for the three coldest months varying from about 5° to 10°C, while a sample of eight in Zambia, that lie between about 1100 and 1700 m, shows mean minima varying from about 7° to 12°C—no considerable difference. It may be added that seasonal migration in birds generally is stronger in Cape Province than in the southern tropics (though not than in the northern tropics; Chapter 13) and, of the thirteen insectivorous species among the families cited in the preceding paragraphs, only two, the nightjar *Caprimulgus pectoralis* and the hornbill *Tockus alboterminatus*, are resident all the year round and not merely breeding visitors. Note also the curious fact that in the hornbills Cape Province possesses only this one mainly insectivorous species compared with seven in the neighbouring territory of South West Africa. The whole question of the seasonality of food and the niches of the local birds in Cape Province, compared with those in Rhodesia and elsewhere, merits field investigation of a kind that is as desirable as it would be laborious.

If now we turn from a geographical to a systematic consideration of the data in Table XIV the first thing to strike us is the paucity of Anatidae in both West Africa and the Sudan, both with very extensive watery habitats, but neither mustering more than seven species, compared with twelve to fourteen in every other territory (except waterless Somaliland, which in this group is not a competitor). The only Anatidae the two vast northern territories possess, apart from the forest species *Pteronetta hartlaubii* and the montane duck *Anas sparsa* in the highlands in the south of the Sudan, are the "geese" *Plectropterus gambensis, Alopochen aegyptiaca, Sarkidiornis melanotos*, two tree-ducks *Dendrocygna bicolor* and *viduata, Nettapus auritus, Anas capensis* (only in a few spots in the Sudan), and *Thalassornis leuconotos* (extending west only to Nigeria). The other four *Anas* spp. of the Ethiopian Region including the pochard "*Aythya*" *erythophthalma*, which are so widespread elsewhere in Africa, are absent from both West Africa and the Sudan, and it is difficult to imagine why this should be so. Incidentally, it may be recalled that the pochard occurs in western South America as well as in eastern and southern Africa. Moreover, the birds composing the two populations are indistinguishable. Because the crossing of the Atlantic is so much shorter north of the Equator it is difficult to believe that this duck has not inhabited West Africa and at no remote date, since no subspeciation has taken place.

It has been suggested that perhaps these African ducks could not stand the competition from the Palaearctic ducks which winter in these territories (Dorst, 1962b); but the only species among these which arrive in large numbers are Garganey *Anas querquedula*, Shoveler *Spatula clypeata* and Pintail *Anas acuta*. It is difficult to believe that the competition from these species would alone have such a devastating effect on the African ducks; and the suggestion is, I feel, weakened by the fact that those African species which are absent from West Africa and Sudan are also absent from nearly all of the Congo, where Palaearctic ducks hardly occur at all. I infer that there must be come fundamental ecological requirement of these African *Anas* species, which are widespread in the east and south (and breed also in Angola), that is missing from West Africa and the Sudan. Incidentally, it may be noted that neither of the African species of harrier *Circus* breeds in West Africa or the Sudan, although Palaearctic harriers are common there in winter.

The heron tribe has a remarkably consistent representation throughout those territories which have water habitats, varying only from thirteen to seventeen species (and twelve in little Gambia). That is, a majority of the nineteen non-forest herons of the Ethiopian Region are present in most parts of it. This is one of the few families in which Cape Province produces one of the highest totals. By contrast, it lacks regular breeding storks except for a very few *C. nigra* in the high north-east, unaccountably, for there is no particular reason for thinking that the stork and heron populations affect each other. Another numerous family in which the territorial totals are still more alike is the Aquilidae, varying only between twenty and twenty-four (except in Gambia and Somaliland), but the proportion these figures represent of the total non-forest species of the Ethiopian Region, which is thirty-five (Table V), is not so high as in the herons.

In group D it is noticeable how the nightjar and bee-eater species fall off steadily with increasing distance from the Equator and this applies also to some passerine families, especially the shrikes and the sunbirds. In the four families named, Cape Province has only three, one, nine and seven non-forest species respectively, compared with four, five, fifteen and eleven in Malawi and eight, five, twenty-eight and fifteen in Kenya. In shrikes Kenya shows an extraordinary predominance, with twenty-eight against twenty in Uganda and fifteen or less in nearly all the other territories except the Sudan (eighteen). This situation is due mainly to an accumulation of shrike species, which must await more information for ecological analysis, in the semi-arid part of the Kenya lowlands.

There are only two passerine families in which Cape Province rates as high as any of the tropical territories, namely the finches and the thrushes. The former is predominantly a family of the Holarctic and the latter is very well represented there, with special links between the Palaearctic and Cape Province in *Monticola*, as mentioned in Chapter 6. The finch situation is of special interest. Of non-forest species there are only three in West Africa, but five to eight (plus up to three montane) species in other territories and ten in Cape Province. Of forest finches there are two montane and one lowland, and thus it seems that in Africa this typically Holarctic family shows a preference for the cooler climates. Hence in some way a temperature factor may be affecting the distribution of the finches. Relations with other seed-eating passerines, referred to below, are uncertain. It may be added that the closely related buntings show a contrast to the finches, for none is montane in tropical Africa and there is no accumulation of species in Cape Province.

Another family in which South Africa ranks high is the larks, for South West Africa has seventeen species and Cape Province nineteen compared with eighteen in the richest tropical territories, Kenya and Somaliland. All four territories contain large areas of dry, very open, terrain, and there is geological evidence that Somaliland and much of south-western Africa have been semi-arid for a long time, while Kenya in this connexion benefits from the proximity of the Somaliland focus of endemism. By contrast, West Africa counts only eleven larks.

The figures for the non-forest weaver birds show unexpected features. Those for the estrildines are rather consistent within the tropics, with West Africa, Sudan, Kenya, Uganda and Zambia varying only from twenty-five to twenty-nine species—which is surprising in view of the smaller ecological diversity of the last territory. In Malawi and Rhodesia the totals decline to twenty and twenty-four, then drop abruptly to only eleven in both South West Africa and Cape Province. The other weavers show a curiously different pattern, richest in Sudan, Kenya and Uganda (forty-two to forty-nine species) but only thirty-three in the next richest territory, Zambia, and slumping to twenty-six, little more than half the Kenya total, in West Africa, and twenty-five to twenty in Malawi, Rhodesia and Cape Province. In fact, the proportion of estrildines to other weavers varies greatly between territories, being about equal in West Africa and Rhodesia, between 3:5 and 4:5 in most of the territories, but dropping to 11:20 in Cape Province. On present information there is no way of rationalizing these results.

Again, on the ecological relations between the various families of birds that are important seed-eaters precise data are lacking, but I think there is not much overlap between them, whatever the situation may be between individual species within a group. Thus, the estrildines and the other weavers, while both predominantly seed-eating, do not overlap much in diet; the estrildines feed mainly on smaller seeds than the other weavers and, unlike most of these, take them on slender-growing grasses or on the ground, where the Ploceinae do not feed though the Passerinae (sparrow subfamily) do. The finches probably take very largely different species of seeds from the "other weavers", but overlap the estrildines slightly more; and the larks, while feeding on the ground, generally do so in different situations from the estrildines. It is interesting to combine the territorial totals of species in these four main seed-eating families of small birds. The result, which gives some indication of the diversity of seed-utilization by passerines in different parts of Africa, is as follows: Kenya, Sudan and Uganda pre-eminent with 102, ninety-one and eighty-seven species respectively; Zambia next with seventy-seven; West Africa sixty-five; Cape Province fifty-eight; and the other main territories forty-five to sixty-two. The paucity of passerine seed-eaters in West Africa is thus shown to be remarkable, in view of its range of habitats and its size. In Africa, so much of which is dominated by monsoon rains, violent local fluctuations of seed supply, with apparent temporary superabundance, are the rule. It is probably true to say that nowhere in Africa has any critical study of the ecology of seed-eating birds in relation to natural foods been attempted, and it will be the more difficult to do in face of these characters in the food supply and the nomadism of many of the species of birds.

Finally, Table XV shows the proportion that each group of birds contributes to the bird fauna of each territory. The passerine share, approaching one half the species, is remarkably consistent. Cape

Table XV

The proportion of the bird fauna contributed, in terms of species, by each group of birds in African territories

Group	W. Afr.	Sudan	Kenya	Uganda	Zambia	Malawi	Rhod.	S.W.A.	C.P.	Range	Mean
A	0·15	0·12	0·14	0·15	0·17	0·18	0·16	0·18	0·20	0·12–0·20	0·16
B	0·11	0·10	0·09	0·09	0·09	0·11	0·10	0·13	0·11	0·09–0·13	0·10
C	0·07	0·07	0·07	0·07	0·06	0·05	0·06	0·08	0·08	0·05–0·08	0·07
D	0·19	0·21	0·19	0·20	0·18	0·20	0·19	0·17	0·14	0·14–0·21	0·19
E	0·46	0·49	0·51	0·48	0·49	0·47	0·48	0·44	0·48	0·44–0·51	0·48

Province is responsible for extreme figures in two groups, the highest in water-birds and the lowest in group D. Cape Province is the only extra-tropical territory in the list, and, as noted above, it is relatively deficient in non-passerine eaters of fruit and large insects, as are bird faunas at higher latitudes. The only other abnormality in Table XV that might call for comment is the low proportion of water-birds in the Sudan, which I find quite baffling. I can only suppose that although its watery habitats make such a grand show on the geographical scale they would on analysis show a lack of ecological variety.

8

Comparisons with Extra-Ethiopian Bird Faunas

COMPARISONS BETWEEN CONTINENTS

There seems to be no comparison of the numbers of bird species in different continents since that of Stresemann (1927–34). Such statistics are of course acutely complicated by the extent to which the concept of the polytypic species is applicable (Stresemann himself recognized only half as many species as Reichenow had done some twenty years earlier), and in this connexion even a thoroughgoing use of the superspecies would also be somewhat unrealistic; but with all their imperfections due to the taxonomic imbroglio Stresemann's figures merit reproduction (Table XVI) and discussion. It may be noted that he did not mention the Oriental Region.

TABLE XVI

Numbers of bird species in different zoological regions

Region	Area (million sq. miles)	Bird species	
		Total no.	No. per million sq. miles
Africa south of Sahara	8	1750	220
South America	7	2500 "at most"	350 at most
Palaearctic	13	1100	80
North America	8	750	90
Australia	3	560	190

The extension of the polytypic species concept has in fact reduced these figures differentially since Stresemann wrote: by only about 5% in Australia, to the 531 of Keast (1961), but by 15% in Africa, to the 1481 arrived at in Chapter 5. This reduced estimate brings the African ratio down to 185 species to 1 million sq. miles, nearly the same as Australia. The reduction in the Palaearctic and Nearctic totals is probably no greater than in Australia. There is no information on a possible revised figure for South America but it is unlikely that it would be anything like halved. Hence it can be accepted that on a unit of area the two north-temperate regions are barely half as rich in species as Africa and Australia, while these are much less rich than South America.

It is of course a commonplace that in most classes of organism species are much more numerous in the tropics than at high latitudes. This applies even when the range of gross habitats, from desert to evergreen forest, is roughly equal in the two areas compared. It applies, moreover, even when the extra-tropical area concerned has not been impoverished by the direct consequences of a Pleistocene glaciation, for, as shown in Chapter 7, the Cape Province of South Africa, between 26° and 35° S., has a poorer bird fauna than Rhodesia, between 16° and 23° S., although it has actually a wider range of habitats. It may, however, be noted that comparisons are sometimes made that from this point of view are unfair. For example, the total of 195 species of birds quoted by Dobzhansky (1959) for New York State against 1148 for Venezuela and 1395 for Columbia is derived from an area of homogeneous climate dominated by temperate woodlands; the South American figures come from areas with a great difference between their wettest and driest parts and with an altitudinal zonation not attained in New York State.

The fundamental reason why low latitudes are so much richer in species than high latitudes is usually attributed to the different natures of the climates. For example, Dobzhansky (1959) has written that in a cold or temperate climate "in order to reproduce, any species must be at least tolerably well adapted to every one of the environments which it regularly meets" and consequently "may be unable to attain maximum efficiency in any one of them". Klopfer and MacArthur (1961) recognize that the tropical increase in number of species "implies a reduction in the size of the exclusive portion of the species niche" and that this "can occur only where climatic stability is such as to assure a fair degree of stability in the availability of the required food and perch sites". Mayr (1963) accepts the same idea when he writes that especially in the humid tropics "seasonal stability and low demands by the physical environment permit strong niche specialization, resulting in great faunal richness" (see also the discussion by Fischer, 1960). The explicit emphasis by Mayr on the humid tropics is important. Without that emphasis and implied reservation it seems to me that statements on this subject give inadequate recognition to the fact that tropical countries under a monsoon climate, with a single rainy season, undergo an enormous seasonal variation in food and vegetation cover, so that conditions over most of the tropics are in fact far less "stable" than is often suggested. Moreover, on the temperate side of the comparison, where birds are concerned allowance needs to be made for their individual mobility and for the opportunities given them to evade ecological differences by migration. On the whole, although no alternative explanation to climatic "stability" presents itself for the greater richness of species at low latitudes, it is less universally acceptable than appears at first sight.

The foregoing considerations do not apply when Africa south of the Sahara and South America are each considered as a whole. Each is mainly tropical but enters the south temperate zone, each has a full ecological range from desert to evergreen forest, both lowland and montane. Hence they are fully comparable and the reason for the big difference in their numbers of bird species is not at first obvious. However, the decisive element in the South American predominance is undoubtedly its abundance of evergreen forest birds (mainly sub-oscine), and this reflects the botanical situation; compared with the tropical American "the African forest is floristically relatively poor and uniform" (Richards, 1952, 1963). Though without special reference to the forest flora, Good (1953) has given some numerical indication of the extent of the difference between the continents in this respect. He has quoted estimates of the number of plant species in Africa as 25000 and in Brazil as 40000. This latter country has an area a little over half that of South America as a whole and perhaps the same proportion of the range of habitats, so that the number of species for the continent can with confidence be put at over 60000, if not nearer 80000. In other words, floristically, South America is some two and a half times as rich as Africa, while it possesses about one and a half times as many species of birds.

I am inclined to find the basis for the present difference in the Pleistocene history. Both continents, of course, shared in full measure the vicissitudes of temperature associated with the glaciations, as indicated in Chapter 3, but owing to the difference in the surface relief of the two continents the distribution of the individual biomes is perhaps likely to have been disturbed by this means more in Africa than in South America. More important, there is, I understand, at present no evidence in South America for great fluctuations in humidity, not directly related to the glaciations but affecting wide areas of the continent, as there is in Africa. In particular, we have, as stated in Chapter 3, the evidence that the main forest mass of Africa, occupying the Congo basin, was broken up as late as the Middle Pleistocene, and after recovering was again somewhat reduced around 12000 years ago; while there are strong grounds for inferring that the other main forest blocks of Africa, those west of the Cameroons, only narrowly escaped elimination at some stage in the Late Pleistocene. These catastrophes must have involved the extinction of many African species, so recently that there would have been only limited opportunities for subsequent evolution of new ones. Hence in Africa the ecological stability postulated as providing the best conditions for the elaboration of biota and the differentiation of niches may have suffered significantly more interruption than it has in South America, at any rate in the richest environment, that of forest.

SOME LOCAL COMPARISONS

The differences between the land-bird fauna of a tropical area and a temperate-zone bird fauna can usefully be illustrated and more sharply focused if we can find areas that to some extent are comparable in their range of habitats and are well documented. I think for this purpose it is worth while to compare on the one hand the Thames Valley with the Gambia and on the other hand "British" Somaliland with Arizona and Morocco. Of these areas, I am helped by personal familiarity with Thames and Gambia and a little experience of Arizona and Morocco.

THAMES AND GAMBIA

For listing purposes, it is convenient to take as the "Gambia Valley" that part which forms the Colony and Protectorate, a strip half a dozen miles wide on both sides of the last 200 miles of the river, following its main bends and widening at the mouth to a distance of about 35 miles between the northernmost and southernmost points on the coast. It is easy to visualize a similar area along the Thames; both rivers have a big estuary and, if the same distances are projected, the eastern front of the Thames area would be a line from about Walton-on-the-Naze to a point in northern Kent. For purposes of comparison it is of course necessary to think of a Thames Valley unpolluted by urbanization and with only sporadic and primitive agriculture, the conditions found by the first Roman invaders. In Gambia urbanization has been negligible so far, but nevertheless a great deal of natural vegetation has been cleared and some allowance must be made for the effects of the recent increase in the human population (see also Cawkell and Moreau, 1963).

The climatic difference between the two areas is striking. The rainfall of around 25 in. in the Thames Valley is well distributed through the year and the extreme range of temperature is in most years from about 30° to −6°C. In Gambia the 45 in. on the coast decreases to about 35 in. in the interior and everywhere the rain is concentrated into five months, beginning in May. The other seven months are rainless, the country becomes sere and partly leafless, and grass-fires are widespread. Nevertheless, some of the leafless trees then bloom, some trees, especially figs (*Ficus* spp.), are loaded with fruit, and the general impression of ecological poverty that one gets from the appearance of the landscape is belied by the number of species of birds that maintain themselves there. The birds in general are not so concentrated in the riverine vegetation and the strips and patches of semi-evergreen forest (which depends on the high water-table) as might have been expected. In most years the extreme range of temperature is from about 40° to about 7°C (in the more variable interior). Thus in Gambia frost is never approached and cold is not an effective climatic control.

In both areas the range of natural habitats is comparatively small. In Gambia the surface soil is everywhere sand and the highest points are sour-looking laterite ridges rising to about 300 ft above the sea. The Thames strip rises to some 800 ft, with much more varied soil: clay, sand, chalk, colite. Nevertheless the Gambia has the more varied vegetation. According to an unpublished list by D. R. Rosevear (kindly communicated to me by E. F. Brewer) in Gambia eighty species of native tree have been recorded, compared with about thirty for Thames.

Both must be credited with extensive mudflats and tidal or seasonal swamps. The main Gambia habitats, in ascending order of importance for birds, are mangroves (with hardly any typical breeding species; Cawkell, 1964), the semi-evergreen woods of the seaward quarter of the valley and the savannas and deciduous woodlands of various subtypes. Clearings in the woodland, which are rarely anything like free of coppice, seem actually more popular with many species of birds than the untouched natural vegetation. The Thames area, relatively undisturbed as it was around 2000 years ago, it is necessary to think of as largely covered with mixed broad-leaved forest and thicket, especially of oak *Quercus* and with few or no conifers (cf. Godwin, 1956). There would be masses of alder *Alnus* in the wetter places. It is difficult to envisage much open country anywhere, apart from the saltings. There may have been some on the poorest sands and on the more windswept higher ground with thin soil on chalk.

In compiling Table XVII I have made no attempt to distinguish species that are locally common from those which are not. So far as Thames is concerned, in order to arrive at a list for the "natural", undisturbed, environment, I have made certain adjustments to the present-day list; in particular I have added about a dozen species, such as Spoonbill *Platalea* and Oriole, still existing in the Low Countries, for which conditions should have been suitable in unspoilt Thames country, but I have brought in no species from much further north or south, nor any dependent on conifers. At the same time, I have taken a chance and have exaggerated the potentialities of the "undeveloped" Thames valley by admitting a few species dependent on man and on cleared spaces, such as starling, swallow and Great Bustard *Otis tarda*; and also the crane *Megalornis grus*, as a species almost certainly banished from the countries surrounding the southern North Sea by the increase in human occupation.

For Gambia the list is as in Table XIV for breeding birds, with the addition of wintering species (see Cawkell and Moreau, 1963). There is a margin of uncertainty about whether a number of species that can be seen in the territory all the year round actually breed there. For purposes of Table XVII, I have been conservative, omitting birds of uncertain status unless they seem common and/or regular in appearance. I should expect intensive observations to add to the list of breeding species. Also it seems that some species not admissible today, such as certain Palaearctic ducks in winter and parasitic cuckoos in summer, would have been admissible fifty years ago. On the whole it can be said that Table XVII probably errs in being too generous to Thames and the reverse to Gambia. But the general picture is certainly sound and such a bias as this makes the results of the comparison all the more striking.

From the statistics in Table XVII it will be seen that in the breeding season the Gambia list is nearly double that of Thames 244:144. This disproportion is greatly increased during the northern winter, when in the local dry season Gambia has lost a few of its breeding species, but has gained more, most from the Palaearctic and a few from the southern edge of the Sahara, so that the figures are 276:82. However, most of the constituent groups of birds depart widely from these proportions. Taking only breeding species, in group A, the water-birds, Thames has about as many as Gambia, 39:38, though nine of the former are only summer visitors; in group B, the raptors etc., Thames has less than half as many species as Gambia, 15:33; in group C, the ground-birds, one-third,

TABLE XVII

Composition of the Gambia and the Thames Valley bird faunas

(R, Resident;   B, Breeding visitor;   N, Non-breeding visitor)

	Gambia			Thames		
	R	B	N	R	B	N
Group A						
Anatidae: Ducks and Geese	5	—	1	10	1	11
Anhingidae: Darters	1	—	—	—	—	—
Ardeidae: Herons	12	—	2	2	3	—
Balearicidae: Cranes	1	—	—	—	1	—
Charadriidae: Plovers	3	—	4	5	—	2
Ciconiidae: Storks	3	—	—	—	—	—
Heliornithidae: Finfoots	1	—	—	—	—	—
Jacanidae: Lilytrotters	1	—	—	—	—	—
Pandionidae: Ospreys	1	—	—	—	—	—
Pelecanidae: Pelicans	1	—	—	—	—	—
Phalacrocoracidae: Cormorants	1	—	—	1	—	—
Podicipitidae: Grebes	1	—	—	3	—	2
Rallidae: Rails	3	—	—	3	2	—
Recurvirostridae: Avocets and Stilts	1	—	1	1	—	—
Scolopacidae: Snipes and Waders	—	—	18	5	1	6
Scopidae: Hamerkops	1	—	—	—	—	—
Threskiornithidae: Ibises and Spoonbills	2	—	—	—	1	—
	38	—	26	30	9	21
Group B						
Aegypiidae: Vultures	4	—	—	—	—	—
Aquilidae: Eagles	19	—	4	6	1	2
Falconidae: Falcons	3	—	1	2	1	1
Sagittariidae: Secretary Bird	1	—	—	—	—	—
Tytonidae and Strigidae: Owls	6	—	—	5	—	—
	33	—	5	13	2	3
Group C						
Burhinidae: Stone Curlews	1	—	—	—	1	—
Glareolidae: Fratincoles and Coursers	2	—	1	—	—	—
Otididae: Bustards	2	—	—	1	—	—
Phasianidae: Game Birds	5	—	1	1	1	—
Pteroclidae: Sand Grouse	2	—	—	—	—	—
Turnicidae: Button Quails	1	—	—	—	—	—
	13	—	2	2	2	—

TABLE XVII (cont.)

	Gambia			Thames		
	R	B	N	R	B	N
Group D						
Alcedinidae: Kingfishers						
Dry-feeding	5	—	—	1	—	—
Water-feeding	3	—	—	—	—	—
Apodidae: Swifts	3	—	—	—	1	—
Bucerotidae: Hornbills	4	—	—	—	—	—
Capitonidae: Barbets	4	—	—	—	—	—
Caprimulgidae: Nightjars	—	2	—	—	1	—
Columbidae: Doves	8	—	—	2	1	—
Coraciidae: Rollers	4	—	—	—	—	—
Cuculidae: Cuckoos	1	4	—	—	1	—
Indicatoridae: Honey Guides	1	—	—	—	—	—
Meropidae: Bee-eaters	2	1	5	—	—	—
Musophagidae: Turacos	3	—	—	—	—	—
Phoeniculidae: Wood-hoopoes	2	—	—	—	—	—
Picidae: Woodpeckers	4	—	—	3	1	—
Psittacidae: Parrots	3	—	—	—	—	—
Upupidae: Hoopoes	1	—	—	—	—	—
	48	7	5	6	5	—
Group E						
Alaudidae: Larks	3	—	—	3	—	1
Campephagidae: Cuckoo Shrikes	2	—	—	—	—	—
Certhiidae: Tree Creepers	—	—	—	1	—	—
Corvidae: Crows	2	—	—	6	—	—
Dicruridae: Drongos	1	—	—	—	—	—
Emberizidae: Buntings	—	—	—	4	—	2
Fringillidae: Finches	2	—	—	7	—	3
Hirundinidae: Swallows	4	1	1	—	3	—
Laniidae: Shrikes	7	—	1	—	1	—
Motacillidae: Wagtails and Pipits	2	—	4	4	3	—
Muscicapidae: Flycatchers	8	—	1	—	1	—
Nectariniidae: Sunbirds	9	—	—	—	—	—
Oriolidae: Orioles	1	—	—	—	1	—
Paridae: Tits	1	—	—	7	—	—
Ploceidae: Weavers						
Estrildines	15	—	—	—	—	—
Others	16	1	—	2	—	—
Prunellidae: Accentors	—	—	—	1	—	—
Pycnonotidae: Bulbuls	3	—	—	—	—	—
Sittidae: Nuthatches	—	—	—	1	—	—
Sturnidae: Starlings	5	1	—	1	—	—
Sylviidae: Warblers and *Regulus*	9	2	7	2	15	—
Timaliidae: Babblers	4	—	—	—	—	—
Troglodytidae: Wrens	—	—	—	1	—	—
Turdidae: Thrushes	5	—	4	6	5	2
Zosteropidae: White-eyes	1	—	—	—	—	—
	100	5	18	46	29	8
Grand totals	232	12	56	97	47	32

4:13; in group D, the other non-passerines, only one-fifth, 11:55; but in the passerines the balance is somewhat redressed, for Thames has two-thirds as many, 75:105, though twenty-nine of the seventy-five are only summer visitors. The groups are worth discussing individually.

Group A, water-birds

Although the totals for the two areas are so similar their composition is very different. In Gambia seventeen families are represented, against only ten in Thames. Also Gambia scores very heavily in resident Ardeidae, with twelve (and also three storks) against a potential five species in Thames at the outside (admitting Purple Heron, Night Heron, Bittern, Little Bittern as well as Grey Heron). Two of the Gambia species, *Ardea melanocephala* and the Cattle Egret *Ardeola ibis*, are typically dry-land feeders for which there is no counterpart in Thames. But, allowing for these, it is not clear why the difference in the water-feeding species against five, should be so great, especially as the size-range of the herons on the two lists is much the same, apart from the outsize *A. goliath* of Gambria.

By contrast, Thames can count eleven breeding ducks and three grebes against five Anatidae and only one grebe (which is rare) in Gambia. Poverty in these families characterizes all West Africa: and in ducks no part of the continent is so poor as West Africa except the Sudan (see Chapter 7).

Group B, raptors and scavengers

Only Gambia has vultures, although the undisturbed Thames, with its wild bovids, deer, swine, wolves, etc., presumably had plenty of carrion. Two causes would contribute to the absence of vultures: one is that in summer the carrion would be to a great extent hidden in the woods, the other is that for many days in each year the up-currents necessary for vultures' soaring would be absent. Of raptors Gambia has twice as many species as Thames at all times of the year. This is doubtless made possible by the fact that apart from its swamps, important also in Thames, Gambia is largely open woodland, with populations not only of birds and rodents, but also of lizards, snakes and large orthoptera accessible from one year's end to the next, whereas in a Thames area dominated by closed woodland the first two would be the only important sources of food at any time of the year and the only ones available at all in the winter. At the same time, a careful study of the extent to which the numerous raptors—and indeed also the four vultures of Gambia—overlap or are segregated ecologically is much to be desired. Gross differences are of course obvious, as between the ecology of a huge Martial Eagle *Polemaetus bellicosus* and the ecology of a kestrel, but as between the kestrels—*Falco ardosiacus* is a resident, *F. tinnunculus* an abundant winter visitor—or even between the kestrels and the Grasshopper Buzzard *Butastur rufipennis*, the ecological relations are not known clearly.

Group C, ground-birds

Here the greater poverty of Thames is prima facie attributable to the shortage of open habitats. The discrepancy between the two areas may indeed well have been greater than the figures suggest, because under primitive conditions the presence of Quail, Partridge and Great Bustard, all of which have been credited to Thames, is questionable. However, this comparison throws into relief the singular fact that the deciduous woodlands of Western Europe seem to have housed no game bird—the pheasant having apparently been introduced by the Romans.

Group D, other non-passerines

In this group the difference is greater than in any other, 55:11, and since most of the species concerned live in trees no facile overall explanation of broad habitat difference can be called in. In the first place, sixteen families appear in Gambia against only six in Thames. It is possible to see how certain differences in the environments contribute to this result. For one thing Gambia has ten non-passerine species that are dependent more or less completely and all the year round on fruit growing on trees—two pigeons and three each of barbets, turacos and parrots—while hornbills also take fruit to a considerable extent. This category has no representative in Thames, and it is evident that Gambia, notwithstanding the length and severity of its dry season, has a continuous supply of fruit, while Thames, like the rest of the North Temperate zone, provides none in most months of the year. Another important category in the Gambian list, unrepresented in Thames except for two summer visitors, one cuckoo and one nightjar, is formed by those species which depend on large insects, the four rollers and three bee-eaters, the five ground-feeding kingfishers and the hoopoe, the two nightjars and the five cuckoos. True, seven of these twenty species, the two nightjars, four of the cuckoos and one of the bee-eaters, disappear from the Gambia during the dry season, but five other species are known only at that time of the year and what are presumably Palaearctic individuals of Great Spotted Cuckoo *Clamator glandarius* also appear in the same season. Clearly, Gambia is able to supply plenty of large insects, especially flying insects, all the year round, as Thames is not.

Of Columbidae the accumulation in Gambia, nine species compared with three in Thames, is indeed remarkable. As already noted, two are fruit eaters, but the other seven, with no considerable range in size, depend on seeds; and this apparent overlap would repay investigation in the field. A final point to be noted in this category is the proliferation of water-feeding kingfishers in Gambia—five species showing a great range in size, compared with a single one in Thames, or for the matter of that in the whole of Europe or most of North America. The size differences in Gambia straightaway indicate ecological separation between several of the species and moreover one of them, *Ceryle rudis*, by fishing independently of a perch, has a foraging range that is all its own.

It will be noted that Gambia has no more Picidae than Thames, but one on the latter list, the Wryneck *Jynx*, does not find its food on trees, and in Gambia the wood hoopoes supplement the woodpeckers as foragers in bark, though they cannot penetrate so deeply.

Group E, passerines

Here the numerical difference between the lists of Gambia and Thames is not so great, 105:75, as in other groups. Important elements in the excess in Gambia are the sunbirds (9:0), bulbuls (3:0), flycatchers (8:1), shrikes (7:1), babblers (4:0), starlings (6:1) and weavers, estrildine and other (32:2). The sunbirds are largely dependent on nectar, which is always available in Gambia, even in the dry season and often on trees that are leafless. The shrikes and babblers are birds that need large insects all the year round and the starlings (of Gambia) depend mainly on fruit; that is, all these three families belong to ecological categories already noted as most conspicuous among the Gambian non-passerines (group D), with no counterparts in Thames. The weavers are to a considerable extent ecological replacements of the finches (2:7) and buntings (0:4), in which Thames excels. But two groups deficient in Gambia for no immediately obvious reason are the tits (1:7) and the corvids (2:6). The latter are represented in Gambia only by the general-purposes scavenger *Corvus albus* and the small "black magpie" (Piacpiac) *Philostomus afer*, which is a ground-feeder. The absence of the nest-robbing corvids, the Magpie *Pica pica*, Jay *Garrulus glandarius* and Crow *Corvus corone*, from Gambia is noteworthy; and there their place is probably taken by snakes and mammalian predators.

SOMALILAND, MOROCCO AND ARIZONA

The essential characters of the three areas are briefly summarized below. It will be seen that they are fairly comparable ecologically, except that Somaliland has the smallest range of habitats. The differences in area, which also operate against Somaliland, are regarded as of minor importance for the present purpose.

SOMALILAND

The country covers an area of 68 000 sq. miles, mostly of sandy plain covered with "wooded steppe" and "subdesert steppe", around 10° N. A range of mountains rises to over 7000 ft but the area above 5000 ft is very small. Rain, exceeding 20 in. a year hardly anywhere and in most areas much less, falls mostly from April to June. Domestic stock are plentiful and until recently also big game. The highest parts of the mountains carry vestigial dry evergreen forest (*Juniperus procera*), but there are only four species of birds locally confined to it; and altitudinal zonation, with replacement of one bird species by another at different elevations, is altogether less general and important than in the other two territories under discussion.

MOROCCO

The area extends over 170 000 sq. miles between 28° and 35° N., with a much wider local variation in rainfall than Somaliland, all coming in the winter. Morocco is largely occupied by a complicated mountain system rising to over 12 000 ft in the Great Atlas. The variation in temperature between winter and summer is much greater than in Somaliland and the habitats range from desert and Mediterranean macchia to broadleaf woodlands of typically temperate trees and open country above the timber-line. The large mammals are now all domestic.

ARIZONA

The state covers 114 000 sq. miles between about 32° and 37° N. Much is high, over 6000 ft, with numerous mountain masses rising to 10 000 ft and the highest to 13 000, but with the general altitude declining practically to sea-level in the south-east. Notwithstanding the broken nature of the country most of it is very dry, averaging perhaps 20 in. in the north-east and as little as 4 in. in the opposite corner, though the high mountains presumably attract more rain, which comes early in the year and also in July and August. The annual and daily ranges of temperature are much greater than in Somaliland. With its ephemeral herbage and its thorny perennials, including its arborescent cacti, of which the Candelabra Euphorbia is the African counterpart, the Arizona plains vegetation is broadly like that of much of Somaliland; but altitudinal zonation is very important and the mountain vegetation culminates in aspen and spruce. As many as one-fifth of all the bird species, mostly passerines, are confined to high altitudes, a much larger proportion than in Somaliland. Large mammals are limited to two species of deer and the domestic stock, now greatly reduced as a result of the calamitous over-grazing which has replaced so much good grassland with inedible mesquite thornbush.

For comparison of the bird faunas (Table XVIII) lists have been extracted from Archer and Godman (1937–61), Heim de Balsac and Mayaud (1962) and Phillips et al. (1964). Water-birds are omitted, since Somaliland is so devoid of their habitat that here valid comparison is impossible. All species of groups B–E are included that have been recorded as either breeding or wintering regularly, even though they may now have ceased to do so. In Arizona and to a lesser extent in Morocco various species change their localities after breeding, but remain within the territorial boundaries, often apparently in reduced numbers; such species have been listed as "resident". It may be noted that the groups into which it is so satisfactory to divide the Old World bird faunas for analysis do not work quite so well when an American bird fauna is being dealt with. For one thing, the non-passerine Trochilidae (humming birds) are the ecological equivalents of the passerine Nectariniidae (sunbirds) of the Old World. Also some of the big species in the passerine, but sub-oscine, Tyrannidae ("American flycatchers") can be

TABLE XVIII

The bird faunas of three dry territories

(R, Resident;   B, Breeding visitor;   N, Non-breeding visitor)

	Somaliland			Morocco			Arizona		
	R	B	N	R	B	N	R	B	N
Group B									
Aegypiidae } vultures	5	—	—	5	—	—	—	—	—
Cathartidae }	—	—	—	—	—	—	2	—	—
Aquilidae	12	—	1	15	—	1	6	6	2
Falconidae	5	—	1	6	—	1	5	—	—
Sagittariidae	1	—	—	—	—	—	—	—	—
Strigidae and Tytonidae	5	—	—	6	1	—	10	1	2
	28	—	2	32	1	2	23	7	4
Group C									
Burhinidae	1	—	—	1	—	—	—	—	—
Glareolidae	3	—	—	2	—	—	—	—	—
Meleagrididae	—	—	—	—	—	—	1	—	—
Otididae	7	—	—	4	—	—	—	—	—
Phasianidae	6	—	—	4	—	—	4	—	—
Pteroclidae	2	—	—	4	—	—	—	—	—
Struthionidae	1	—	—	1	—	—	—	—	—
Tetraonidae	—	—	—	—	—	—	1	—	—
Turnicidae	1	—	—	1	—	—	—	—	—
	21	—	—	17	—	—	6	—	—
Group D									
Alcedinidae (dry-feeding)	2	—	—	—	—	—	—	—	—
Apodidae	3	—	—	1	3	—	1	—	—
Bucerotidae	3	—	—	—	—	—	—	—	—
Capitonidae	4	—	—	—	—	—	—	—	—
Caprimulgidae	4	—	—	2	—	—	1	3	—
Coliidae	1	—	—	—	—	—	—	—	—
Columbidae	12	—	—	4	—	—	3	2	—
Coraciidae	1	1	—	—	1	—	—	—	—
Cuculidae	3	1	—	1	1	—	2	—	—
Indicatoridae	1	1	—	—	—	—	—	—	—
Meropidae	3	1	—	—	2	—	—	—	—
Musophagidae	1	—	—	—	—	—	—	—	—
Phoeniculidae	2	—	—	—	—	—	—	—	—
Picidae	3	—	—	2	—	—	10	—	—
Psittacidae	1	—	—	—	—	—	—	—	—
Trochilidae	—	—	—	—	—	—	1	6	1
Trogonidae	1	—	—	—	—	—	—	1	—
Upupidae	1	—	—	—	1	—	—	—	—
	46	4	—	10	8	—	18	12	1

TABLE XVIII (cont.)

	Somaliland			Morocco			Arizona		
	R	B	N	R	B	N	R	B	N
Group E									
Sub-oscines									
Cotingidae	—	—	—	—	—	—	—	1	—
Tyrannidae	—	—	—	—	—	—	9	11	1
Oscines									
Alaudidae	18	—	—	12	1	—	1	—	—
Bombycillidae	—	—	—	—	—	—	1	—	—
Certhiidae	—	—	—	1	—	—	1	—	—
Cinclidae	—	—	—	1	—	—	1	—	—
Corvidae	4	—	—	6	—	—	8	—	—
Dicruridae	1	—	—	—	—	—	—	—	—
Emberizidae	3	—	—	4	—	—	26	6	10
Fringillidae	5	—	—	9	—	—			
Hirundinidae	2	—	—	3	4	—	—	5	1
Icteridae	—	—	—	—	—	—	8	3	1
Laniidae	11	—	2	2	1	—	1	—	—
Mimidae	—	—	—	—	—	—	6	1	—
Motacillidae	4	—	2	3	1	2	1	—	—
Muscicapidae	5	—	—	—	2	—	—	—	—
Nectariniidae	5	—	1	—	—	—	—	—	—
Oriolidae	—	—	—	—	1	—	—	—	—
Paridae	2	—	—	3	—	—	6	—	—
Parulidae	—	—	—	—	—	—	5	7	2
Ploceidae									
Estrildinae	8	—	—	—	—	—	—	—	—
Others	14	—	—	3	—	—	—	—	—
Prunellidae	—	—	—	1	—	—	—	—	—
Pycnonotidae	1	—	—	1	—	—	—	—	—
Sittidae	—	—	—	1	—	—	3	—	—
Sturnidae	11	—	—	1	—	—	—	—	—
Sylviidae	16	—	7	8	10	2	5	—	—
Thraupidae	—	—	—	—	—	—	2	1	—
Timaliidae	2	—	—	1	—	—	—	—	—
Troglodytidae	—	—	—	1	—	—	6	—	1
Turdidae	6	—	6	12	5	4	6	1	—
Vireonidae	—	—	—	—	—	—	3	2	—
Zosteropidae	1	—	—	—	—	—	—	—	—
	119	—	18	73	25	8	99	38	16
Grand total	214	4	20	132	34	10	146	57	21

regarded as the ecological equivalents of the non-passerine Coraciidae (rollers) and Meropidae (bee-eaters) rather than of the passerine Muscicapidae. It may be added that the opposite numbers of the Sylviidae include both the Vireonidae and the Parulidae (which are treated as a subfamily of the Thraupidae by Mayr and Amadon (1951).

It will be seen from Table XVIII that Somaliland with 218 breeding species, i.e. comprising both residents and breeding visitors, is much richer than Morocco with 166, but not much richer than Arizona, with 203. However, when allowance is made for the much more important degree of altitudinal replacement in Arizona this territory must evidently have appreciably less rich bird communities than Somaliland. The pre-eminence of this tropical area comes out much more strongly if we look at the figures as they are during the northern winter. Somaliland, losing four breeding visitors but gaining twenty winter visitors, then accommodates 234 species. (The number of winter visitors is actually smaller than might have been expected, for it compares with thirty species for Gambia (see above) and no less than seventy for Kenya, as shown in Chapter 14.) In both Morocco and Arizona the turnover of species is much greater than in Somaliland, for the Moroccan list is reduced to 142 species and that for Arizona to 167—to respectively 0·6 and 0·7 of the Somaliland total. Thus it is clear that, taking the year as a whole, the tropical territory, though the smallest and apparently ecologically the most limited, provides the largest number of niches for birds.

When the different sections of the bird faunas are considered it will be seen in the first place that the three totals in group B are very near each other, more so than in any other group. Morocco is actually just the richest of the three in this group, and one is reminded that among the African territories examined in Chapter 7, this is the one group in which the extra-tropical Cape Province holds its own with those of lower latitudes in the Ethiopian Region. Arizona owes its relatively favourable position in this comparison to its abundance of owls, which more than outweighs the fact that it has only two vultures against five in each of the other territories. All three territories show a great range of body-size in their owls so that on this ground alone much ecological segregation is indicated.

In group C, the ground-birds, Somaliland is the richest with twenty-one species, including seven bustards and six phasianids, while Morocco also has an excellent total, seventeen, including the remarkable number of five sand-grouse. By contrast, Arizona has only six species in this group, made up of Turkey, which is a forest bird, the high-altitude Blue Grouse and four phasianids, one of which had been "grazed-out" by the beginning of this century. Indeed, only three of Arizona's ground-birds belong to the sort of open country inhabited by most of Somaliland's twenty-one species; and in particular there seems to be no American equivalent to the coursers (Glareolidae) or the bustards (Otididae) or the sand-grouse (Pteroclidae).

In group D, the other non-passerine birds, Somaliland is again pre-eminent, with fifty species, compared to only eighteen in Morocco and thirty in Arizona. Actually, for the present purpose this latter number might be regarded as only twenty-three, since the seven Trochildae are, as already noted, the ecological equivalent of the passerine sunbirds, but on the other hand an uncertain number of the sub-oscine Tyrannidae must be regarded as taking the ecological place of bee-eaters and rollers. In the pre-eminence of Somaliland in this group a most important element is provided by those families and genera which are dependent on large insects and on fruit, for the latter of which categories there is no equivalent in the temperate-zone areas. But perhaps the most striking single feature of the Somaliland list is the dozen Columbidae, compared with only four in Morocco and five in Arizona. Allowing for the two montane species associated with forest, *Columba arquatrix* and *Streptopelia lugens* and the other frugivorous species, *Treron waalia*, this leaves nine seed-eaters, even more than in Gambia, as mentioned above, and an example of potential ecological overlap equally inviting investigation. In the Arizona list the comparative wealth of woodpeckers is very noteworthy; with ten species, they account for one-third of all the group D species there and they total twice as many as in Morocco and Somaliland combined. The fact that four of the ten species are thoroughly montane helps somewhat to explain this.

In the passerines Somaliland, with 119 breeding species, has fewer than Arizona (139), but a good many more than Morocco (98). Again Arizona scores heavily because so many species replace each other altitudinally. The Somaliland total owes much to the proliferation of four families, the larks (Alaudidae) with eighteen species, the shrikes (Laniidae) with eleven, the starlings (Sturnidae) also with eleven, and the weavers, estrildine and others, with twenty-two. In Africa the first of these families is typical of dry open country, which so dominates Somaliland, and the other three families have numerous representatives elsewhere in the dry parts of Africa. Between them these four families account for sixty-two species, compared with only twenty in Morocco and two in Arizona. Here there is only a single lark, *Eremophila bilopha*, and a single shrike *Lanius ludovicianus*, the former Holarctic. The place of shrikes in Arizona is perhaps to some extent taken by large members of the Tyrannidae. The lark has been, as it were, borrowed by Arizona from the tundra and it is difficult to imagine why nothing comparable is typical of the American dry country. To some extent the niches of the larks and of the estrildines in Somaliland may be occupied in Arizona by Fringillidae/Emberizidae.

This is no place to discuss the peculiarities of the American bird fauna, but certain features, especially the paucity of group C birds (see below) typical of dry areas, might perhaps result from the vicissitudes of the Pleistocene. Antevs (1954) and Martin and Harrell (1957) have independent evidence that, with reduced temperature and increased rainfall, the vegetation zones were lowered in south-western U.S.A. by some 3000 ft during the Last Glaciation. This must have involved a great reduction in the area of arid country and in its summer temperatures. Still, according to the map of Kendeigh (1961, p. 285), reconstructing the vegetation of North America during a glacial maximum, "desert" persisted in a strip from Death Valley southwards down the east side of the Gulf of California.

Morocco has nearly as many larks as Somaliland and of all three territories the largest number of Hirundinidae, of Sylviidae and Turdidae. By far the biggest predominance is in the last family, with ten more than in Arizona and eleven more than in Somaliland. The Moroccan Turdidae range from such birds as the Mistle Thrush *Turdus viscivorus* and Blue Rock Thrush *Monticola solitarius* to a series of dry-country chats *Oenanthe* spp. (whose interrelations have been described by Brosset (1961)). It is difficult to suggest to what extent ecological equivalents to many of these Moroccan Turdidae occur in the other territories. It is no easier to understand the relative abundance of Corvidae and Hirundinidae in Morocco, but when we come to the Sylviidae, which are nearly as abundant in Somaliland as in Morocco, the great deficiency of them in Arizona is doubtless made up there by the Parulidae.

It is also interesting to note how differently parasitic birds are represented in the three territories. Somaliland has most, with four cuckoos (one dependent on Corvidae and starlings), two honey-guides (on barbets and woodpeckers) and three Viduinae (on estrildine weavers, of which Somaliland has eight species but Morocco and Arizona none). By contrast with the nine parasites of Somaliland, Morocco has only one, the "European" Cuckoo *Cuculus canorus*,* and Arizona two, the cow-birds, *Molothrus ater* and *Tangarius aeneus*. Clearly, the Somaliland bird fauna carries a much more varied burden of parasitism than those of the other two territories.

GENERAL COMMENT

The foregoing comparisons between bird faunas of the temperate zone and of tropical Africa are preliminary and very tentative. They raise many unanswered questions about the structures of the bird faunas concerned, about the ecological equivalence of different groups of birds in different circumstances and about the extent to which niches are subdivided. In each of the comparisons made, the tropical areas, Gambia and Somaliland, are shown to have richer bird faunas than the temperate-zone territories used for comparison, even during the northern summer, and very markedly during the northern winter, even though this is the dry season in the northern tropics, with what appears to be contemporary ecological impoverishment. In the summer Arizona indeed runs Somaliland close, but not if allowance is made for its greater altitudinal replacement. The most important element in the pre-eminence of the tropical areas is the group D birds, those dependent on fruit and on large insects all the year round. Also in the dry country of Somaliland the complicated exploitation of its resources by a wealth of larks, doves and game birds is very remarkable.

* Another cuckoo that parasitizes Corvidae in Somaliland, *Clamator glandarius*, would be expected to breed in Morocco, since hosts are there and this cuckoo breeds in Algeria and Tunisia, as well as in Iberia, but there is still no breeding record for Morocco (Heim de Balsac and Mayaud, 1962).

9

The Ethiopian Lowland Forest Bird Fauna

This chapter deals with the African version of what Darlington (1957) has called "the true, very old, incredibly rich, intricately organized, unforgettably majestic rain forest of the continental, lowland tropics". In fact, as already mentioned, it appears that the African lowland forest is by no means so rich as the South American or the Malaysian; the reason may be sought in the vicissitudes it has undergone in the latter part of the Pleistocene, as described on geological evidence in Chapter 3. In times that on the evolutionary scale are very recent, the distribution of the lowland forest in Africa has differed greatly from what it is today and its range has been greatly reduced; and it is interesting to consider the existing distribution of the forest species of birds on that background.

As shown in Fig. 3, the lowland forest today consists mainly of two great blocks, one in the western part of West Africa, the other in the eastern part of West Africa and most of the Congo, with some relatively tiny areas elsewhere. The first block dominates much of Sierra Leone, Liberia, Côte d'Ivoire and Ghana, disappearing on the west in Portuguese Guinea and on the east in Togo. It is convenient to restrict the term "Upper Guinea Forest" to this block, as has been done by Chapin (1932) and others. Its length from east to west is nearly 1000 miles and its variable depth inland from the coast averages about 150 miles. However, the "actual forest" in this area has been estimated as occupying only about 17 million hectares = 70 000 sq. miles (quoted in Phillips, 1959) and much of that is under secondary growth in various stages following agricultural occupation.

The Upper Guinea forest is isolated from the forest belt of Nigeria on the east by 150 miles, conveniently called the Dahomey Gap, through which savanna reaches the coast. The Nigerian forests run east from the neighbourhood of Lagos across the lower Niger and through the corridors, totalling only about 50 miles in width, between the montane areas of the Cameroons and the corner of the Gulf of Guinea, to the great mass of the Congo forest. Because of this constriction and because the Nigerian forests have by some authors been included in the "Upper Guinea Forest" and by others in the "Lower Guinea Forest" I think it preferable to keep them separate under their own name. Moreover, owing to the density of the human population in this part of Africa the clearing has been so extensive, that, although on the vegetation map the Nigerian forest block is depicted as occupying nearly 50 000 sq.

miles, it appears that hardly any primary forest remains (cf. Richards, 1939), while fully grown tall secondary occurs only in patches. East of the Cameroons the great Lower Guinea (Congo) forest extends east continuously to the foot of the mountains forming the eastern rim of the Congo basin, so that this block, which is extremely irregular on its southern edge, averages about 1300 miles from west to east by 500 from north to south. In the south-west it extends discontinuously across the lower Congo into northern Angola, with vestigial patches along the Angola scarp all the way to about 11° S. East of the main Congo block and of the Rift Lakes, Albert, Edward, Kivu and Tanganyika, there are small outliers of Congo-type forest in the extreme south of the Sudan, of which the Lotti is the best known, in Uganda, especially the Budongo and Bugoma forests east of Lake Albert, and just inside the western edge of Kenya. There are also some vestiges of lowland ever-green forest on the east side of Lake Tanganyika.

Completely cut off from these outliers of the Guinea forests by the arid country of northern Kenya, by the mass of the Kenya highlands and by the dry interior of Tanganyika, barriers 300–500 miles in width, lie the vestigial coastal forests of East Africa. Naturally of a much poorer and drier type than the Guinea forests, they have suffered greatly in the last hundred years of "development", so that several species of birds discovered in them in the 1880's have not been found there since. A strip of "coastal forest–savanna mosaic" is mapped all through Kenya and Tanganyika, the "best" forest patches being the Sokoke-Arabuko in Kenya and the Pugu forest near Dar es Salaam, while there are very small but biologically important patches 50–100 miles inland, at the foot of the Uluguru and the Usambara Mountains. For convenience I shall refer to these collectively as the "East African coastal forests". Their isolation from the Guinea forests is complete today, for geographi-cally intermediate patches of semi-evergreen forest, especially at the foot of Kilimanjaro and near the foot of Mt Kenya, are of a different character, dependent on ground water and devoid of typical evergreen-forest birds.

South of Tanganyika, Malawi and the Zambezi Valley have frag-mentary lowland forests; and, as noted in Chapter 5, Zambia abounds in *mushitu*, patches of a few acres in which a few typically evergreen-forest species of birds persist. More important, because forming the "tropical corridor" of South African zoogeography, a narrow strip of "coastal forest–savanna mosaic" starts again just outside the tropics, at 25° S., and runs round as far as the Knysna forest in the middle of the Cape Province coast.

Beginning our examination of the lowland forest bird faunas in the north-west, it is illuminating to compare the Upper Guinea (Sierra Leone–Ghana) forest with the furthest removed part of the Lower Guinea forest, the north-eastern. For the present purpose this is arbi-trarily taken as east of about 26° E. and north of about 2° S., an area of about 250 × 300 miles containing the comparatively well-explored basins of the Ituri and Semliki. This area, though arbitrary, includes half a dozen species that are practically endemic to it, namely, the swift *Chaetura melanopygia*, the pigeon *Columba unicincta*, the thrush *Geokichla oberlaenderi*, the bulbul *Phyllastrephus lorenzi* and two of the most distinctive of all the weavers, *Ploceus aureonucha* and *P. flavipes*. On the east side of the Semliki River, the district of Bwamba (worked by the van Somerens, 1949) at the north end of Ruwenzori and just inside Uganda, forms the eastern limit of the continuous Lower Guinea (Congo) forest.

The forest bird fauna of Upper Guinea numbers 182 species*; this includes nine endemic species, five of which have vicariant species somewhere to the east of the Dahomey Gap, the pairs being the phasianids *Agelastes meleagridis/niger*, the flycatchers *Artomyias ussheri/fuliginosa*, the non-parasitic cuckoos *Centropus leucogaster/anselli*, the babblers *Picathartes gymnocephalus/oreas* and the estrildines *Parmoptila rubrifrons/woodhousei*. Of the five genera concerned, only *Centropus* is represented in the Guinea forests by any species other than the pair named. The other Upper Guinea endemics consist of the warbler *Apalis sharpei*, the woodpecker *Campethera maculosa*, the babbler *Malacocincla rufescens* and the fishing owl *Scotopelia ussheri*. The Upper Guinea total of 182 species compares with 212 for the north-east Congo (as compiled from maps in Schouteden, 1948–56). Thus the Upper Guinea total, from the detached end of what must always have been, on the continental scale, a long narrow ecological peninsula, cannot be regarded as particularly impoverished, for it numbers 86% of the total for north-east Congo, which is not isolated in any way from the rest of the 600000 sq. miles of the Congo forest. But the whole of this vast area counts only 266 typically forest species, as compiled from the records in Chapin (1932). This total is remarkably unimpressive when put against the total of 269 resident species, which, however, include a few belonging to clearings, recorded by Slud (1960) on no more than 2 sq. miles of forest in Costa Rica. It would be interesting if someone with adequate knowledge could give corresponding figures for a block of South American forest.

As mentioned before, I would ascribe the comparative poverty of the African list to the disruptions that the forest biome, superficially of such immense stability, has suffered, especially in the last hundred thousand years or so. First, the redistribution of Kalahari sand over most of the present Congo forest area towards the end of the mid-Pleistocene indicates that the forest must have been vastly reduced and pushed into edges and corners, presumably with concurrent extinctions. The resultant isolation of surviving populations would have been conducive to speciation, and of course highly effective if it continued long enough.

---

* This figure is not quite in agreement with that of Marchant (1954), in part because of taxonomic rearrangements, in part because personal judgments differ on whether a few species should be regarded as forest birds or not.

Such a fragmentation could go far to explain the situation in the *Malimbus* weavers as described by Moreau (1958a). The nine closely allied species are almost entirely confined to the Guinea forests, with great geographical overlap. Apparently four to six of the species occur together over most parts of the forested area from Sierra Leone to the eastern side of the Congo basin. It is far easier to envisage the evolution precedent to the present situation as taking place in a series of isolated forests rather than in a block such as the present one, which, apart from the Dahomey Gap (and recent human depredations), is continuous.

After this dry period forest presumably re-established itself widely over the present basin but during the glacial maximum montane conditions would have encroached from all sides, reducing the area of lowland forest to perhaps half what it now is, as can be inferred from the position of the 500 m contour in Fig. 2. A similar contraction, alternating with expansion, presumably occurred with every glaciation as a result of the temperature changes and irrespective of the predominance of humidity or aridity. However, since the Last Glaciation we know that at least the south-east of the Congo basin has suffered again from aridity, around 11 000 years ago, as discussed in Chapter 3, with consequent reduction of forest in that part.

The Upper Guinea forest, being practically all within 500 m of sea-level, would have felt no decisive effects from the Pleistocene changes in temperature but it has suffered great changes in area as a consequence of the changes in the extension of Ethiopian vegetation northwards, whether it was bounded by desert, as now, or by Palaearctic vegetation as apparently during the Neolithic. When the extension was at its widest, the area of the Upper Guinea forest may well have been four times what it is now, and when the Sahara came 300 miles further south than it does now the forest area must have been greatly reduced. No doubt the Upper Guinea block was fragmented, as must be postulated from the ranges of forest mammals (Booth, 1958), and on theoretical grounds it might be inferred that it was almost eliminated by the southward movement of the drier zones towards the Gulf of Guinea; but the existence of the endemic species referred to above shows with virtual certainty that forest to some extent persisted. Those five species which have close relatives on the opposite side of the Dahomey Gap might conceivably have evolved since the Late Pleistocene restriction and subsequent recovery of the Upper Guinea forest, but the other four species are unlikely to have done so. Such violent vicissitudes in the range of the biome might be expected to result in much extinction, and it is all the more surprising that the Upper Guinea forest bird fauna should be so comparable in number of species with that of the Congo block.

No less than 157 species (86%) of the Upper Guinea forest bird fauna appear also in the north-east Congo list, 1500 miles away. Hence, the west-to-east continuity is great, in parallel with that of the savanna zone and of the steppe zone to the north (see Chapter 10). This continuity in the range of so many forest birds is the more noteworthy because throughout the Pleistocene the constriction of the lowland at the corner of the Gulf of Guinea would always have been about as narrow as it is now. Certainly during the last glaciation the eustatic lowering of the sea-level by some 300 ft would have turned the Fernando Poo strait and the mangrove creeks on the shore of the continent into dry land, but concurrently the montane areas bordering the lowland corridor would have been enlarging and so encroaching on it. Also at this stage, which persisted in full intensity for some thousands of years and began to give way only some 18000 years ago, this lowland corridor round the corner of the Gulf of Guinea would have greatly reduced significance because, as shown by the contours in Fig. 2, it could at that time give access to nothing more than the narrow coastal strip of Gabon. Access to the lowland centre of the Congo basin was cut off because, the lower limit of montane conditions having come down to about 500–600 m a.s.l., the comparatively low northern and western rims of the Congo basin, though not high, would have become a montane barrier. Hence the present continuity of range of so many species from Upper Guinea to the north-eastern corner of the Congo forest has presumably been established, or more likely re-established, since the glaciation. To this general impression of continuous ranges there are, however, a number of exceptions which will be discussed later.

Meanwhile it is interesting to consider the status of the forest bird fauna in Nigeria, which is geographically intermediate. Bounded as the Nigerian forest is, by the Dahomey Gap on the west and the Cameroons constriction on the east (Fig. 25), its bird fauna might be expected to be to some extent individual and to a considerable extent homogeneous. This is not confirmed by the investigation of Marchant (1954). He used the available collections from Lagos (around 3° 30′ E.), Ondo-Benin (also west of the Niger and around 5–5° 5′ E.), Owerri (east of the Niger and at about 7° E.) and Kumba (in the Cameroons gap and at about 9° 9′ E.) (see Fig. 25, p. 58). Omitting those species found through Nigeria and beyond without subspecific change, he showed that round Lagos fifty-two out of eighty-one birds (0·64) belong to Upper Guinea subspecies, in Ondo-Benin forty out of eighty-six (0·42), in Owerri twenty-nine out of 103 (0·28) and in Kumba twelve out of 126 (0·10). Although, as noted above, a small minority of Upper Guinea species (9) do not appear in the Nigerian forests, the fact that so many of the birds at Lagos are subspecifically identical with those west of the Dahomey Gap shows that though this is prominent on the vegetation map today it cannot have been an effective barrier for long. Indeed, it seems to depend on no topographic or orographic feature but on a very localized cold up-welling (Brooks and Mirlees, 1932), which might be of little permanence. East of Lagos, considering that intervals between collecting stations are only about 100–150 miles, the variation in the birds is remarkable at the subspecific level, but it is fairly regular. A more marked discontinuity might perhaps have been expected between Benin and Owerri, separated as they are by the lower Niger River. From the subspecific ranges of the forest primates Booth (1958) found reason to think that to these animals the river had been a significant barrier and, although of course it would not have equal potentialities where birds are concerned, it may, however, well have been more formidable at intervals in the recent past than it is now, as discussed earlier (Moreau, 1963).

While most of the species common to the Upper Guinea forest and the forest in the north-east of the Congo appear to have as continuous a range between these two areas as the vegetation permits, twenty-four other species appear to be absent from intermediate stretches of the Guinea forests, the gaps being from 600 to over 1500 miles in width. Examples are the following, longitudes being approximate: *Scoptelus castaneiceps* 13–26° E., *Smithornis capensis* and *S. sharpei*, both 13–28° E., *Erythropygia leucosticta* 2° W. to 23° E., *Malimbus coronatus* 13–22° E., *Prinia bairdii* 13–25° E., *Malacocincla puveli* 13–29° E. No doubt a proportion of these cases are merely due to inadequate collecting, but, as Chapin (*in litt.*) has agreed, it is improbable that this applies to all; and when we remember that this north-eastern corner of the Lower Guinea forest has six endemic species it is clear that a genuine problem of distribution exists. It might be suggested that these restricted endemics and these discontinuities of range are a legacy of the mid-Pleistocene fragmentation of the Congo and restriction of its species to the seaward and the eastern edges of the basin. But these effects are likely to have been obliterated by the subsequent effects of the glaciation, with the prolonged concentration of lowland life in the centre of the basin, unless some independent lowland refuge persisted on the north-eastern periphery of the area throughout the glaciation.

Actually, such a refuge, though a very small one, may perhaps have been available in the Semliki Valley, which, with the district of Bwamba just inside Uganda, at the northern foot of Ruwenzori, forms the extreme north-eastern extension of the Congo forest. This valley is at the top of the Albertine Rift Valley, which communicates with the extensive Sudan lowland. Since the surface of Lake Albert stands at 618 m and much of the Semliki Valley is below 700 m (see Fig. 33), an area of perhaps 1000 sq. miles is near enough to the critical lower level of about 500–600 m reached by the montane during the glaciation to provide a possible refuge for strictly lowland species. However, only one of the half-dozen north-east Congo endemics, *Columba unicincta*, has been found in Bwamba today (van Someren and van Someren, 1949). Moreover, with the single exception of a bulbul, *Phyllastrephus (baumanni) hypochlorus*, all the birds of the lowland forests east of the Albertine Rift, those of Uganda, western Kenya and the southern edge of the Sudan, isolated as they are, belong to species and also to subspecies found west of the

Semliki, in the Congo. These facts as a whole do not point to the persistence of an isolated Semliki–Bwamba refuge during the glaciation which would have been a focus of subspecific differentiation, but, except for *Ph. hypochlorus*, rather to a post-glacial colonization from west of the Albertine Rift as montane conditions evacuated the Lake Victoria basin. Further, if a Semliki–Bwamba refuge were not operative, the north-east Congo must have been reoccupied by lowland species ex-

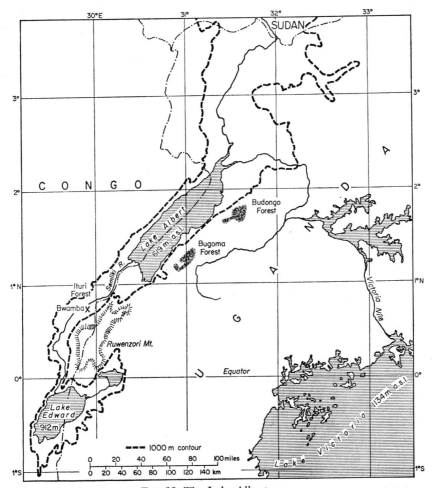

FIG. 33. The Lake Albert area.

panding from the centre of the basin as the montane conditions receded, and their spread across the Lake Victoria basin must have been a later stage of the same movement, and hence considerably after the end of the glaciation. In the result, the discussion has come full circle and we are no nearer to the solution of the problem why the north-eastern Congo harbours a group of endemics and also two dozen other species that reappear on the opposite side of the Congo basin, but not in the middle of it.

It may be mentioned incidentally that what is one of the most note-worthy birds in the lowland forest, the African Peacock *Afropavo congensis*, only described in 1936, is now known to have an extensive range in the Congo, from at least 18° to 28° E. It is unlike any of the other African Galliformes but has, as its name implies, a considerable re-semblance to the peacocks of the Oriental Region. Unless at some stage the common ancestor inhabited much drier vegetation than the present-day peacocks, it is difficult to postulate extension of the necessary vege-tation across the gap between the ranges of the African and Asiatic birds at any stage in the Pleistocene and presumably the connexion must be sought earlier. As mentioned in Chapter 6, a somewhat similar but even more acute problem is raised by two birds that are known only from montane forest on the eastern edge of the Congo and have close rela-tives in India.

A large proportion of the Congo forest birds extend south to the southern part of the basin, and some of them beyond, into northern Angola, while eastwards, about half of them occur in one or more of the forests of Uganda (see Chapter 15), some of them ranging past Lake Victoria to the western foot of the Kenya highlands. However, no more than fourteen species (7%) reappear on the other side of the Kenya highlands, in the East African coastal forests, and half of these are ex-ceptionally widespread birds occurring sporadically in suitable vegeta-tion in the Rhodesias and right down into South Africa. In fact, the ultimate forest of Africa, the Knysna, close to the southern coast of Cape Province at 34° S., is inhabited by four of the Congo species, along with three that are typical of the East African coastal forests (but not the Congo), eight that are typically montane within the tropics, and five South African endemics, in addition to wide-ranging raptor and other species, some of which in the tropics are not typically forest species at all (see Chapter 15 for another reference to this bird community).

The East African coastal forests, exiguous and broken up as they are today, count only thirty-eight species of birds for the whole length of some 600 miles from the Juba River to the Rovuma; and this total includes two species, the thrush *Geokichla guttata* and the warbler *Apalis chariessa*, which have not been found on the East African coast since their original discovery some eighty years ago.* (They are not extinct; the main stronghold of *chariessa* is in Malawi and the thrush survives there and south to Natal.) Moreover, some of the thirty-eight species seem today to be quite narrowly localized within the coastal strip. Much the richest section is the most humid, that on the seaward foot-hills of the East Usambara Mountains, some 40 miles inland, which house thirty-two of the species.

Several of the East African coastal species, which do not extend south of about 6° S., are of special zoogeographical interest, because they are

* Since this was written this thrush has been found again on the Kenya coast (A. Forbes-Watson, *in litt.*).

represented also in the Congo and Uganda, with the barrier of the high and/or dry interior of Kenya and Tanganyika intervening. A good example is the turdine *Neocossyphus rufus* (Fig. 34), with a gap of 600 miles in its distribution. A babbler, *Malacocincla* (=*Illadopsis*) *rufipennis*, has much the same range and in both these species the East African

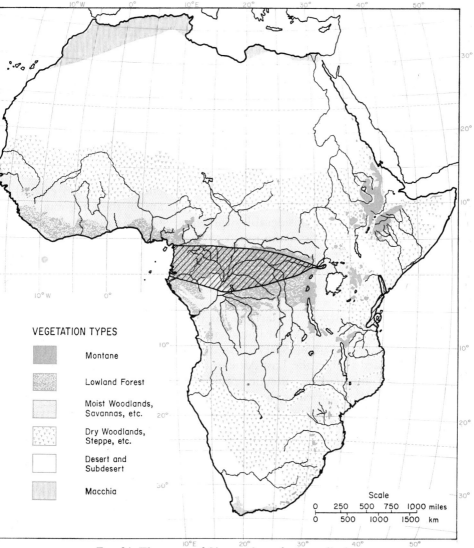

FIG. 34. The range of *Neocossyphus rufus* (a turdine).

population is subspecifically distinct. Confined to the low-level forests of the Usambaras there are two more comparable birds, *Bubo poensis*, with its West African and East African populations little differentiated (Moreau, 1964a), and an estrildine *Spermophaga ruficapilla*. The Usambara form *cana* of this weaver is so distinct that Chapin thought it might be classed as a different species—it is much more of a forest bird, living mainly on the hard seed of the peculiar grass *Olyra latifolia,* than its western relative. The same uncertainty about the degree of relationship applies to the weaver *Ploceus golandi* of the coastal forest in Kenya, which has been suggested as conspecific with *P. weynsi* of the Congo. Certainly the *Spermophaga* populations and the *Ploceus* populations can be accepted as forming superspecies and perhaps the same applies to the East African coastal *Macrosphenus kretschmeri* with *M. flavicans* or *M. concolor* of Uganda. These are enigmatic birds which have been classified recently as babblers, as bulbuls and finally as warblers.

Here, then, we have three cases in which West African species reappear on or near the East African coast with only subspecific differences and two or three in which differentiation has gone further.* Two other birds may be mentioned in this connexion. Among the turacos of the *corythaix* group—whether members superspecies or of a polytypic species (Moreau, 1958b)—*fischeri* of the East African coast from the Juba River to 6° S. is most like *schutti* of Uganda and the Congo, and resembles it more than it does *livingstonii,* which occurs south of *fischeri* in coastal Tanganyika. Finally, a less cogent case than any of the foregoing, because the bird is a considerable wanderer while the others are thoroughly sedentary, the Palm-nut Vulture *Gypohierax angolensis* has its only East African location in the coastal zone (and on Pemba Island) in association with the typically West African oil-palm *Elaeis guineensis,* which is however not quite certainly native there (see discussion in Moreau, 1933).

Whether we accept the last case as valid or not, these birds in conjunction pose a difficult distributional problem, which is reinforced by data from other groups, especially butterflies. Of these Carcasson (1964) has recorded that in the East African coastal forests the relatively poor fauna is comprised of endemic species or very distinct subspecies belonging to West African genera. It seems therefore necessary to postulate that in the past there has been a lowland forest connexion between the Congo and the East African coast. In fact the two levels of differentiation amongst the birds cited may mean that a connexion was established on two separate occasions. It is not easy to envisage such a connexion. The lowest corridor between the two areas is that past Lake Rudolf, but it is so excessively dry today that under anything like the present atmospheric circulation and under any increase in rainfall that can reasonably be suggested a lowland evergreen forest connexion by this route seems impossible. South of the Rudolf gap the Kenya highlands are interposed and they are so high that even at an interglacial maximum of temperature they would not be warm enough for lowland forest. Hence the most likely line of connexion between the Congo forests

---

* Since this was written another example has been added by the discovery on the Kenya coast of a "new" owl, *Otus ireneae,* that forms a superspecies with *O. icterorhynchus* of West Africa and the Congo (Ripley, 1965).

and those of the East African coast is through Uganda and the interior of Tanganyika, where much of the surface is about 4000 ft but little above 5000 ft. Here too, however, the climate is dry but not so excessively as in the Rudolf gap, and in the north of Tanganyika the long dry season is interrupted by the "short rains". Both increased rainfall and further mitigation of the dry seasons must be postulated to bring evergreen forest across this area, but not during a glaciation because the contemporary temperatures would be too low. In fact we can form no idea of when the necessary conditions can have occurred.

Discussing the butterflies, which provide more data than do the birds, Carcasson has concluded that, while no broad-front connexion has been possible between the Congo and east-coast forests since the uplifting of the East African plateau, details of distribution and differentiation suggest to him that there has been a brief and narrow connexion in the course of the Pleistocene, probably through the gap between Lakes Nyasa and Tanganyika. Here the difficulty is that the intervening country is high—over the montane limit today. Conceivably it would have been warm enough during the hypsithermal but that was no more than around 6000 years ago, and again much heavier annual rain, together with a briefer dry season than the present, would have been needed to clothe this col between the lakes with lowland evergreen forest. Actually this very same piece of country must be postulated to have served as a north-and-south corridor for very dry-country, acacia, species of birds and mammals at some stage of the Late Pleistocene (see Chapter 10), which makes the problem of the Congo–east-coast forest connexion more difficult still.

10

The Lowland Non-forest Bird Fauna

As indicated in Chapter 3, it is necessary to envisage that during the most severe stages of the glaciations and certainly during the last phase of severity, around 25000–18000 years ago, animals and plants intolerant of montane conditions would have been restricted to tropical Africa west of the Cameroons, to the southern Sudan (with a prolongation up the Albertine Rift), the middle of the Congo basin and a rim round the coast, with of course ramifications up the main river valleys, such as the Zambezi. This geographical arrangement would have been highly conducive to speciation, especially for non-forest birds if, as is possible, the coastal strip was here and there occupied by evergreen forest. In the 18000 years since the conditions associated with the glaciation began to relax, all those species which had been cooped up within the narrow limits indicated above would have extended their ranges, mainly from the edges of the continent towards the interior (see Fig. 27, p. 99); in the course of it individual species would meet others that are closely related; then either their expansion would be blocked or a mutual ecological adjustment would follow. The latter would lead to co-existence and an enrichment of the bird community concerned.

The concepts outlined above have their difficulties. Unless it is accepted that many of the existing species that are typically lowland have been evolved to that level in the last 18000 years or so, it has to be accepted that as many species as exist today, including about 500 passerines (Table V, Chapter 5), were packed into something like one-sixth of the area they now occupy. Moreover, half of this reduced area was in West Africa, which today is poor in non-forest species, as discussed in Chapter 6, and especially in endemics, as will be shown below. Hence in the remaining lowlands of the continent the crowding of species involved in the situation outlined above was much greater than the figures of relative area would indicate. Perhaps some species classed as typically lowland today were not so limited in the past. In any case it is probable that the average geographical range of individual species was far smaller than it is today and that there was much vicariance among closely related species.

The habitats occupied by the lowland non-forest birds are of course far more varied than those of the other ecological categories designated in Chapter 5, namely the montane birds and the forest birds. Bird habitats have formally been listed for a few areas of Africa, especially Malawi (Benson, 1953), Zambia (Benson and White, 1957), Eritrea

(Smith, 1957) and the Cape Province (Winterbottom and Skead, 1962). For the present purpose it is useful to recapitulate in a different form some of the botanical information already given, and it suffices to list main habitats as shown below. It may be recalled that in African ecological usage "savanna" is a grassland with trees, so that "savanna" intergrades on one side into woodland, on another into wooded steppe and on a third into open grassland.*

A. Permanent watery habitats, including beds of *Phragmites* and papyrus.

B. Flood plains. As they dry after their temporary inundation they become covered with a dense growth of herbage, mostly heavily seeding grasses, among and near which vast numbers of weavers (Ploceidae) breed.

C. Deciduous woodland and wooded savanna, divisible by the dominant trees into Acacia, Mopane, Brachystegia (in the south; the related (*Isoberlinia* in the north) and High-grass–Short-tree savannas. (These are the "Woodlands undifferentiated, relatively moist types" of Keay (1959), which are arranged here in order of increasing humidity.) So far as Fig. 3 is concerned, *Acacia* is the typical tree genus of the "Wooded Steppe" areas and occurs over much of the "Wooded Savanna"; Mopane Woodlands form areas, too small to depict on that scale, in the "Wooded Savanna" (dominated by *Brachystegia* spp.) south of the Equator. Bird communities of sample areas of these three woodland types are listed and discussed in Chapter 15. Finally Long-grass–Short-tree Savanna occupies the most humid part of the "Wooded Savanna", in a narrow belt round the edge of the Guinea forests.

From the birds' point of view each of the woodland/savanna types is separated into two sub-habitats, the trees and the underlying herbage. This latter is nearly everywhere seasonal, largely disappearing after the rains. Then it is burnt in those (less arid) areas where it rises from perennial rootstocks and grows more rank; and in the arid areas, where the herbage is ephemeral, it dies to very small proportions.

D. Riparian woodland (Figs. 35 and 36). As a line or a narrow belt of trees this appears along drainage channels, whether they usually carry water or not; the trees are bigger and the component species are different from those of the surrounding savanna or steppe.

E. Open country with short grass, low bushes and fewer trees (including the Karoo of South Africa).

F. Rocky slopes and cliffs.

G. Mangroves. These seem to be of little importance to African birds (Cawkell, 1964), less for example than in Australia (personal information), but more study is required.

H. Cultivation. By far the greater part of cultivated land in Africa is not "clean"; it has residual trees, "weeds" and regenerating natural growth, especially of a bushy nature. "Cultivation" of this type is attractive to many birds, apart from those which batten on the crops themselves.

* With especially this group of vegetation types in mind, Emlen (1956) has elaborated a scheme for describing the bird habitats by means of quantitative data.

FIG. 35. Vegetation along Grumiti River and floodplain grassland, Serengeti National Park, Tanganyika. The dark spots are anthills with peculiar vegetation. Photo: P. J. Greenway.

FIG. 36. Gallery forest (riverine vegetation), Aloma Plateau, south-eastern Sudan. Photo: A. C. Hoyle.

Most parts of tropical Africa of course offer a mosaic of habitats. Apart from the fact that watery places, riparian woodland or cultivation are liable to appear in all but the most arid areas, the components of the mosaic vary a good deal. For example, Zambia is composed mainly of flood-plains and of woodlands among which acacia, mopane and brachystegia are represented, with subsidiary areas of baikiaea and cryptosepalum woodlands and semi-evergreen thickets on termite mounds. On the other hand the dominant components of Somaliland are acacia steppe and subdesert steppe, in places with rocky hillsides.

Attention has been given in the past to dividing the Ethiopian Region into avifaunal "districts". Chapin (1923) designated seventeen (reproduced as Fig. 37), including five devoted to forest and montane areas,

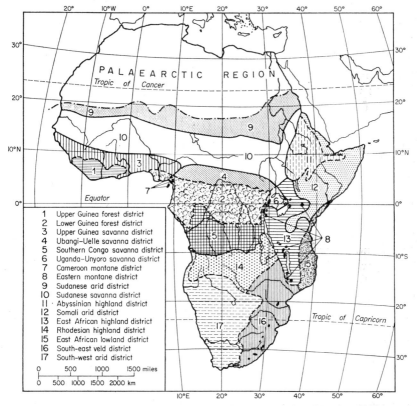

1 Upper Guinea forest district
2 Lower Guinea forest district
3 Upper Guinea savanna district
4 Ubangi–Uelle savanna district
5 Southern Congo savanna district
6 Uganda–Unyoro savanna district
7 Cameroon montane district
8 Eastern montane district
9 Sudanese arid district
10 Sudanese savanna district
11 Abyssinian highland district
12 Somali arid district
13 East African highland district
14 Rhodesian highland district
15 East African lowland district
16 South-east veld district
17 South-west arid district

FIG. 37. Chapin's Ethiopian avifaunal divisions. (After Chapin, 1923.)

which in this book are covered in Chapters 9, 11 and 12. The birds of the remaining twelve "districts" are thus left to be covered in the present chapter. To some extent their boundaries are in rough agreement with the main blocks of vegetation types in Keay's (1959) map. For example, the Sudanese and Somali Arid Districts of Chapin together more or less occupy the "Acacia-Commiphora Wooded

Steppe" plus the "Subdesert Steppe" of northern tropical Africa, as shown in Fig. 3; but on the other hand the mutual boundaries of avifaunal districts 13, 14, 15 and 16 do not have counterparts on the vegetation map. South of the Zambezi subdivision has been carried much further, for within Chapin's two "Veld" districts, 16 and 17, twenty-three "biological subdivisions" have been mapped (McLachlan and Liversidge, 1957). A weakness of these various "districts" is that no lists or numerical evaluation of their distinctive characters have been published, but it is clear so far as birds are concerned that the various "districts" are of very unequal status, as are the "subdivisions". By contrast with the subdividing that has been done, Benson and Irwin (1966) have found that there is a case for merging the Rhodesian and East African Highland Districts (13 and 14 in Fig. 37), since the proportion of avifaunal difference is not high.

The difficulties of delineating avifaunal "districts" closely or consistently are particularly well shown in the northern tropics, where the ecological pattern is relatively simple, with rainfall decreasing and vegetation becoming more sparse from south to north, until the Sahara desert is reached at between 15° and 20° N. In conformity, various species of birds are typical of the drier north but not of the more humid south of the northern tropics, and vice versa. Such changes are, however, most difficult to map satisfactorily. Chapin designated latitudinal belts north of the Equator (in ascending order of humidity) as the Sudanese Arid, Sudanese Savanna and Guinea Savanna + Ubangi–Uelle Savanna. Lynes (1925) on personal experience in the Sudan, agreed in the main with Chapin's map, but proposed some modifications and in support gave lists for his area. In particular he would increase the width of the Sudanese Arid District, bringing its southern boundary down to 10° N. and thus reducing the Sudanese Savanna to a very narrow strip indeed, which in fact merely covers transition from Sudanese Arid to Guinea + Ubangi–Uelle Savanna. Moreover Elgood (personal communication) has noted in travelling from south to north in Nigeria that typically northern "arid" species are first encountered on comparatively barren slopes, while typically southern species appear sporadically north of their main range in spots that are exceptionally humid for the locality. Also it may be noted that some relatively "arid" species, such as the lark *Mirafra buckleyi*, appear near Lagos on the Gulf of Guinea, some 200 miles south of their main Nigerian range, presumably as a result of working east along the coast from the comparatively dry "Dahomey Gap".

In the face of such difficulties as these it is more fruitful not to use the conventional subdivisions into "districts" etc., but to consider the lowland non-forest birds *de novo*. As a preliminary, it is convenient briefly to dispose of the raptors (group B) and the water-birds (group A). The non-forest species in both these groups have on average very extensive ranges, and hardly any of them are confined to restricted areas of

Africa. As shown in Table XIV (pp. 130–133), thirty lowland non-forest raptors and forty-four water-birds are common to the extremities of the Ethiopian Region, West Africa and Cape Province, out of totals of sixty-eight and ninety-six belonging to those groups. The proportions are 0·44 and 0·46 respectively, compared with only 0·11 in the passerines; and it may be recalled that although one of these areas is tropical and the other not, both have a very wide range of comparable habitats.

As classified in Table V (pp. 84–87), about one-eighth of both water-birds and raptors etc. are today tolerant of montane conditions, but although the proportion is the same it seems possible that the present extension of the raptor ranges is due to post-glacial dispersal to a greater extent than is the water-bird extension. The dominants of many water-bird habitats today are papyrus, *Phragmites* reeds and *Typha* bulrushes. In the inner tropics today all of these extend upwards into the montane zone though in Zambia and Malawi papyrus is not found above about 4000 ft (C. W. Benson, *in litt.*). It can be inferred then that at the height of a glaciation such a great water-bird focus as Lake Victoria would retain its essential features and perhaps much the same water-birds as at present, though Lake Bangweulu, for example, would lack papyrus, and no doubt lose some species in consequence. It may be added that there are about a dozen passerines that are dependent on swamp vegetation. Such are *Cisticola carruthersi* and the shrike *Laniarius mfumbiri*, closely associated with papyrus, and three species of reed-warblers *Acrocephalus*.

Leaving aside the group A and group B birds, it is of much interest to make some analysis of the endemism in different parts of Africa among non-forest birds that can be regarded as typically lowland (Table XIX). In the first place, 178 species are more or less limited to northern tropical Africa, the strip between the Equator and the Sahara, extending over some 14° of latitude in West Africa but more than 20° in the east. This total does not include several birds that were until recently regarded as distinct species limited to the northern tropics but are now treated as conspecific with birds of the southern tropics, thus making cases of discontinuous range, which are discussed separately below. Nor does the total of 178 species include a number that are unknown in Africa south of the Equator but are shared by northern tropical Africa with areas to the north and east—the Sahara (as several species of sandgrouse Pteroclidae), the Palaearctic Region (as *Athene noctua*) or India (as the parrakeet *Psittacula krameri*).

These 178 northern tropical endemics are readily divisible into two ecological groups and two geographical. In the first place, the ecological pattern is mainly latitudinal, as already indicated (Fig. 3); throughout West Africa and east nearly to the Nile the transition from desert-steppe to the most humid savanna is compressed within a strip some 700 miles deep. Further east, in the Sudan and the Nile basin, the equatorial forest disappears, while the series of vegetation belts, arid to

Table XIX

Endemism of non-forest species in different parts of Africa

Family	Ethiopian Region	Somali Arid endemics	Other northern tropical endemics	Endemics of South Africa and S.W.A.	Brachystegia endemics
		Number of species			
Group C					
Otididae	16	3	1	4	—
Phasianidae	25	2	4	3	—
Pteroclidae	10	—	—	2	—
		5	5	9	—
Group D					
Bucerotidae	12	3	1	2	1
Capitonidae	23	4	2	—	2
Caprimulgidae	14	3	3	1	—
Coliidae	6	1	1	1	—
Columbidae	15	2	4	—	—
Coraciidae	6	—	1	—	—
Meropidae	11	1	1	—	—
Musophagidae	9	1	3	—	—
Phoeniculidae	6	1	—	—	—
Picidae	13	—	1	1	2
Psittacidae	13	1	1	2	—
		17	18	6	5

Number of lowland non-forest species

Group E					
Alaudidae	44	10	6	15	—
Corvidae	6	—	1	—	—
Emberizidae	8	1	1	1	1
Fringillidae	14	3	1	5	—
Hirundidae	19	—	3	—	1
Laniidae	30	8	2	3	—
Motacillidae	18	3	—	3	1
Muscicapidae	27	3	4	5	4
Nectariniidae	29	6	4	2	2
Paridae	9	1	2	2	—
Ploceidae					
Estrildinae	43	3	8	1	2
Others	69	12	7	4	—
Promeropidae	1	—	—	1	—
Pycnonotidae	12	1	1	3	2
Sturnidae	29	8	5	4	—
Sylviidae					
Cisticolae	25	2	2	1	1
Others	49	5	5	10	3
Timaliidae	17	1	4	2	—
Turdidae	41	3	7	12	2
Zosteropidae	2	1	—	1	—
	—	71	63	75	19
Grand total	—	93	86	90	24

humid, bends south round the north-east of Lake Victoria. At the same time arid country, dominated by subdesert steppe and acacia steppe, becomes proportionately far more important than in the west and centre of the continent. It surrounds the Abyssinian highlands, fills the Horn of Africa and extends south through eastern Kenya to as far as central Tanganyika.

The ranges of the birds peculiar to the northern tropics are either (1) essentially latitudinal, following the vegetation belts part or the whole of the way across the continent from the neighbourhood of Senegal to the neighbourhood of the Red Sea or Gulf of Aden (a distance of some 4000 miles and a total area of some 3 million sq. miles), or (2) belonging to the eastern end of the northern tropics, dominated by acacia steppe, essentially from Eritrea and Somalia to Kenya and beyond, covering about 500000 sq. miles. These two types of range, one essentially east–west, the other north–south, are in fact the basis of the distinction between Chapin's series of latitudinal belts and the Somali Arid District of Fig. 37. Of course, virtually no two species have exactly the same geographical ranges; in particularly, a few of the "latitudinal" species tend to bend south into Uganda and some of the Somali species follow the acacia steppe south throughout its extension across the Equator into Tanganyika. But the decision whether to count a species as of latitudinal or Somali range is easy in practically all cases. The representation of individual families in the two types of endemics is shown in Table XIX.

For purposes of discussion it can be assumed that most of the species now found as endemics of the two types designated were evolved there but this will doubtless not apply to all. Some of the existing ranges are doubtless relics of more extensive ones. For example, *Scoptelus aterrimus*, with a latitudinal range from Senegal to Eritrea, has an isolated population in Angola, and *Francolinus albogularis* of Senegal to the Cameroons reappears in south-eastern Congo and again in eastern Angola (Fig. 38). Presumably these species have extended in recent times in a belt round the Congo forest. Under present vegetational conditions similar ranges cannot exist for species typical of subdesert, acacia steppe or dry savanna, because these vegetation types do not occur close to the southern edge of the Congo forest; but conditions may have been significantly different on the occasions when aridity extended north from the Kalahari dry focus, as described in Chapter 3 and as discussed further below. Actually, of the 178 endemic northern tropical species only a very small proportion, less than a score altogether, belong to humid habitats or the most southerly and humid latitudinal belt. Examples are the turdines *Cossypha albicapilla* and *C. niveicapilla*, *Hypergerus atriceps* (variously classified as a babbler and a warbler) and the penduline tit *Remiz* (=*Anthoscopus*) *flavifrons*. One species, the warbler *Drymocichla incana*, extending from northern Cameroon to Uganda, is peculiar in being a swamp bird. Incidentally it seems that, on number of endemic species, there can be little positive basis for designating the Guinea Savanna or the Ubangi-Uelle districts. The great majority of the north-tropical endemic birds belong in fact to the drier habitats, especially acacia savanna, acacia steppe and subdesert.

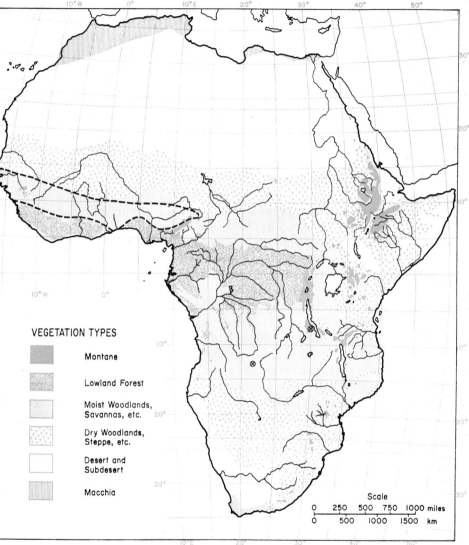

VEGETATION TYPES

Montane

Lowland Forest

Moist Woodlands, Savannas, etc.

Dry Woodlands, Steppe, etc.

Desert and Subdesert

Macchia

Scale

| 0 | 250 | 500 | 750 | 1000 miles |
| 0 | 500 | 1000 | 1500 | km |

FIG. 38. The range of *Francolinus albogularis* (a bush fowl). – – – –, Approximate range in West Africa; ⊗, occurrences elsewhere.

The endemic status of the non-forest bird fauna of West Africa (west of the Cameroons) is in some respects anomalous. As stressed in earlier chapters, this section of the continent forms an ecological peninsula, bounded on the south and west by the sea, on the north by desert during some periods, by Mediterranean vegetation during others. At its eastern end communication with the rest of tropical Africa has, for lowland non-forest birds, always been subject to impediments of one sort and another. The corner of the Gulf of Guinea, in the former British Cameroons, has presumably been covered permanently with very wet evergreen forest and northwards from here a line of highlands, of which the Bamenda–Banso are the most important, runs towards Lake Chad. The effectiveness of this line of highlands as a barrier for the kind of birds under discussion will have increased whenever the climate became wetter or cooler, as during a northward swing of the equatorial rainbelt or a glaciation. The Lake Chad depression must itself also have been an appreciable barrier to east–west communication during the period of its great extension down to some 8000 years ago (Chapter 3), when it was some 600 miles from south to north and over 200 miles wide. On the whole, then, the geography and ecological history of West Africa should, one would think, have been conducive to a good level of endem-ism in non-forest birds. To make an assessment one may take all those species which appear to have their eastern limit west of the Chad barrier or near it, and all those species which show a subspecific difference some-where near this line. Actually, published information on ranges is likely if anything to exaggerate the importance of this line, because the country east of it as far as Darfur is not well known ornithologically and more "endemic" West African species and subspecies may in fact range further east than at present recognized. Even so, on present evidence barely a dozen species qualify as West African non-forest endemics. One comes nowhere near the Chad line, namely the savanna starling *Lam-protornis iris*, Portuguese Guinea–Côte d'Ivoire; the others extend east to Nigeria or the Cameroons. Conversely, about half a dozen birds from the Sudan find their western limit near the Chad line. Special interest attaches to those cases in which an endemic West African species is re-placed east of the Chad line by a member of the same superspecies, namely *Francolinus bicalcaratus/icterorhynchus*, the grey turacos *Crinifer piscator/zonurus*, the grass warblers *Cisticola rufa/troglodytes* and the fly-catchers *Batis senegalensis/orientalis*. Possibly the situation as between the turacos *Musophaga violacea* and *M. rossae* may be similar, but further inquiry is needed (Moreau, 1958b). These few cases strongly suggest speciation on the two sides of the Chad line as a result of the isolation it has facilitated. At the same time there are only about eighteen birds which are represented by different subspecies on the two sides of the same boundary. It is possible that cases of this nature were more numer-ous about the end of the glaciation and subsequently, while Mega-Chad was still in being; but that, as the present situation, which presents less impediment to east-and-west communication, established itself in the last few thousand years, some subspecies that formerly were limited to east and west respectively of the line of the Lake, not having attained genetical isolation, have so far intermingled as to lose their distinctive characters.

The comparative poverty of West Africa in endemics may be compared to its poverty in number of non-forest bird species and of plant species, as discussed in Chapter 6. Clearly its non-forest bird fauna is merely a somewhat impoverished version of the latitudinal bird fauna. On the other hand, this differs sharply in composition from the Somali bird fauna, which in one-sixth of the area possesses nearly as many endemics, practically all adapted to dry country. This surely means that in some respects the Somali focus of aridity has been more undisturbed than the latitudinal dry belt. As noted in Chapter 3, Clark concluded that the climate of Somaliland had not been better than semi-arid at any time during the Pleistocene. There is no reason to suppose that a latitudinal dry belt ever ceased to exist, at any rate in its eastern half, but at least in West Africa it must have undergone great vicissitudes. At one stage late in the Pleistocene, when the Sahara carried Mediterranean vegetation south to the latitude of Lake Chad, the belt of dry Ethiopian vegetation may well have been narrowed, and this may have taken place again in West Africa during the period when the desert dunes advanced for 300 miles south of their present condition and compressed the whole series of vegetation zones towards the Gulf of Guinea.

The marked difference between the latitudinal and Somali bird faunas, as judged by their endemics, has presumably been facilitated by the fact that for lowland dry-country organisms inter-communication is constricted by the highlands of Abyssinia. Round these, one narrow and circuitous strip of dry lowland runs up the Eritrean coast and another past Lake Rudolf. This, however, is barely 200 miles wide and during the glaciation could have been further reduced by the fact that part of it, just west of Lake Rudolf, is high enough to have been "montane". As shown in Table XIX, the Somali area excels especially in endemic larks, shrikes, Ploceine weavers and starlings, but is inferior in several families, especially bustards, swallows and estrildines.

Another main focus of endemism in Africa lies in the dry southwestern area. If one takes the area comprised in the 900000 sq. miles of South West Africa and the Union of South Africa as a whole, it possesses about ninety-two endemic species of non-forest birds. The majority of these belong to the driest part, which covers the Karoo, Bechuanaland and South West Africa, from which elements extend eastwards into the drier parts of the Transvaal and northwards up the arid coast of Angola, just as Somali elements extend south into the dry acacia country of Tanganyika. The coincidence of the dry bird fauna with the extension of the acacia steppe, as shown in the vegetation map of Africa (Keay, 1959) simplified in Fig. 3, is marked. Several other of the endemics, such as the turdine *Oenanthe bifasciata*, belong to rather dry hilly country or, like the pipit *Macronyx capensis*, to open grassy country. Three species, the two sugar-birds *Promerops* and the sunbird *Nectarinia* (*Anabathmis*) *violacea*, have a special association with proteas (which are not an important element in the vegetation elsewhere in the continent)

and presumably depend greatly on their nectar. No species are endemic
to the eastern lowlands of South Africa, which are relatively humid; as
various authors have pointed out, they are inhabited by birds whose
ranges are otherwise tropical. It is also noteworthy that, although the
south-west winter-rains corner of the Cape Province is so rich in plant
endemics (e.g. Weimarck, 1941), no species of birds are similarly
limited.

As in the case of the Somali arid focus of endemism, that of south-
western Africa has doubtless been favoured by the fact that aridity has
been an extremely persistent feature of this part of the continent—
probably for much longer than in Somalia, far back beyond the Pleistocene, as noted in Chapter 3; a manifestation of this is the remarkable
succulent flora endemic to the Karoo and such a vegetable fantasy as
Welwitschia mirabilis.

The distribution of the endemic species between families in south-
western Africa and in the two parts of the northern tropics is shown in
Table XIX. Only one important family, the cuckoos (Cuculidae), is
unrepresented; all the cuckoos that frequent the dry areas of Africa also
occupy the more humid, and they resemble the water-birds and the
raptors in the exceptionally high proportion of species that extend from
West Africa to the Cape. It will be seen that each of the dry areas has
certain specialities. South Africa is outstandingly strong in group C,
but the northern tropical areas score low because they share some of
their species with areas outside the Ethiopian Region. The poverty of
South Africa in group D endemics is remarkable and is another aspect
of the reduction in strength of this group as a whole as the tropics are
left behind, as discussed in Chapter 7. Again, South Africa is weak
in endemics of the Laniidae (shrikes), Sturnidae (starlings) and
weavers (both ploceine and estrildine), but exceptionally strong in
Alaudidae (larks), Fringillidae (finches), Sylviidae (warblers) and
Turdidae (thrushes). The Somali arid focus excels in shrikes, sunbirds,
ploceine weavers and starlings but is markedly inferior to the latitudinal
north tropical area in estrildines. It is not easy to suggest reasons for
any of these predominances. It may, however, be remarked that the
figures for endemic lark species are high, fifteen for South Africa and ten
for the Somali arid, though only six for the rest of the northern tropics.
This gives a total of thirty-one species out of the forty-four for the whole
of Africa south of the Sahara. Such a proportion, nearly three-quarters,
is not reached in any other family and this situation is presumably
because the areas in question offer more than any other part of Africa
the open lightly vegetated country to which larks as a family are
adapted. Similarly, and for the same understandable reasons, the
Otididae (bustards) score high, with a total of eight endemics in the dry
areas out of sixteen species for the Ethiopian Region as a whole. Yet the
same proportion, six out of twelve, is reached by the Bucerotidae (hornbills) whose ecology is very different from that of the larks and bustards.

We may now turn to cases of discontinuous range. In so far as each species is typically associated with certain types of vegetation and not others, it is the rule of course for ranges to be to some extent discontinuous. This is especially marked in birds restricted to forest and montane habitats, as discussed in other chapters; but there are also many cases (see especially Benson and White, 1962) among lowland non-forest birds. Omitting those in which the circumstances are not very clear and those in which the gaps in range are relatively small, less than 400 or 500 miles, forty-four cases remain. Some of these depend of course on a taxonomic decision, not always undisputed, whether two isolated populations should be regarded as conspecific or not. A better basis for a discussion of this nature would naturally be the superspecies, but for this a comprehensive review does not yet exist, though work upon it has started by Hall and Moreau. When it is completed, the cases to be considered will obviously be more numerous than they are at present. Only one gap in range, the absence of the Ostrich *Struthio camelus* from a belt of southern Africa comprising Zambia and Malawi, seems to be historically very recent indeed and probably due to human interference.

A few spectacular cases of discontinuous range do not coincide with the vegetation types or anything else that is obvious. For example, *Francolinus streptophorus* (Fig. 39) is known from savanna north and south-west of Lake Victoria, then reappears 1500 miles to the west, in Cameroon, where it is extremely rare. The Maccoa Duck *Oxyura maccoa*, with a total range from Cape Province to Abyssinia, shows a gap of over 1200 miles, between south-western Rhodesia and northern Tanganyika. *Euplectes gierowii*, a typical bishop (ploceine) of rank grass, has one population in Angola, another in southern Kenya and northern Tanganyika, and a third in the area Abyssinia–eastern Congo. Most remarkable of all, the lark *Mirafra ruddi*, so distinctive as to have been regarded as forming a monotypic genus (*Heteromirafra*), has one population confined to grassland in an area of less than 100 sq. miles on the western border of former British Somaliland and the other to grassland 2500 miles away in the area traversed by the border between Cape Province and Transvaal (Fig. 39). All three species are presumably alike in being represented by relicts of a former vastly more continuous population. (One is reminded of the White Rhinoceros *Ceratotherium simum*, reduced in modern times to one population in the West Nile Province of northern Uganda and another in Zululand, but known from fossils to have been flourishing as far afield as Morocco only a few thousand years ago.) Another case of wide discontinuity, that of the bee-eater *Merops nubicus*, may have a different origin. The nominate birds breed across the northern tropics from Senegal to Somalia, then migrate south as far as coastal Tanganyika and the northern Congo. The southern birds (*nubicoides*) breed from southern Angola to Malawi and Natal, thereafter spreading for the off-season widely over the Congo and Tanganyika (Fig. 40). Here an alternative hypothesis is tenable,

that either population might have originated from migrants belonging
to the other.

All the remaining forty species that show major discontinuities in
their ranges are typically birds of dry open country, acacia steppe or

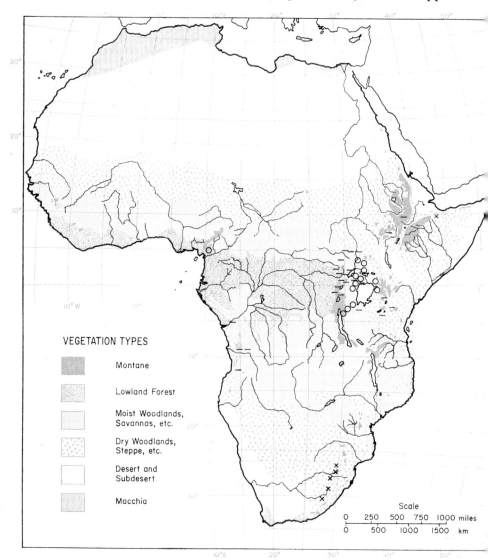

VEGETATION TYPES

Montane

Lowland Forest

Moist Woodlands,
Savannas, etc.

Dry Woodlands,
Steppe, etc.

Desert and
Subdesert

Macchia

Scale

0   250   500   750   1000 miles

0   500   1000   1500   km

Fɪɢ. 39. Three examples of unexplained discontinuous ranges: *Francolinus streptophorus* (a bush fowl, ◯), *Euplectes gierowii* (a weaver, —), and *Mirafra ruddi* (a lark, ×).

acacia savanna, and in most of these cases the gap seems to have some
connexion with vegetation type: for in thirty-three of them the gap
comprises or includes either southern Tanganyika or the whole of Tanganyika, in twenty-one of these part or whole of Zambia is included in

the gap, and in five other cases the gap occurs in Zambia with or without Rhodesia also. An extreme example is that of the weaver *Ploceus rubiginosus* (Fig. 41), with one population from Eritrea to central Tanganyika and the other in South West Africa. Benson and White

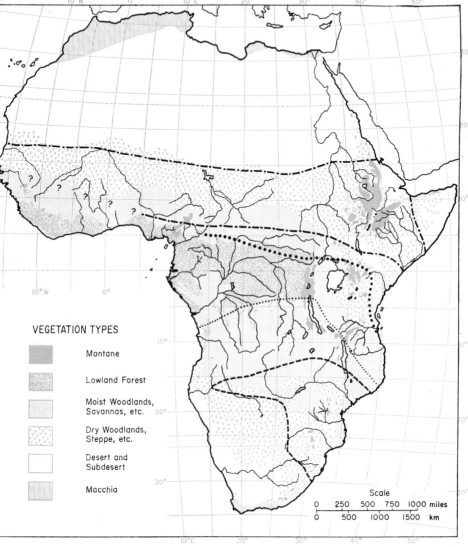

FIG. 40. The ranges of the Carmine Bee-eaters *Merops n. nubicus* and *nubicoides*. *Merops n. nubicus*: —·—·, breeding limits; ····, off-season limit; *Merops n. nubicoides*: — — — —, breeding limits; ····, off-season limit.

(1962) cite also several mammals that conform to this pattern. In nearly all cases the discontinuity includes the high ground on the Tanganyika–Zambia border and at least part of the block of brachystegia woodland that interposes between the north-eastern acacia steppe with its focus in Somalia and that with its focus in south-western Africa. Now, as discussed in Chapter 3, there is reason to suppose that a dry spell affected this part of Africa as recently as around 11 000 years ago and of sufficient severity to break up the brachystegia block and replace

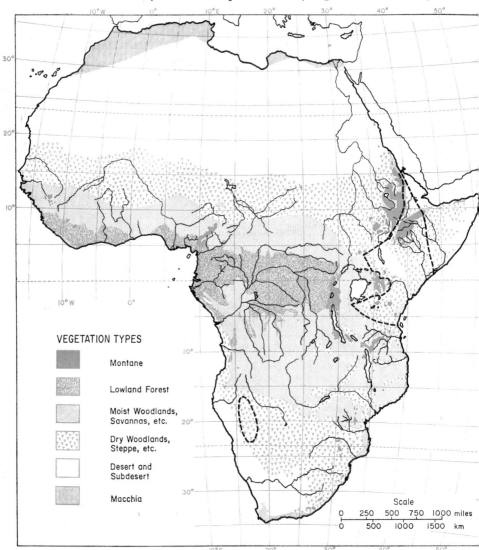

VEGETATION TYPES

Montane

Lowland Forest

Moist Woodlands, Savannas, etc.

Dry Woodlands, Steppe, etc.

Desert and Subdesert

Macchia

Scale

0   250   500   750   1000 miles

0   500   1000   1500   km

FIG. 41. An extreme example of discontinuous range in birds of dry, acacia, country; *Ploceus rubiginosus* (a weaver).

nearly all of it by drier and more open types of vegetation, as shown in Fig. 23. This would have left a gap of only some 300 miles between the acacia steppe of Tanganyika and that of Rhodesia and the southern Congo; and the gap would have been filled with open woodland, perhaps including mopane, acacia and also brachystegia, which would present the minimum of impediment to the movement of dry-country arrivals from either side of the gap. Hence, for all the species under discussion something approaching a continuous range can be envisaged for a period little more than 10 000 years ago. It may be added that in three cases the present gap in range is occupied by a closely allied species that is not typically associated with acacia, namely, *Phoeniculus damarensis* (a wood hoopoe) replaced by *P. purpureus* between Kenya and South West Africa, *Parus afer* (a tit) by *P. griseiventris* between northern Tanganyika and Rhodesia, and *Eremopterix leucotis* (a lark) by *leucopareia* between the southern part of Zambia and central Tanganyika. In these three cases then, the presumptive effect of vegetation type in limiting and splitting ranges is likely to be operating through, or be reinforced by, interspecific competition.

A faunistic situation in Angola that may well be the legacy of the same dry period has been described by Hall (1960b). Because of its aspect, facing the sea, a zone along the north–south scarp and for the most part only about 50 miles broad has probably maintained through recent climatic vicissitudes a climate more humid than either the coastal strip or most of the interior of Angola. Faunistic differences between the areas to the east and the west of this narrow zone suggest that there has in the recent past been a much wider gap between the populations concerned, presumably due to the desiccation of the interior. At the same time the escarpment zone itself accommodates several endemic or near-endemic species that are independent of forest (e.g. *Colius castonotus*, *Cisticola bulliens*, the flycatcher *Batis minulla* and the shrike *Prionops gabela*).

Today Angola east of the Scarp is dominated by brachystegia woodland that extends east across the south-eastern Congo and Zambia to the Indian Ocean, an area of about 1 million sq. miles with only minor interruptions. The birds of this great block of a well-defined vegetation type have been discussed by Benson and Irwin (1966). They conclude that the bird fauna of the brachystegia belt consists of 112 species; of these eighty-nine species range outside the belt and many of them are widespread in Africa south of the Sahara. Twenty-three species are "virtually endemic to Brachystegia", almost exactly one-fifth. Their distribution by families is as shown in Table XIX. It will be noticed that the brachystegia block has only about one-quarter as many endemics as the other three sections of Africa included in that Table, but the comparison is not altogether a fair one, since the brachystegia habitat is by definition one of more or less homogeneous woodland of a certain type, whereas each of the other three is more varied. In fact, as Benson and White remark, the species of birds endemic to brachystegia woodland indicates that this vegetation type has been in existence for a considerable period. However, the proportion of endemics is not so high as in some of the areas under discussion above. For example, former British Somaliland, situated in the Somali arid focus, possesses 185 breeding species of groups C–E, and of these fifty-four are Somali endemics, nearly one-third, compared with the one-fifth of endemics in brachystegia. This might indicate that the Somali arid focus has provided a field for bird evolution for a longer period, at least without major disturbance, than has the brachystegia belt; and this would agree with the inferences, set out above, derived from geological data.

11

The Montane Forest Bird Fauna

As will have been gathered from references earlier in this book and as is demonstrated by the figures in Table V, the montane forest birds of the Ethiopian Region are sharply distinct from those of the lowland forest, and in the tropics they are typically met with at elevations of over 5000 ft, though lower close to the sea and of course in sub-tropical South Africa. As shown in Fig. 2, the relief of Africa is such that the areas high enough for montane conditions to be developed are scattered, and the montane forests are even more discontinuous as a result of human depredations, clearing and burning. These forests form, then, ecological islands (see Fig. 42 for those of East Africa, Fig. 43 for a more general view, with groupings to be explained below); many of them are within sight of one another, but such an intervisible group may be separated from the next one by a wide stretch of lower country—as much as 300 miles between the montane forests on the eastern side of Uganda and those on the western, and over 1200 miles between the latter and those of the Cameroons. The intervening areas are covered with a variety of vegetation that ranges from lowland evergreen forest to subdesert steppe and is all foreign to the typically montane species. Nevertheless, a large proportion of the montane birds (and of plants) reappear in even the most widely separated areas—a situation that is, moreover, exactly paralleled in the butterflies (Carcasson, 1964). Statistics will be given below and meanwhile, as examples of different types of discontinuous range, three species may be cited. *Pogoncichla stellata*, a small turdine of the forest floor, extends sporadically from the mountains on the Sudan–Uganda border and Ruwenzori (but not Abyssinia) through eastern Africa to the forests of South Africa, there reaching approximately sea-level (Fig. 44). The warbler *Phylloscopus* (=*Seicercus*) *umbrovirens* has its most southerly station on the Uluguru Mountains of central Tanganyika and, via east Congo, Kenya and Abyssinia, reaches the relict juniper forests in the mountains of Somaliland and reappears on the mountains of south-western Arabia (Fig. 45). The small babbler *Alcippe abyssinicus* occupies mountain forest in eastern Africa from Abyssinia to south of Lake Nyasa and, unlike the species already cited, reappears in western Africa, in western Angola and on the mountains of the Cameroons (Fig. 46).

Notwithstanding the distances by which some of the montane forests are isolated, most of them resemble continental rather than oceanic islands in one important respect—their isolation is, on the scale of evolutionary time, very recent. As will have been gathered from the description of Pleistocene events in Chapter 3, during the last glacial phase and up to about 18000 years ago, the lower limit of montane conditions would have been reduced from about 1500 m to 500 m a.s.l., so that, as can be seen from Fig. 2, most of the present isolated montane

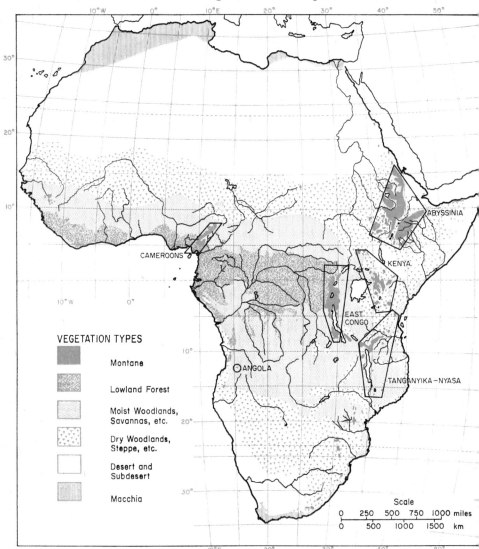

FIG. 43. Montane forest groups of Africa.

areas would have been in communication with each other. To what extent this applied to the forests themselves is uncertain, since we do not know, and can probably never know accurately, the distribution of forest over the huge and continuous montane block. However, in so far as any gaps between the forests were climatically montane, the degree of isolation between them was minimal; and it is certain that, the post-glacial rise in temperature being as it was, the present conditions of isolation would not have been established until after 10 000 years ago. In these circumstances the populations of most of the existing montane forests are essentially relicts occupying and providing refuges in ex-

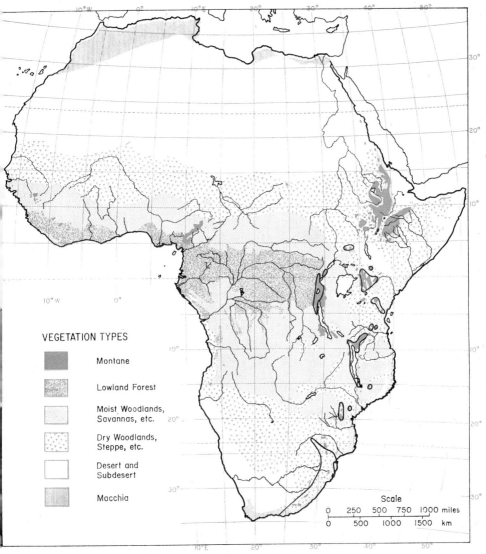

VEGETATION TYPES

- Montane
- Lowland Forest
- Moist Woodlands, Savannas, etc.
- Dry Woodlands, Steppe, etc.
- Desert and Subdesert
- Macchia

Scale

| 0 | 250 | 500 | 750 | 1000 miles |
| 0 | 500 | 1000 | 1500 | km |

FIG. 44. The range of *Pogonocichla stellata* (a montane turdine).

ceptionally favourable parts of the recently continuous montane area. (As discussed by Hall and Moreau (1962), such refuges today support many of the rarest bird species in Africa, with extremely limited ranges and with in some cases populations that are likely to total less than 2000.) On the other hand, there is no doubt that a small minority (less than half a dozen) of the forests isolated on mountain tops today would also have been so, though more narrowly, during a glaciation. Such a forest must therefore have been colonized by the same sort of dispersal as colonizes an oceanic island, and the birds must have reached there

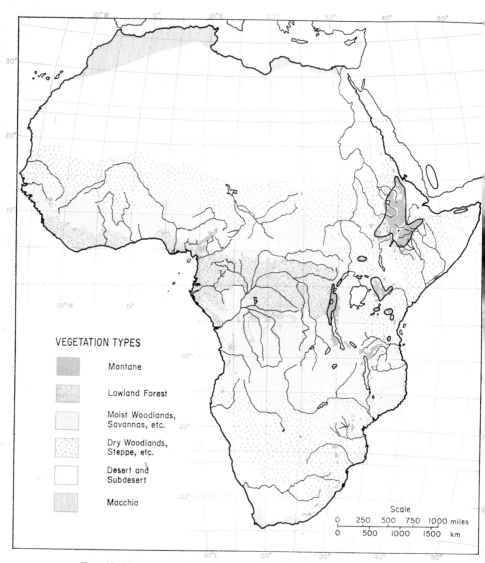

FIG. 45. The range of *Phylloscopus umbrovirens* (a montane warbler).

as a result of individual activity; but in perhaps only two of the forests in question, namely, Kulal and Marsabit in northern Kenya, which are mentioned separately below, would the isolation have been as much as 30 miles of lower and drier country quite unsuitable for forest species.

The conditions, under which most montane forests were at one time

in communication and at another were isolated, have recurred repeatedly during the Pleistocene, in accord with the successive alternations of glacial and interglacial periods. During an interglacial, when
world temperatures were higher than they are now, the lower boundary
of the montane environment would have been at a greater altitude above

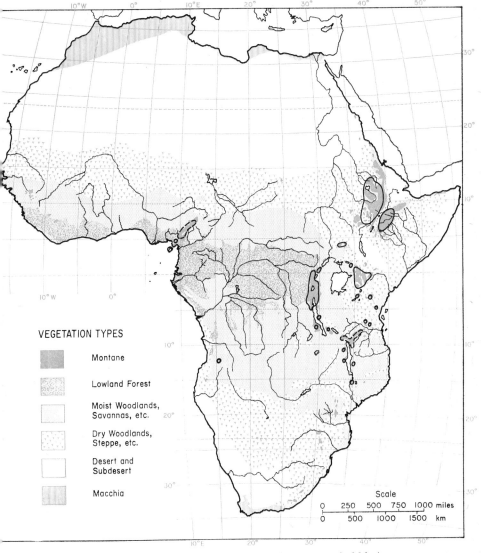

VEGETATION TYPES

Montane

Lowland Forest

Moist Woodlands,
Savannas, etc.

Dry Woodlands,
Steppe, etc.

Desert and
Subdesert

Macchia

Scale

| 0 | 250 | 500 | 750 | 1000 miles |
| 0 | 500 | 1000 | 1500 | km |

FIG. 46. The range of *Alcippe abyssinicus* (a montane babbler).

the sea than it is now, so that the montane areas would have been more
restricted. But there would have been virtually no destruction of the
forests by human agency—for the last interglacial ended more than
70 000 years ago—and hence in all probability the forest refuges would
have been at least as numerous as they are now.

While most of the existing distribution of montane forest birds can undoubtedly be relict, not all of it can be, and it is interesting to consider the evidence for active colonizing by such birds. Actually it is conflicting. On the one hand montane forest birds of any species have very rarely been collected at any distance away from their typical habitat, and during many years in Tanganyika my own observations confirmed this generalization. Yet in a few days in Kenya in 1964 Myles North showed me a forest barbet *Pogoniulus bilineatus* in acacia scrub by Lake Naivasha and a forest turaco *T. hartlaubi* in big acacia trees near Lake Nakuru. Although in the Rift Valley, these birds were altitudinally well into the montane zone, at about 6000 ft, and they were midway across the gap of about a dozen miles which separates the forests east and west of the Rift. Since they are both birds dependent on fruit, which is not supplied by acacias, they were probably in transit. It should be noted moreover that both are birds of the forest canopy, and hence likely to be at least temporarily more adaptable than birds of the forest interior, with its cool dark eco-climate.

The avifaunal situation on Marsabit (see Fig. 42), an old volcano in northern Kenya which carries the most isolated mountain forest in Africa, is in accord with these casual observations on the mobility of forest birds. The Marsabit forest is surrounded by extremely dry country, desert steppe or actual desert, and situated 80 miles from the nearest possible sources of montane forest birds, namely Kulal Mountain and the Matthews Range. Moreover, there is no reason to suppose that this isolation was greatly diminished even at the height of a glaciation. It is perhaps significant that widespread forest mammals belonging to species which it is particularly difficult to imagine crossing hot dry country, namely tree hyraxes *Dendrohyrax*, Sykes' (Blue) Monkey *Cercopithecus mitis* and Colobus *C. polykomos*, are all absent from Marsabit today (D. M. Minter, *in litt.*; R. H. Carcasson, *in litt.*). It is true that this mountain is somewhat peculiar ecologically; both altitude (1695 m) and rainfall are abnormally low for montane forest, so that it must depend to an unusual extent on mist. Also, although there is a rich variety of trees (B. Verdcourt, *in litt.*), two of the most typical montane elements, *Podocarpus* and *Juniperus* appear to be absent. But these peculiarities do not affect the significance of Marsabit for the purpose of the present discussion.

For information on the birds of Marsabit we have incidental records in van Someren (1922, 1932) and Jackson and Sclater (1938), supplemented by unpublished information given me by R. H. Carcasson, Stuart Keith and J. G. Williams from their personal experience. It is evident that seven montane forest species occur there at the present day, namely, *Coracina caesia* (cuckoo shrike), *Muscicapa* (=*Alseonax*) *adusta* (flycatcher), *Onychognathus walleri* (starling), *Phyllastrephus placidus* (bulbul, the only bird of the forest interior, rather than of the canopy or edge, on this list), *Phylloscopus* (=*Seicercus*) *umbrovirens* (warbler), *Turdus* *abyssinicus* (thrush) and *Zosterops senegalensis jacksoni* (white-eye).* Only the flycatcher is represented by an endemic subspecies. It is, then, virtually certain that the other six species have not merely crossed 80 miles of "bad" country at some time but that they continue to do so with sufficient frequency to prevent subspeciation. This is indeed a surprising conclusion but it is in perfect agreement with that reached for butterflies by Carcasson (personal communication).

The other particularly isolated montane forest of northern Kenya, that on Kulal (2381 m) near the southern end of Lake Rudolf, and like that of Marsabit almost certainly never connected with any other, provides evidence of the colonizing ability of a few other species over some 30 miles of low arid country. There is no reason to suppose that the Kulal bird fauna is fully known, but the forest, amounting to some 40 sq. miles (Verdcourt, 1963), is inhabited by several species not established on Marsabit, namely, a dove *Aplopelia larvata*, a sunbird *Nectarinia* (=*Cinnyris*) *mediocris*, the thrushes *Geokichla piaggiae* and *Pogonocichla stellata*, a bulbul *Andropadus* (=*Arizelocichla*) *tephrolaema*, a flycatcher *Melaenornis* (=*Dioptrornis*) *chocolatinus* and *Tauraco hartlaubi*, (J. G. Williams *in litt.*; Stuart Keith *in litt.*). Again there seems to be no subspeciation in the Kulal forest birds except in the Zosterops.

One other area provides unequivocal evidence of the ability of certain montane forest species to cross a much greater distance than even that isolating Marsabit. The mountains of the Yemen are inhabited by the otherwise African (and Abyssinian) *Streptopelia lugens* (dove) and the same species of *Turdus* and *Phylloscopus* as have reached Marsabit. Of these two the Yemen populations form endemic subspecies. A gap of at least 200 miles separates them from any possible habitat in Africa; a flight across the southern end of the Red Sea or the western end of the Gulf of Aden and over the arid coastal lowlands on both sides is inescapable and the distance between montane stations will not have been much reduced during the glaciation. Here, then, we have proof that three montane forest species have been able to colonize across a gap of some 200 miles. Actually, more species may have achieved this but owing to the destruction of their habitat in the Yemen mountains have left no trace.

In conflict with the foregoing evidence for the mobility of montane forest birds we have the evidence for subspeciation on mountains in southern Kenya and northern Tanganyika. As mentioned on p. 206 below, several of the birds on the Chyulu Mountains, much less isolated than either Kulal or Marsabit, are subspecifically distinct, while in three species, actually including *Turdus abyssinicus*, which has not differentiated

* There are in addition one or two forest-edge species, such as *Anthreptes collaris*, which are not typically montane. Two species that have been recorded from Marsabit in the past have not been found there in recent years, namely, *Apalis cinerea* (warbler) and *Tauraco hartlaubi*, the latter recorded allusively, rather than in precise terms, by van Someren (1932). No specimen of either bird from Marsabit can be traced in museums. Either the records were due to error or the populations were transitory.

on Marsabit, there is repeated subspeciation between montane forests separated by gaps of low dry country no more than 10–20 miles wide. In any case, whatever the conflict of evidence of this kind, two things are clear. One is that such positive evidence as there is for the colonizing ability of montane forest birds relates to birds of the canopy or the edges, far more than to birds of the interior. The other is that any demonstration of ability to cross a gap of 80 miles or even, as with the Yemen, of 200 miles, is of little relevance to the case of the forests of the Cameroons and Angola, which are separated by some 1200 miles from their nearest neighbours. It is necessary to assume that the bird faunas of those forests, especially of their interior, owe their partial identity with others in the centre and east of Africa, not to individual dispersal but to the changed ecological conditions of the glaciation. In other, less remotely isolated, montane forests no doubt both factors will have operated in varying degrees.

In view of the history and the present extent to which individual montane forests are isolated, it is of interest to examine their bird faunas with particular reference to their degrees of mutual relationship and of endemism. It would of course be useful for comparative purposes if we could devise a numerical index of isolation. It would, however, depend on several factors, which may be reduced to (1) width of the isolating gap, (2) its altitude (as an indicator firstly of temperature and secondly of post-glacial duration of isolation), (3) nature of the vegetation of the gap (lowland evergreen forest presumably providing the least formidable barrier to montane forest birds, and desert the most). However, for each factor the values would have to be ascribed on an arbitrary scale and the results of combining such values could hardly have much validity.

We now know much about most of these bird faunas, but few of them so thoroughly that negative evidence can be regarded as really conclusive. Hence the emphasis in the following discussion will be on the positive records, with little reference to the apparent absences of species from stations where they might have been expected. Also, for the present purpose it seems unnecessary—and it is impracticable—to present a complete list of what has been recorded for each of the many isolated montane forests and I shall therefore deal primarily with geographical sections as delimited in Fig. 43, namely (a) the Cameroons group,* (b) the mountains on the eastern rim of the Congo basin, from west of Lake Albert through Ruwenzori and the Kivu mountains to the west side of Lake Tanganyika, with the addition of Mt Kungwe on the east side, (c) the montane areas of Kenya, (d) the Abyssinian plateau, (e) the highlands of Angola, (f) the montane areas of Tanganyika–Nyasa,

* West of the Cameroons group the only localities that approach or slightly exceed the 1500 m lower limit of the montane zone are the Futa Jallon plateau in the Guinea Republic, the Bintumane and Nimba group on its south-eastern border and the rather dry Jos plateau of Nigeria (see Fig. 25). None of them has any area of typical montane forest or has yielded any species of bird typical of it.

(g) the montane areas south of the Zambezi. Such treatment is in a sense an oversimplification, but the object is to show the main significant features of the montane species' distribution. Since we are dealing with isolated populations, taxonomic difficulties involving discrimination between specific and subspecific rank repeatedly occur. I have, as before, consistently used the broadest polytypic concepts of species that have been put forward and I have used my own judgement in allocating birds to superspecies where this course seems justified. From the discussion that follows I have excluded species that are lowland in any considerable part of their range. In the result, about three-quarters of the species under discussion are passerines and of the others many are such individually mobile birds as pigeons and raptors. The treatment in this chapter is zoogeographical; the bird faunas of a number of individual montane forests are discussed from the ecological point of view in Chapter 15.

The Montane Forest Bird Fauna of the Cameroons

Systematic presentation of the data and much of the original exploration of these montane forests are due to Serle (1950, 1954, 1957). Subsequently Eisentraut (1963) has discussed the birds of Cameroon Mountain, along with all the other vertebrates and has given useful descriptions of the habitats. As shown in Fig. 47, from an area at over 3000 ft a.s.l. arise the small montane patches of Manenguba Mountain, Kupe Mountain, the Obudu plateau and the much more extensive montane area of the Bamenda–Banso highlands stretching away north towards Lake Chad. During the glaciation all these montane patches would have been continuous and would have been in communication, though only narrowly by way of the northern rim of the Congo basin, with the mountains of central and eastern Africa, as shown in Fig. 2. Just separated from this montane group, today and also during the glaciation, are the Rumpi Hills to the west, and more widely, by some 50 miles, Mt Cameroon itself (13 350 ft; 4072 m) to the south. Even at the height of the glaciation the montane forest of Mt Cameroon would probably have been isolated,* though by less than 30 miles of lower ground; and this has probably at all times been covered with evergreen forest, the vegetation type that presumably presents a less formidable barrier to montane forest birds than other types do.

Mts Cameroon and Manenguba are volcanoes, the former still intermittently active, while the neighbouring montane areas are all of ancient crystalline rock. On Mt Cameroon, although nearly twice the height of the others, "fires, both natural and man-made, and the extra porous

---

* On the south-east face of Mt Cameroon montane conditions come so abnormally low (see p. 81) that during the glaciation they might even have reached sea-level, but this may well not apply to the northern, inland side of the mountain. But at the same time the Cameroon montane might have been in communication with that of Fernando Poo island (see Chapter 16), which biologically is a member of the Cameroon montane archipelago.

lava soil are perhaps the outstanding ecological characters" (Keay, 1955), and the montane forest fades out at the abnormally low altitude of 2100 m (about 7000 ft). Because of these conditions and because the Bamenda–Banso highlands are of greater area and are more in contact with the corridor along the northern rim of the Congo basin, one might

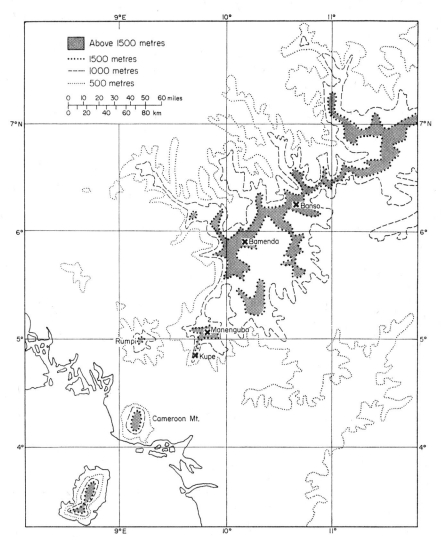

FIG. 47. The montane areas of the Cameroons.

expect these highlands to be richer in species than the other montane areas in the Cameroons. This is indeed realized strikingly for plants, but less so for birds. The combined forests of the Kupe-Manenguba-Obudu-Bamenda-Banso block certainly contain half a dozen species that do not occur on the far smaller forest on the upper slopes of Mt

Cameroon, but the latter possesses two species not found on the bigger block.

The extent to which the bird fauna of the Cameroons montane archipelago as a whole is shared with the nearest other montane archipelago, that on the eastern edge of the Congo basin, is shown in Table XX, taxonomic categories being distinguished. It will be seen that well over half the species 0·14+0·49) of the Cameroons bird fauna are common to the two areas, which are separated by over 1200 miles of country everywhere so low that only at the height of the glaciation was a montane causeway between them developed, and that a narrow one (Fig. 2). The remainder of the Cameroons species in question are all endemic, but half of these belong to superspecies represented also on the East Congo mountains. Moreover, of the shared species six are subspecifically identical in these widely separated areas, though none of them can be regarded as potentially very mobile. They consist of the apparently very sedentary *Heterotrogon vittatum*, the babbler *Malacocincla* ("*Alethe*") *poliothorax*, which is a bird of the forest floor, the insectivorous *Ploceus insignis* (which has differentiated on Fernando Poo), the estrildine *Cryptospiza reichenovii* and two warblers *Apalis pulchra* and *A. cinerea*. This last case might be delusive. The Mt Cameroon population, which has always been isolated and has most likely been derived from the Bamenda block, rather than the other way about, is indistinguishable from the East African population, while the Bamenda population is different, although it is on the causeway from East Africa. The only character involved is a most labile one, intensity of melanin; so that the identity of the Mt Cameroon birds with the East African may be due to secondary convergence.

The several levels of affinity involved, as shown in Table XX, tempt one to suggest that here we have evidence for at least two periods of communication with the east; probably those species which are shared, and almost certainly those which show no subspecific distinction, have had a continuous range round the northern rim of the Congo basin as recently as the last phase of the Last Glaciation, while the differentiated representatives of superspecies may derive from stocks that reached their present isolated stations during an earlier phase or an earlier glaciation. It is of course impossible to say which of the species concerned evolved in the east and spread west and which the other way about. Nor, as shown in Fig. 2, is it possible to rule out communication southwards by way of the country above 500 m, which extends, except for very narrow gaps, through the interior of Gabon and along the eastern rim of the Congo basin to the highlands of Angola (see below). There is one direct suggestion of intercommunication between the Cameroon montane forests and Angola; the endemic turaco of the Bamenda highlands, *T. bannermani* (not found on Mt Cameroon itself), is much more like the endemic *T. erythrolophus* of Angola (which is not, however, montane) than any other turaco.

Table XX

Affinities of montane forest bird faunas

	Cameroons		Tanganyika-Nyasa		Kenya		Abyssinia	
	No.	Proportion of total	No.	Proportion of total	No.	Proportion of total	No.	Proportion of total
Subspecies shared	6	0·14	7	0·11	20	0·42	8	0·42
Other species shared	21	0·49	27	0·42	12	0·25	7	0·37
Endemic species that are members of superspecies shared	8	0·19	6	0·10	3	0·06	1	0·05
Other endemic species	8	0·19	12	0·19	1	0·02	3	0·16
Other species	—	—	12	0·19	11	0·26	—	—
	43		64		47		19	

(with East Congo montane — Cameroons, Tanganyika-Nyasa, Kenya; with Kenya montane — Abyssinia)

Because of its extreme isolation and the fact that its communication with other montane areas has always been at best tenuous, in its circumstances the Cameroons montane bird fauna resembles a continental-island bird fauna less than does that of any other montane area. Mt Cameroon and to a lesser degree the Bamenda block are so situated that they are not likely to be serving as refuges for any wide area and hence their endemic species are more likely to have been evolved locally than to be relicts. They include two monotypic genera, the warblers *Poliolais* and *Urolais*, hitherto generally accepted but not impressive (Hall and Moreau, 1962). But even without these the Cameroons archipelago scores heavily in endemics at the specific level, as will be seen from Table XX. In fact, if an endemics "score" is calculated for them on the same scale as for the bird faunas of the marine islands round Africa (Chapter 19), this land-based ecological archipelago ranks equal to the highest of them (3·5); and even if the two genera are discounted, with 2·7 it scores almost the highest.

THE MONTANE BIRD FAUNAS OF THE EAST CONGO

These forests form a disconnected series some 600 miles long, with its northernmost member on the west side of Lake Albert, nearly 200 miles from the next, on Ruwenzori. Thereafter, a succession of forests follow, with no gap exceeding about 50 miles, on the mountains overlooking Lake Edward, on the Kivu highlands and volcanoes, on the blocks overlooking the Ruzizi Valley and the northern end of Lake Tanganyika on both sides; and finally, after rather wider gaps, the Kabobo and Marungu montanes on the west, the Kungwe on the east, bank of the Lake. Since the lowest gaps—and even the surface of Lake Tanganyika—exceed 700 m (2500 ft) a.s.l., all these montane forests could have formed a single block during the glaciation. Because the area to the east in the basin of Lake Victoria (surface 3717 ft a.s.l.) is so high as well as largely humid today, communication between this East Congo series and the forest of the Kenya highlands, now separated by 300 miles, would have been unimpeded not only throughout the last phase of the Last Glaciation but also for some time after, probably to about 13 000 years ago.

The total number of typically montane forest species in the East Congo mountains is sixty-three, compared with forty-three in Cameroons and forty-seven in Kenya (see below). Such differences might be expected from the much greater extension of the East Congo forests and their geographically central position. This has made them accessible as a refuge from all directions as the total montane area of Africa shrank with the post-glacial rise in temperature. The proportion of endemic species in East Congo, 21/63, is nearly as high as in the Cameroon montane forests, which are so much more isolated.

Few species are known from one end to the other of the East Congo series and some have not been found in intermediately located montane forests where they "ought" to occur, even though collecting might seem to be adequate. For the present purpose the ranges of individual species need not be referred to, except a few that merit special comment. In the first place two of the endemics, and those in their way among the most remarkable birds in Africa, are known only from a small mountainous area north-west of Lake Tanganyika; they are *Pseudocalyptomena graueri* and the owl *Phodilus prigoginei*. *Pseudocalyptomena*, forming a monotypic genus, is one of the only six sub-oscines in Africa and is well distinguished from all of them. As its name implies, its nearest relatives may be in the genus *Calyptomena* of south-eastern Asia. The owl is remarkably like the Indian *P. badius* and forms the only other species in this genus. As discussed on p. 117, the existence of these birds so far from relatives implies extinction of their common stock over vast areas; and it is impossible to formulate a hypothesis as to why just this ecological island of a few hundred square miles on the eastern edge of the Congo should have served as the refuge in Africa for both these genera.

From the ecological point of view the most noteworthy birds are the four *Cryptospiza* species, which between them comprise a well-marked genus of estrildine weavers strictly confined to the montane forests of tropical Africa. Their distribution is outlined in Fig. 48: *C. salvadorii*, Abyssinia, East Congo mountains and Kenya to northern Tanganyika; *reichenovii*, south of that to Malawi and Rhodesia, then on East Congo mountains and far to the west, Angola, the Cameroons and Fernando Poo; *jacksoni* and *shelleyi*, both on the East Congo mountains but nowhere else. All four occur on Ruwenzori and in some other localities but in varying degrees of abundance. The remarkable thing is that, although the species differ somewhat in size, all depend on the seeds of forest grasses and herbs, a food supply that they probably have to themselves except where the fringillid *Linurgus olivaceus* occurs.* This striking degree of apparent ecological overlap demands investigation, which would, however, be excessively difficult owing to the birds' elusive habits. To account for the co-existence of the species in the Ruwenzori and other forests it seems necessary to postulate multiple "double invasion" (see especially Mayr, 1942, p. 173), a phenomenon well known in oceanic islands. As mentioned elsewhere, it appears that if on two separate occasions colonists from the same stock reach an island the later comers may be able to establish their strain permanently and maintain its identity provided that the earlier arrivals have diverged far enough genetically for interbreeding to be impossible and far enough ecologically for competition not to be so severe as to preclude co-existence. How far the latter condition has been satisfied on Ruwenzori we do not know; evidence on the food eaten is too vague to be worth anything. However, assuming, as we must, that two species could not have evolved

* Since this was written J. G. Williams (*in litt.*) has found the seeds of the same species of balsam *Impatiens* in the stomachs of all four cryptospizas when he was collecting them in the Impenetrable Forest of south-west Uganda.

in the Ruwenzori forest, which is continuous, then three others have appeared there from ouside. While "double invasions" of oceanic islands are not uncommon (see African examples in Chapter 19), multiple invasions certainly are; one example that occurs to me is, however, the three of *Zosterops lateralis* stock from Australia into Norfolk Island.

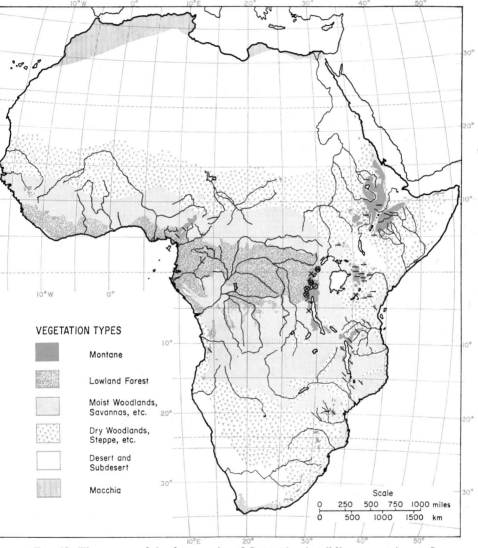

FIG. 48. The ranges of the four species of *Cryptospiza* (estrildine weavers). ⌢, *C. jacksoni*; ⌣, *C. reichenovii*; —, *C. salvadorii*; ◯, *C. shelleyi*.

If I had to suggest the sequence of *Cryptospiza* evolution I would presume that some forest in the centrally placed East Congo series saw the origin of the genus; that *jacksoni*, *shelleyi* and the common ancestor of *reichenovii* and *salvadorii* each evolved in isolation; that the last two differentiated on mountains in different parts of eastern Africa; and that at one time and another they all four invaded some part of each other's territory. For the evolutionary process conditions would have been most favourable during one or more interglacials, when the isolation of the mountain forests from each other would have been greatest; and glacial periods, when isolation broke down, would have facilitated the mutual invasions.

Before leaving the East Congo group I should draw attention to its outlier on the east side of Lake Tanganyika, the forest on Mt Kungwe which was explored ornithologically for the first time, and very effectively as was shown by a later expedition, by the African collector Salimu Asumani, working under my instructions at long range in 1940 and 1941 (Moreau, 1943; Ulfstrand and Lamprey, 1960). The Mt Kungwe forest is isolated on the north by the low Kigoma–Malagarasi River gap, some 40 miles wide, from the Urundi highlands (on the east side of the head of Lake Tanganyika), which connect with the Kivu highlands and have been deforested only in recent historical times. On the south the Mt Kungwe forest is isolated by the low Karema gap, some 50 miles wide, from the Ufipa montane area, which carries relict forests, between Lake Tanganyika and Lake Rukwa. These again are separated from those round the head of Lake Nyasa by barely 100 miles, nearly all above the critical altitude of 1500 m. This, as well as the Karema and Kigoma gaps, is clothed with savanna and brachystegia woodland, but owing to its higher altitude would be a less important barrier. Eastwards Kungwe is separated from any other montane area by some 400 miles of the interior of Tanganyika, but the whole of this area, as well as the lower gaps north and south of Kungwe, would have had a montane climate during the glaciation.

The Kungwe montane forest bird fauna consists of twenty-eight species, only one of which, the warbler *Phylloscopus* (=*Seicercus*) *ruficapilla*, is not found in the East Congo group of forests further north, and in only ten of the twenty-seven species are the Kungwe populations subspecifically distinct. By contrast, only twelve of the Kungwe species are known in the mountain forests around Lake Nyasa and in eight of those the subspecies are different. Clearly then, in birds the affinities of Kungwe with the East Congo group are much the stronger, and the Kigoma gap has been of minor importance. To the south of Kungwe the bird fauna of the vestigial forests on the Ufipa plateau differs strikingly. Of its sixteen species only eight are shared with Kungwe (though eleven are shared with East Congo, on the far side of Kungwe from Ufipa), and three of the eight are subspecifically distinct. In the other direction, eastwards, the Ufipa forests have eleven of their species in common with the Nyasa forests, with five of them showing subspecific differences. These figures indicate that, unless the vestigial nature of the Ufipa forests and their fauna has produced a misleading impression, their affinities are with Nyasa to the east rather than with Kungwe to the north; and consequently that the Karema gap has been much more effective than the Kigoma gap as a faunal barrier, although from Fig. 42 little difference in this respect would have been expected. Ufipa in fact cannot be allocated to either the East Congo or the Tanganyika-Nyasa group of montane forests. It may be added that the faunal relations of the Kungwe and Ufipa montane forests are much the same in butterflies as they are in birds according to Carcasson (1964), except that he is inclined to allocate Ufipa to the Tanganyika-Nyasa group.

THE MONTANE FOREST BIRD FAUNAS OF KENYA

In Kenya the main montane forest area, almost continuous in recent historical times, runs from Mt Elgon south to the Tanganyika border (Loliondo) and east to Mt Kenya. The area is divided by the Rift Valley but, as this is barely 20 miles wide in places and for 100 miles its floor is at over 1500 m a.s.l., it can have been no obstacle during the glaciation. Faunistically the montane forests as far to the north as the Uganda–Sudan border (on the Imatong and Didinga Mountains) belong to the Kenya area, with which they are connected by a discontinuous arc of montane forests running north from Mt Elgon (Thomas, 1943); so too belong the montane forests of Chyulu, Kilimanjaro, Mt Meru, the Crater Highlands of northern Tanganyika, and the intermediate Monduli, Essimingor, Gelai and Ketumbeine volcanoes. The biota of all these forests merely reproduce that of Kenya in a depauperate form, and all of them would presumably have been in communication during the glaciation since the country intervening between them is all more than 1000 m a.s.l. These favourable conditions do not apply to the mountains detached to the north of the Kenya highlands—the Matthews Range, Mt Nyiro, Kulal and Marsabit, each of which is surrounded by acacia steppe, which degenerates in places to desert steppe and even full desert. The bird faunas of the last two mountain forests, which are peculiarly isolated, have already been discussed at the beginning of this chapter.

It is remarkable that the entire Kenya group, though it includes numerous montane forests that are isolated today, contains only a single endemic species, *Francolinus jacksoni* (and that belongs to a superspecies represented in montane forests as remote as the Cameroons, Abyssinia and Angola). Moreover within the Kenya group as a whole subspecific variation is limited—far more so than in the Tanganyika–Nyasa group discussed later—even in its most isolated components, namely Kulal and Marsabit. In this respect the Chyulu range in southern Kenya, investigated by van Someren (1939), shows more differentiation, although it is very young and much less isolated (see discussion by Moreau, 1963). Rising to about 2300 m, it is due to recent vulcanism and is perhaps less than 40000 years old in its entirety; its montane forest is separated by only about 40 miles from the forest girdle of Kilimanjaro, with which it was doubtless connected during the glaciation and probably for long after. Nevertheless, there is clearly much incipient subspeciation among the Chyulu birds, both in the forest and outside it, for van Someren described as endemic to the range altogether twenty-six new subspecies, although out of the twenty-one passerines Mackworth-Praed and Grant (1960) have recognized only six of them (as well as two non-passerine subspecies) and White (1960–63) only two.

Taking the Kenya group as a whole, the affinities of the montane forest birds with East Congo are remarkably close. Two-thirds of the forty-seven Kenya species reappear in East Congo, as shown in Table XX. (Actually fifteen of them reappear in the Cameroons.) Further, twenty out of the thirty-two species common to the main Kenya highlands and to East Congo have not diverged subspecifically in these two areas. In fact the proportion of shared subspecies is three times as great as that between Cameroons and East Congo. Two factors will have contributed to this result: one is that the Cameroons montane forests are four times as remote from the East Congo as are those of Kenya. The other is that, owing to the greater height of the country between Kenya and East Congo, it would have remained suitable for montane biomes several thousand years longer, before communication was broken. It will be noted from Table XX that one-quarter of all the Kenya montane species are neither endemics nor shared with East Congo; this is because they extend southwards through Tanganyika.

A problem is posed by a series of small mountains, especially Mutha, Mugongo and Endau, each rising to about 1400 m and capped with forest, isolated to the east of the main Kenya highlands. They are unexplored ornithologically except Endau, which is unaccountably poor in forest birds (Tennent, 1964), notwithstanding its area of over 10 sq. miles of forest dominated by montane tree species, e.g. *Podocarpus gracilior*, *Piptadenia buchananii*, *Olea chrysophylla*, with some admixture of lowland species. Whereas the Nairobi forests, which are the nearest to Endau in the main highlands, contain about forty-five species of which most are typically montane, the Endau forest houses only eleven forest species, only three of which are typically montane. Forest turacos, hornbills, weavers, thrushes (*Turdus*), drongos, bulbuls and *Viridibucco* and *Pogoniulus* barbets, whether lowland or montane, are all missing. This situation is the more remarkable since in Malawi mountains of the same height with less than 1 sq. mile of forest sustain a typical bird fauna (see below and also Chapter 15). It might be suggested that in Kenya, though not so far from the Equator as southern Malawi, the rise in temperature during an interglacial or even an interstadial might have destroyed the montane forest on Endau. How then does it come about that today montane tree species are so much better established than montane forest birds? Perhaps the process described on p. 5

is to be invoked; but in any case it seems that for whatever reason, the invasion of Endau by individual birds is less active than that of the more isolated marsabit discussed earlier in this chapter.

THE MONTANE FOREST BIRD FAUNA OF
ABYSSINIA

It has been comparatively easy to rationalize the relationships between the montane forest bird faunas of the Cameroons, the East Congo and Kenya highlands, but the Abyssinian situation is baffling. It is true that our knowledge of the Abyssinian bird fauna is unsystematized and the ranges of individual species are difficult to define, but nevertheless the general impression to be gathered from the existing information is not likely at this date to be seriously misleading. The physical conditions seem suitable for a rich forest bird fauna. The extensive Abyssinian plateau carries large areas of forest in much of which the dominant forest trees of the East African highlands are present (Logan, 1946); though there is considerable variety, for, as Gillett (1955) has remarked, "the flora of southern Ethiopia is much more like that of Kenya and Uganda than it is like that of northern Ethiopia". Also, in terms of altitude and distance, the isolation of the Abyssinian forests is prima facie not very unfavourable. A belt of country at over 500 m a.s.l. on both sides of Lake Rudolf connects the main massifs of Abyssinia, the Imatong Mountains and Kenya, and at the height of the glaciation may have carried montane vegetation, perhaps in the west even forest. Using, as before, the 1500 m or the 5000 ft contour as the lower montane limit today, the southernmost point of the Abyssinian montane, at Mega, is only some 200 miles from the northernmost point of the main Kenya highlands, and moreover the Matthews Range, Mt Nyiro and Mt Kulal supply intermediate stepping-stones on which montane forest exists. The gaps, of which the longest is the 120 miles between Kulal and Mega, are, however, ecologically very severe, for they consist of successive belts of desert, subdesert steppe and acacia steppe.

So far as we know, the extensive montane forests of Abyssinia all together house only nineteen species of birds, classified as in Table XX, compared with forty-three in the Cameroons, sixty-three in East Congo and forty-seven in Kenya. Odder still, the whole bulbul family (the Pycnonotidae) and the genus *Apalis*, each of which contributes more than one species to all the other montane forest bird faunas, are entirely absent from those of the Abyssinian plateau. This situation is all the more anomalous because the smaller forests clothing the top of the Imatong groups of mountains on the Sudan–Uganda border, already referred to as connected discontinuously with Mt Elgon, have a bird fauna of normal composition consisting of twenty-four species (all shared with Kenya) which is actually larger than the Abyssinian. It would be interesting to ascertain whether the reduced number of species that are present have individually more extensive niches in the Abyssinian forests than elsewhere.

Another unexpected feature of the Abyssinian forest bird fauna is that, while the non-forest bird fauna of the Abyssinian plateau shows high endemism (see Chapter 12), only three out of its nineteen forest species are endemic—a babbler *Parophasma galinieri* (which everyone has retained as a monotypic genus) and two turacos. One of these, *T. ruspolii*, may well be the rarest bird of Africa, for it is known only from a patch of *Juniperus* forest of about 10 sq. miles and is hardly likely to occur anywhere else (Moreau, 1958b; Hall and Moreau, 1962). Practically all the other species in this Abyssinian bird fauna are shared with Kenya, the possible exception being *Oriolus monachus*, provisionally accepted as belonging to the same superspecies as Kenya orioles (the taxonomy of this genus being controversial). Thus this bird fauna as a whole suggests that Abyssinia has by no means been the centre of evolution for montane forest birds that it seems suitable to be, and that very recently on the evolutionary time-scale there has been interchange of species with Kenya—probably from Kenya rather than the other way about.

It is true that the not very considerable gaps between the montane stations on the way from Kenya are so dry that they are unlikely to have been bridged by suitable vegetation at even the height of the glaciation. On the other hand, many montane forest species of tree certainly got across and it is not clear why they are not accompanied by a more adequate assortment of the associated bird fauna, especially as Marsabit Mountain, which is isolated by 80 miles of acacia steppe and subdesert steppe, carries a number of Kenya montane birds in its forest cap. The lack of bulbuls and *Apalis* warblers in the Abyssinian forests is particularly baffling, because it is impossible to believe in an ecological reason, effective at the present day, for the absence of them all.

The impoverished state of the Abyssinian forest bird fauna may be comparatively recent on the evolutionary time-scale. A few pages earlier it was recalled that *Phodilus* and *Pseudocalyptomena* of the East Congo forests have their nearest allies in India and beyond. The same applies to the small warbler "*Artisornis*" *metopias* of montane forests in Tanganyika, for it is undoubtedly a tailor-bird *Orthotomus* (Hall and Moreau, 1962), another Oriental genus. Furthermore, of the three species under discussion two, the first and the last, are remarkably like their remote eastern relatives. Since the sundering distances are so great and since most forest birds are so sedentary it is necessary to postulate, whatever the palaeoclimatological difficulties, that the ancestral stock of these birds occupied a chain of geographically intermediate stations, though not necessarily all at once. From its situation and its circumstances it is inconceivable that Abyssinia was not one of them and that at some period it did not serve as a staging post for these species, just as it has doubtless been the means by which the two montane forest birds

Turdus abyssinicus and the warbler *Phylloscopus ambrovirens* found their way between Africa and the mountains of the Yemen. Hence it seems that a catastrophe must at some stage have overtaken the Abyssinian forest bird fauna, though not so total as to eliminate its endemic *Parophasma* and turacos. The area is much too extensive all to have been blasted simultaneously by the vulcanicity to which parts of the plateau have been subject. Any catastrophe must surely have been climatic; cold could merely have driven the montane biota lower down the slopes; an annual rainfall so low or a dry season so unmitigated as to be intolerable to evergreen forest growth remains a possibility. Are we then to believe that the evergreen forest flora of Abyssinia survived when most of the associated birds succumbed or that it derives from a subsequent re-colonization? The latter hypothesis is favoured by the high proportion of bird subspecies shared with Kenya. One would like to have a thorough analysis of other animals from this point of view.

The Montane Forest Bird Fauna of Angola

Something like 20000 sq. miles of Angola reaches 1500 m a.s.l., but only very small areas of this reach 2000 m (Fig. 49), and montane evergreen forest is practically confined to vestiges on Mts Moco, Mombolo and Soque. These are all within a few miles of 12° S., 15° E., isolated to the north by some 1200 miles of lowland forest and savanna from the montane forests of the Cameroons and to the east by 1000 miles of mainly brachystegia country from those of the East Congo. Notwithstanding their very small size and their remoteness, these relict Angolan forests hold sixteen species which are usually regarded as typically montane forest birds. Two other such species have been recorded, *Cryptospiza reichenovii* and *Ploceus insignis* (Hall, 1960a; Traylor, 1963) at Gabela,* 10° 51' S., 14° 19' E., on the Angola scarp, some of which is still under forest, though "only a shadow of what it once was" (Shaw, 1947). Here the temperature/altitude gradient is steep, as on other slopes facing the sea, as already noted for Usambara and Mt Cameroon; the mean annual temperature at Gabela meteorological station, 1093 m a.s.l., is only about 19°C, which in most of tropical Africa is about the temperature expected at nearly 1500 m, the lower limit of the montane biomes. Hence it may be inferred that in the recent past montane forest birds had a much more extensive range in Angola than they have at present, especially in a long strip from Sa de Bandeira in the south to the Moco area and also along the top of the scarp. The fact that such remote and relict forests as those of Angola hold eighteen montane species throws into relief once more the poverty of the Abyssinian forests, with only nineteen species.

* Gabela is the type-locality of two endemic species, the shrike *Prionops gabela* and the turdine *Sheppardia gabela* (which forms a superspecies with *S. cyornithopsis*). Both appear to be associated with the forest there, the shrike being definitely lowland, the turdine probably lowland (Traylor, 1963).

The eighteen species in Angola comprise three endemic species, which are members of widespread superspecies, nine species of which the Angola representatives are subspecifically distinct and six in which they are not. Thus, compared with the Cameroons forests, which are isolated by about the same distance, the Angolan montane forests lack endemic species that are not members of superspecies. Though they have the same proportion of subspecific endemics, on the method of estimating endemism applied to marine islands in Chapter 19, the

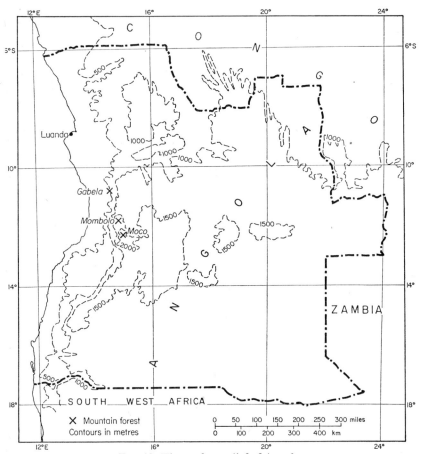

FIG. 49. The surface relief of Angola.

Angolan group scores only 1·0, compared with at least 2·7 for the Cameroons (see p. 200). In one important respect, however, the conditions of isolation are different in Angola, for in this part of Africa ecological specificity has somewhat broken down, as mentioned in Chapter 5. Thus, whereas none of the birds of the Cameroons or the East Congo montane forests have been found in the intervening lowland forests or savanna, half the typically montane species that appear in the Angolan montane forest relics do so sporadically also in fringing forest or mere evergreen thicket at lower altitudes either elsewhere in Angola or in Zambia (cf. Benson and White, 1957). In these parts of Africa, then, they have shown themselves more adaptable than elsewhere; and since the country intervening between the Angolan relicts and the montane forests of East Congo and Malawi is all above 1000 m, from the temperature point of view montane conditions could have continued to exist across eastern Angola, western Rhodesia and the Katanga until well after the last glacial maximum, in fact until some 13000 years ago. Also, patches along the Congo–Zambezi watershed slightly exceed 1500 m, the longest gap between such montane areas being one of about 150 miles. It is, however, not easy to envisage an exchange of forest birds across this fairly high ground because it lies in the monsoon belt and undergoes a long dry season. At present it is mostly covered with brachystegia (deciduous) woodland and, as indicated in Chapter 3, as late as about 10000 years ago and for some thousands of years before that, the area from Angola to Katanga and western Rhodesia was actually much drier than it is now. Hence the montane forest connexion must have been realized either before this dry period or after it, within the last 10000 years. Moreover, evergreen forest here could become general only if the dry season was mitigated. As noted in Chapter 3, meteorological theory is divided as to whether such mitigation might be expected at the height of a glacial period or at the height of an interglacial. To the latter the hypsithermal period of around 6000 years ago approximated in temperature; but this factor in itself would have hindered rather than facilitated the spread of montane forest, which required conditions cooler than the present, not warmer. On these grounds it looks as if the period of intercommunication for the montane forest birds of Angola must be sought earlier, during the Last Glaciation, 25000–18000 years ago. But in that case it is not clear how certain montane forest birds came to survive the subsequent arid period and appear as they do today in a few spots of ground-water forest (*mushitu*), as described in Chapter 5, to some extent favourable but not typical for this category of bird.

The Tanganyika–Nyasa Montane Bird Faunas

Returning to the east side of the continent, we have to deal with an area of great complexity, possessing many separate montane forests (Fig. 42), most of them on isolated massifs rising to between 5000 and 10000 ft a.s.l. Actually "Tanganyika–Nyasa" does not exactly fit the scope of this section, which is intended to cover the area, about 800 miles long, from east of Kilimanjaro south to the Zambezi, omitting those Tanganyikan mountains farther west (Kilimanjaro to Crater Highlands, and Kungwe) already referred to in earlier sections. "Tanganyika–Nyasa" in fact starts in the north with the very small forests on Kasigau and the so-called Teita Hills in the south-eastern corner of Kenya, about 40 miles from Kilimanjaro. (Teita Hills is a ridiculous diminutive for imposing masses that reach 7000 ft a.s.l.; the surviving forests total only about 3 sq. miles but hold the remains of a typical montane bird fauna.) Just over the Tanganyika border rise the North Pare, South Pare, West Usambara and East Usambara Mountains. The North Pares are separated from Kilimanjaro by the 25-mile Kahe gap with some ground-water forest even today, and the East Usambaras drop sharply on their seaward side into coastal forest and savanna. Otherwise each of these massifs is surrounded by dry acacia steppe and is isolated, though narrowly, to an extent that would not have been eliminated during the glaciation. All these massifs are of ancient crystalline rock (gneiss), apparently the remains of raised blocks that have been carved by circum-erosion over an immense period. They rise grandly from the low ground, showing in places enormous precipices of bare rock.

Southwards the Usambaras face over 100 miles of *Acacia*, *Brachystegia* savanna and *Brachylaena* (dry evergreen) woodland, which was probably replaced by humid forest at the height of the glaciation, but most of it not montane. Southwards again a great tilt-block, with raised edge facing towards the Indian Ocean and the monsoons, runs from the Nguru Mountains in the north until it merges with the volcanic outpourings of Mt Rungwe and the Poroto Mountains at the head of Lake Nyasa. Catching much rain, this block still has on its seaward edge heavy forest, which was doubtless continuous until recently. To the east of the tilt-block, at about 7° S., the Uluguru Mountains, another massif of gneiss, are isolated by some 60 miles of savanna, much of which is too low to have been clothed in montane forest at any stage of the glaciation; yet the Uluguru montane forests are inhabited by a full and typical bird fauna. Finally, Malawi is filled with a constellation of isolated mountains—the biggest, Mlanje (10 000 ft)—most of them within 50 miles from each other; and there are a few more to the east, in Portuguese territory. Most of the montane forests in this area have probably always been separated from each other, though only narrowly.

Without presenting a complete schedule of the bird faunas of this multiplicity of montane forests, which would be out of proportion here (and anyway some of them are not fully explored), it is difficult to do justice to the remarkable situation in this part of Africa. On the one hand, across all the discontinuities of habitat a large proportion of the species range widely. For example, of thirty-four species (not all purely montane) found in the 3 sq. miles of forest on Cholo Mountain in southern Malawi by Benson (1948), twenty-nine inhabit also the Usambara Mountains 800 miles away to the north. On the other hand, individual montane forests show much more individuality than do members of the Kenya group, and not only in birds. For example, the Usambaras are the centre of the genus *Saintpaulia* ("African Violet"), all twenty species of which come from forests within the limited area from the Teitas to the Ngurus.

The Tanganyika–Nyasa montane forest group is separated from the nearest main element in the Kenya group, Kilimanjaro, by only about 25 miles and from the main part of the East Congo group, round the head of Lake Tanganyika by 400–500 miles. Since the intervening country, the interior of Tanganyika, all lies at more than 900 m a.s.l. and much of it above 1200 m, it was presumably capable of carrying montane vegetation until well after the glaciation, but most of it has so poor a rainfall today that it may be doubted whether actual forest was widespread there, even at the height of the glaciation.

The status of the Tanganyika–Nyasa montane forest bird fauna and its relation to the East Congo are shown in Table XX. It will be seen that the fauna is rich, with sixty-four species, and its endemism is high—eighteen endemic species, only six of them members of superspecies. Actually, several of them are very restricted in range, known from only one or two stations (see examples below). Thus at the specific level the proportion of endemics, 0·29, is nearly as great as that in the far more isolated Cameroons. Furthermore, of the thirty-four species common to East Congo and Tanganyika–Nyasa, four-fifths shows subspecific differences between the two areas, a proportion as high as between East Congo and the much more remote Cameroons. It seems necessary to suggest that the ecological barriers between Tanganyia–Nyasa and East Congo have been exceptionally effective. As discussed above, between these two groups there are on the east side of Lake Tanganyika two stepping stones, the Kungwe block nearer to East Congo, which it favours in its bird fauna, and the Ufipa block, nearer to the mountains round Lake Nyasa but avifaunistically not close to them. Evidently both the 40-mile gap between Ufipa and Kungwe and the 100 miles intervening between Ufipa and the mountains round Lake Nyasa have barred the movement of evergreen forest birds for a considerable period. Since this latter gap is crossed by a corridor of high ground, much of it at about the critical level of 1500 m, it can only have been effective if it was very dry. This is the complement of another postulate we have had to make when describing discontinuous ranges of a number of dry-country animals (Chapter 10), namely, that at one time a north-and-south connexion existed from Kenya through Tanganyika to Rhodesia that was dry enough to facilitate the movement of mammals and birds belonging to the acacia biome. Here then the postulates arrived at by induction from the present-day distribution of each of two different bird faunas coincide; and there is indeed, as noted in Chapter 3, reason to think that around 10000 years ago the rainfall in this part of Africa may have been little more than half what it is now.

We may turn now to consider the two blocks in the Tanganyika groups, Usambara and Uluguru, which would have been at least narrowly isolated at even the most marked stage of the glaciation. They hold in fact some of the most noteworthy endemics: the Ulugurus a shrike *Malaconotus alius*, with no near relative in its genus, and a warbler

Scepomycter winifredae, which forms a "good" monotypic genus; the Usambaras an endemic turdine, *Alethe montana*, the endemic weaver *Ploceus nicolli*, perhaps conspecific with *olivaceiceps*, and two species, *Apalis moreaui* (warbler) and *Phyllastrephus orostruthus* (bulbul), otherwise known only from Unangu and Namuli forests respectively, far away in Portuguese territory east of Lake Nyasa. Again, the Teita forests, less than 50 miles from those of Kilimanjaro and of course within easy sight, hold, besides two strongly marked endemic subspecies, a thrush *Turdus helleri* that replaces the *T. abyssinicus* of all the neighbouring montane forests but is so strongly differentiated that it would strike most people as belonging to the same superspecies rather than to the same species.

It may be noted incidentally that the forests on these relatively small mountains of ancient crystalline rock, especially those on the Usambaras and the Ulugurus, carry bird faunas that are richer in species and in endemics than are the forests on most volcanic mountains, even the greatest of them, Kilimanjaro (see also Chapter 15). Carcasson (1964) has noted the same thing for butterflies and from my personal knowledge I can say that it applies also to a very different class of organism, the epiphytic orchids. One would naturally assume that the ultimate reason for this difference is that the volcanic mountains are younger and have not been able to offer the same stability of conditions on their surfaces as have the other mountains in question; but the greatest volcanic mountains are old enough for striking endemic species of plants to have been evolved in their vegetation zone above the timber-line (see, for example, Hedberg, 1951) which is moreover the area in propinquity to, and most vulnerable to, volcanic eruptions.

A few birds, such as the turdines *Modulatrix* (=*Illadopsis*) *stictigula* and *Sheppardia sharpei*, occur in a number of forests from Usambara to Malawi, but nowhere else. In some ways more remarkable are those like the cuckoo *Cercococcyx montanus* found in the Usambaras, the Ulugurus and Malawi and also in the East Congo mountains to Ruwenzori, though absent (and in the case of this noisy cuckoo certainly not overlooked) from Kenya and Kilimanjaro. There are other, individual, oddities of distribution besides some already mentioned. For example, the green oriole *Oriolus chlorocephalus* ranges from Usambara to Malawi and reappears in the forest vestiges on Lolkissale Mountain, away to the south-west of Kilimanjaro (but not on Kilimanjaro itself). Again, *Alethe* (=*Bessonornis*) *anomala* extends from Malawi to Uluguru, then reappears 250 miles away to the north-west, in the Mbulu forests at the southern end of the Crater Highlands block. Much of the intervening country is now very dry but the topography is such that it is comparatively easy to envisage a series of stations, if not a continuous chain of forests, across this part of Tanganyika for long after the glacial maximum. As a final example, *Phylloscopus* (=*Seicercus*) *umbrovirens* comes south throughout the Kenya group, Kilimanjaro and also the Pare forests, apparently misses the Usambaras (which are occupied by a different, southern, congener, *ruficapilla*), but reappears on the Ulu-gurus (Fig. 17). Here *ruficapilla* is found also, but the scanty evidence points to ecological separation, with *umbrovirens* perhaps confined to a very small area at the highest elevations (about 7000–8000 ft) on the dry side of this much-dissected ancient massif, and *ruficapilla* more generally distributed in the rest of its forests.

An especially noteworthy feature is the high degree of subspecific differentiation in the isolated forests in the neighbourhood of the Kenya–Tanganyika border, a differentiation without parallel elsewhere and one for which I am unable to suggest an explanation. In the various isolated mountain forests in an area about 150 by 50 miles, that includes the Usambaras, the Pares, the Teitas and Kilimanjaro, five good subspecies of montane forest *Zosterops* (*senegalensis, sensu lato*) have developed, four of the turdine *Pogonocichla stellata*, three of the warbler *Apalis thoracica* and three members of the *Turdus abyssinicus* superspecies. This is a degree of differentiation, within an area of less than 8000 sq. miles—less than the area south of the Thames from Kent to Wiltshire and Dorset-shire—which would call for remark in a marine archipelago and is not indeed surpassed in the islands off Africa (Chapters 16, 18). It is true that because the gaps intervening between the forests concerned are so low and are mostly of acacia steppe, they can hardly have been filled with montane vegetation even at the height of the glaciation. Neverthe-less, the gaps are all so extremely narrow that the unequivocal evidence of reduced gene-flow far exceeds expectation and is at variance with the much lower prevalence of subspeciation within the whole Kenya group from the Imatongs to Kilimanjaro, where geographical conditions for subspeciation would seem more favourable, above all on Marsabit.

Montane Forest Bird Faunas South of the Zambezi

South of the Zambezi typically montane forests are few. They are practically confined to a discontinuous line along the high eastern edge of Zambia, with an outlier, Mt Gorongoza (in Portuguese territory), studied by Pinto (1959), to some relict forests nearly 300 miles away in the Woodbush area of the northern Transvaal, and fragments in the interior of Natal. With increasing latitude the temperature at the same altitude above the sea drops; as Liversidge (1959) has remarked, south of the Limpopo "the clear cut isolation of montane areas rapidly dwindles" and eventually in the Knysna Forest of Cape Province, some 1500 miles south of the Zambezi, the few birds that are typical of mon-tane forest in the tropics are found little above sea-level, a situation paralleled in South America.

Relict as these forests south of the Zambezi are, it is at first sight sur-prising how few species of birds they contain—less than half those occupying the montane forests of Malawi. Certainly during the access of aridity in these latitudes around 10000 years ago the isolation of these southern forests would have been even more marked and the forests themselves were presumably still more restricted in area. Only one species has a claim to have originated south of the Zambezi, namely the turdine *Pogonocichla swynnertoni*. This is of special ecological interest because it is limited to Gorongoza and neighbouring forests on the eastern border of Rhodesia. There, in an area of less than 100 sq. miles, it co-exists with its only congener, *P. stellata*, another insectivorous ground-feeder, but with a great discontinuous range from the northern end of Kenya to the Cape Province (Fig. 41). Concerning these two species, see Oatley (1966).

GENERAL

The foregoing discussion has shown the extent to which the same species of birds appear in montane forests isolated from each other by stretches of country that is too hot and usually also too dry for them. Yet four-fifths of the bird species in the montane forests of Malawi reappear in Usambara, 700 miles to the north; two-thirds of those in the Kenya group reappear on the East Congo mountains and so do those of the Cameroons group, which are four times as widely isolated, without any intermediate montane stations whatsoever. It is true that, as shown by the discussion of the montane bird fauna of Marsabit above, several forest species must frequently cross what seems to have been a persistent gap of never less than 50 miles and one of the greatest ecological severity. Moreover, a much greater variety of species, including a good proportion of those native to the lowest stratum of the forest and hence habituated to the full darkness and humidity of its ecoclimate, must be postulated to have been interchanged between a number of montane forests, notably between Cameroon Mountain, and the Bamenda block (in the broad sense used on p. 198) and between some of the Malawi mountains, over gaps that have never been less than 20–30 miles. But this evidence of short-distance colonization by forest species cannot explain the high proportion of species shared by the Cameroons and the East Congo montane forests, now isolated by over 1200 miles. It cannot be doubted that in this case the most potent factor in present distribution, and one that has been operable in varying degrees all over Africa, is the freedom of intercommunication that prevailed between these now isolated forests during the last glaciation and in some instances until 10000 years ago or less. However, for some unexplained reason this freedom, which Fig. 2 shows should have extended also to Abyssinia, does not seem to have had the expected results there, and the poverty of the montane forest bird fauna there can probably be taken as showing that the post-glacial gap of some 300 miles from the Kenya highlands, though possessing a few intermediate stations, is too great often to be crossed as a result of individual dispersal by forest birds.

It may be added that, when the geographical ranges of individual species are reviewed, it is found that a certain major kind of discontinuity is absent. None of the Cameroon montane species that appear also in Kenya fail to do so in the intervening montane forests of East Congo, and none of the Abyssinian birds occur in Tanganyika if not also in Kenya. But on a smaller scale discontinuities are numerous, as instanced in the Tanganyika–Nyasa section above. Most of these cases are presumably the result of extinction in individual montane forests in recent times; and this is a fate to which small isolated populations, such as many of those under discussion, are known to be prone even when the habitat is not grossly disturbed.

As regards endemism, as well as distribution, both local evolution and accentuation of relict status contribute to the situation we see today and it is impossible to disentangle their roles. It seems likely, however, that local evolution has been a more important factor in the Cameroons group than elsewhere, since that group has been on the whole more isolated than any other. It is interesting that its endemism, as high as in any of the African marine islands, is far higher than in the next most isolated group, the Angolan montane forests, equally far removed from their nearest neighbours to the east, but with what has evidently been less severe ecological isolation. By contrast, the centrally situated East Congo montane forests have as high a degree of endemism as those of the Cameroons, while that of the Kenya group is very low and that of the Tanganyika–Nyasa group much higher. Perhaps the most striking feature is the intense subspeciation that has been achieved by some birds inhabiting forests, often within sight of one another, close to the Kenya–Tanganyika border, while on the northern outliers of the Kenya group, especially Kulal and Marsabit, isolated by greater distances and worse terrain, even subspeciation is almost absent. Finally, to the east of the Kenya group, another small mountain, Endau, with what appears to be a forest quite suitable for montane birds, is almost devoid of them.

The Montane Non-forest Bird Faunas

Like the montane forest, the montane non-forest environment of tropical Africa has of course a sporadic distribution, being confined to highlands, for the most part over 5000 ft. But unlike the montane forest, the non-forest consists of two separated layers, the upper one, the so-called afroalpine moorland, being above the timber-line, which on some mountains is as high as 10 000 ft. Thus, while all montane stations are isolated from each other by lower ground, the afroalpine areas are further isolated by the ring of montane forest from which they emerge. Because of this additional isolation, the distribution of the afroalpine species poses a special problem which needs to be discussed separately from that of the other montane species. For up-to-date discussions of the afroalpine flora, see Hedberg (1961, 1962, 1964).

The Habitat

In the lower, more widespread, montane zone the non-forest vegetation is of several types, all of which probably owe much to human interference, by burning, cultivation and grazing. Some highland areas are covered with an open savanna, scattered with trees, for example with *Acacia spirocarpa* in parts of Abyssinia, *Akocanthera* in parts of Kenya, *Brachystegia* in the highest part of Angola. The montane grassland nowhere has the tall upright growth of the lowlands; it is often well-grazed, forming a turf of a kind not found in tropical Africa at low altitudes, or it may have clumps of coarse grass, as the *Eleusine jaegeri* of Kenya and northern Tanganyika, which is inhabited by the specialized montane weavers, *Euplectes jacksoni* and *progne*. Where the country has been overgrazed and the soil eroded it tends to be covered with a mass of dense woody herbage and low bushes, e.g. *Crotolaria, Clerodendron, Leonotis, Lippia, Solanum, Tarconanthus* and *Vernonia*. Another widespread type of vegetation is a dense low growth, much like rank European heathland and dominated by ericaceous plants, *Phillipia* and *Blaeria*, together with *Smithia* and often with scattered small trees, perhaps of *Myrica, Lasiosiphon* or *Agauria*. This type of vegetation occurs not only on the mountains of eastern Africa, but also on the Bamenda highlands of the Cameroons, though not on Mt Cameroon itself (Keay, 1955). *Pteridium* (bracken) is common, together with *Rubus* (bramble) and rank *Hypericum*. Vegetation of these kinds may be interspersed with islands of montane forest, where this has not all been cleared, and on some mountains its upper limit is formed by the forest cap or, on the greater mountains such as Kilimanjaro, by the forest girdle. As will be seen from Fig. 2, the biggest continuous block is in the Abyssinian highlands, and the next in Kenya and northern Tanganyika. As recently as about 18 000 years ago montane conditions extended continuously from Abyssinia to South Africa, but there is no means of knowing which areas would have been covered with forest and which not. Nor is it easy to envisage the nature of the non-forest montane vegetation before it was subjected to wholesale interference by man. Botanical opinion inclines to the belief that it consisted mainly of deciduous or semi-deciduous woodland, open enough to have plenty of ground-cover, especially of bushes and woody herbage, no doubt including most of the species mentioned above.

THE BIRD FAUNA OF THE MAIN MONTANE AREAS

It will be seen from the figures given in Table V (p. 84) that the birds peopling the montane non-forest habitats are in a sense less specialized than those of the montane forest. In the latter there are three times as many birds peculiar to that habitat as there are birds that live in both lowland and montane forest (120:39), whereas in the non-forest the proportions are nearly equal, 74:69, three of the 74 being confined to the afroalpine. This difference is presumably due to the fact that the montane forest is a relatively homogeneous and undisturbed habitat in which it has been possible for more saturated local communities to develop. The historically recent changes in the non-forest will have disrupted the natural habitats and probably at the same time have produced a greater diversity. This may have caused some extinction among the original non-forest montane species but concurrently it will have opened opportunities for the lowland species with which each montane area is surrounded and is in immediate contact.

Table V shows that the number of typically montane non-forest species of birds is less than one-tenth the number of typically lowland. In one group, the "other non-passerines", the proportion is far lower, with only eight montane species compared with 164 lowland (and in addition about five species which frequent montane as well as lowland). This is in part due to the paucity of tree-growth in the montane non-forest, where for example hornbills and turacos would not find much that was suitable for them and woodpeckers would have only very limited opportunities. This factor cannot, however, be invoked to account for the most striking absences among the passerines—no typically montane lark, nor bunting, nor shrike, nor babbler—for which some types of the montane non-forest environment would seem suitable. By contrast, finches (Fringillidae), thrushes (Turdidae) and also sunbirds (Nectarinidae) are relatively better represented in the non-forest montane environment.

Of sunbirds that do not seem to be in any way dependent on forest we have, apart from the typically afroalpine *Nectarinia johnstoni*, discussed further below, *N. famosa* (Abyssinia–Cape Province), *N. tacazze* (Abyssinia–northern Tanganyika), *N.* (=*Drepanorhynchus*) *reichenowi* (Kenya and northern Tanganyika), *N. kilimensis* (Kenya, Rhodesia and Angola), *N. purpureiventris* (mountains of eastern Congo), *N. bocagei* (Angola highlands), *Nectarinia afra* (=*Cinnyris afer*) (Ruwenzori to Cape Province) and *Nectarinia preussi* (=*Cinnyris reichenowi*) (Cameroon Mountain and mountains of Eastern Congo). In some East African localities three or four of these sunbirds can be seen in the closest proximity, apparently using the same species of flowers, and it would be interesting to work on their food relationships. The fact that there should be so many species, all practically confined to between about 4000 and 9000 ft in the tropics, suggests that conditions at some stage must have been extremely favourable to the evolution of these sunbirds and that there must have been several centres for it, but it is not easy to speculate usefully on the conditions under which it took place. As noted above, during the cool spell of a glaciation and prior to the excessive use of fire by the human population, enormous areas of Africa presumably carried such species of woody herbs as *Leonotis* and *Crotolaria*, which are favourite sources of food for the montane sunbirds. Mere distance would have had a certain isolating effect, but this necessary prelude to speciation would have been provided more adequately during non-glacial stages, when the montane block was broken into islands, as it is now.

Of the seventy-four montane non-forest specialists, it may be recalled that a few are shared with the Palaearctic: two grebes *Podiceps* spp., Lammergeier *Gypaetus*, Alpine Swift *Apus melba*, Quail *C. coturnix* and Chough *P. pyrrhocorax*. To the non-forest montane bird fauna as a whole the most extensive existing montane block, that of Abyssinia, makes the biggest contribution; it accommodates forty-seven typically montane non-forest species, including twenty-one endemics, and also the Chough which is found nowhere else south of the Sahara.* All the non-endemics (except the Chough) are shared with Kenya and some of them with other mountains as well, but there are no examples of such ranges as, for example, Ruwenzori and Abyssinia, omitting Kenya. (To do full justice to the peculiarity of Abyssinia in non-forest birds, mention should also be made of the odd monotypic corvine *Zavattariornis* (which some have thought perhaps a starling) and the endemic swallow *Hirundo megaensis*, both restricted, inexplicably so far as obvious ecological reasons go, to a small part of the southern edge of the Abyssinian plateau. They fail to qualify as typically "montane" because their limits of altitude are too low, both species living around 4000 ft.) It is surprising to find that extra-tropical South Africa makes nothing like so important a contribution to the non-forest montane bird fauna of the

---

* It may be mentioned that the very isolated Gebel Marra massif, rising to over 3000 m at about 13°N. 24°E., in the western Sudan, carries no typically montane nor Palaearctic species of bird, although there is an area of 40 sq. miles above 2400 m where the vegetation has affinities with the Palaearctic and with the Abyssinian plateau some 800 miles to the east.

tropics as does Abyssinia. South Africa has three endemics, an ibis and two finches, confined to the mountains of the Drakensberg and Basutoland, and shares with the Abyssinian montane (as well as East Africa) eleven species which can consequently not be reckoned as characteristically either South African or Abyssinian; but there are only ten species occurring in South Africa as well as the tropics which fail to reach Abyssinia. Moreover, thirteen of the seventy-four non-forest montane species are found only on the mountains of the inner tropics, without representatives in either Abyssinia or South Africa. Hence, on the whole, unless present-day distributions are deceptive (which is quite possible as a result of local extinctions), most montane non-forest birds have originated on either the Abyssinian plateau or the other tropical highlands, but few in South Africa.

At the same time it is a puzzling fact that, while the list of typically montane non-forest species includes no less than nine sunbirds, the Abyssinian montane has only two, and neither of them endemic. Another point is that three of the Abyssinian endemics have obvious vicariants in Kenya, *Corvus crassirostris/albicollis*, *Macronyx flavicollis/sharpei* and *Onychognathus albirostris/tenuirostris*, which are yet different enough to be universally accepted as "good" species. These facts, in conjunction with the high endemism in Abyssinia where other non-forest birds are concerned, suggest that barriers to interchange between the Abyssinian plateau and the montane areas to the south and west have been severe and have been effective for a long time A factor not to be neglected, however, is that the Abyssinian plateau offers a much more extensive unforested area of varied conditions than any other part of tropical Africa; its present high predominance in number of non-forest endemics may be in part a post-glaciation phenomenon and due to its acting as a more effective faunal refuge, while more extinction took place elsewhere. Those three pairs of vicariant species, just quoted, could conceivably be the result of divergence in the course of the last 18000 years.

Before leaving the non-forest bird fauna of Abyssinia, some specific peculiarities may be recalled. First, it is the only place south of the Sahara where the Red-billed Chough exists (as a slightly differentiated subspecies); and it is not difficult to derive this bird from the Palaearctic during the Last Glaciation, by way of the chain of mountains between the Nile and the Red Sea, which can be presumed to have had a climate at that time like that inhabited by this Chough in Morocco today. Another special Palaearctic affinity is the owl *Asio abyssinicus*, which is thought to be conspecific with *A. otus*. Of the other Abyssinian endemics the most remarkable is *Cyanochen cyanopterus*, the Blue-winged Goose, with nothing at all like it in Africa and probably its least remote relative in South America (*Chloephaga*). Aside from the endemic families of Africa, this goose is perhaps the bird most isolated, in the taxonomic sense, of all those on the African continent.

It has been shown above that the high status of Abyssinia as a home for non-forest montane species is in complete contrast to its low status in respect of forest species. On the other hand, the position in the Cameroon highlands is the converse of this. There the montane forest bird fauna is rich and shows strong East African affinities (Chapter 11), but by contrast few such non-forest birds occur. There is one notable endemic, the aberrant zosteropid *Speirops melanocephala*, which belongs to the upper edge of the montane forest on Mt Cameroon and to the few square miles above that where arboreal vegetation occurs sporadically. It is the only member of its otherwise insular genus on the African continent, but has a near relative on the oceanic island of São Tomé, way out in the Gulf of Guinea. It may be an apparently unique example of an insular bird successfully establishing itself on the mainland; its Cameroon environment is, however, itself an ecological island comparable in its isolation to a marine island. The situation is complicated by the fact that the *Speirops* of geographically intermediate Fernando Poo, the bird fauna of which is otherwise wholly that of Cameroon Mountain, belongs to a different species (see discussion in Moreau, 1957).

The only non-forest montane birds in the Cameroon highlands with East African representation are the owl *Tyto capensis* and the scrub haunting warblers *Bradypterus cinnamomeus*, *Cisticola brunnescens* and *C. discolor*, if the last is regarded as conspecific with *C. hunteri*. This suggests that the corridor of montane conditions developed along the northern rim of the Congo basin during the glaciation was too heavily forested for most non-forest birds to use it. It is true that the total area of non-forest habitats in the Cameroon highlands is quite small, but nevertheless several species, that are not typically or consistently montane elsewhere in their range, have emphatically become so in the Cameroon highlands: *Cisticola robusta*, the bishop *Euplectes capensis* (which ranges up to 10 000 ft, according to Eisentraut (1963)), and the pipits *Anthus novaeseelandiae* and *A. similis*. (In the grasslands of the Bamenda plateau the last two are actually accompanied by a third pipit, *Anthus leucophrys*.) Even with the addition of the endemic *Speirops*, the total representation of non-forest species on the Cameroon highland is, however, poor. It may be added that those highlands further west which approach the montane limit, the Jos plateau, Nimba, Bintumane and the other peaks on the Sierra Leone border, or the Futa Jallon massif, do not support any typically montane birds, though the last has quite a large area of apparently suitable country at over 4000 ft.

THE AFROALPINE MOORLAND

We may now turn to a discussion of the afroalpine moorland, which is developed only above the timber-line, from 3500 to 4100 m upwards on different mountains but nowhere reaches higher than about 5000 m. The physical conditions have been excellently described by Hedberg

(1964) and by Salt (1954) with special reference to Kilimanjaro. Apart from the reduced air and oxygen pressure they stress the intense solar radiation, the dry conditions, the rapid changes in temperature and the prevalence of nocturnal frosts lasting for many hours, "summer every day and winter every night" throughout the year. Ecologically the afro-alpine is the counterpart of the páramo of South America (Hedberg (1964) and see Dorst (1957) for an ornithologist's view-point), but whereas that formation has a narrow continuous range of immense length along the Andes and a considerable bird fauna of its own, in tropical Africa there are only a few detached areas where it can occur: on the highest parts of the Abyssinian plateau; on Mt Elgon, Mt Kenya and the Aberdare range in Kenya; on Kilimanjaro, Mt Meru and the Crater Highlands in northern Tanganyika; on Ruwenzori and some of the Kivu mountains on the eastern edge of the Congo basin. As will be shown below, the endemic birds of the afroalpine areas are few indeed. One other mountain in Africa that is fully high enough for an afroalpine zone, Mt Cameroon (13 000 ft), an intermittently active volcano, is exceptional; its montane forest gives way to grassland as low as 7000 ft, presumably because the soil at the higher altitudes is so immature (Keay, 1955). The degree to which the moorland is isolated by forest varies: isolation seems to be complete on Ruwenzori, Elgon, the Aberdares and Kilimanjaro; on Mt Kenya and Mt Meru the forest ring is broken on the dry, north-east side, so that a corridor of bush and grassland connects the moorland with the open country below the forest, while on the Crater Highlands and the Abyssinian plateau there appears to be a transition of open-country vegetation from some 6000 or 7000 ft upwards without any forest barrier at all. In East Africa the largest area of afroalpine vegetation is that of Mt Kenya, some 500 sq. miles, including the barren ground of the highest altitudes. The Abyssinian plateau carries several areas as big as this above the timber-line, but unfortunately no general description of this country that can be used from an ornithological point of view seems to exist. Some good impressions of the Abyssinian highlands by a biologist are, however, given in Scott (1952, 1955).

Many people familiar with Africa will think of the afroalpine moorland as the oddest and most exciting landscape of the continent; but with only sixty to 175 species on individual mountains (Hedberg, 1964) the floras are not rich. Where forest is present one emerges from its upper edges, dominated by gnarled *Hagenia* trees heavy with *Usnea* lichen or by giant heath *Erica arborea*, on to a bushy area that on Kilimanjaro, Mt Kenya and Ruwenzori leads towards the ultimate screes and the glaciers of the central peaks. Shortly, tussock grasses (*Carex*) and *Luzula* rushes predominate. The ground is often boggy and in places there are swamps and tarns, partly seasonal. Proteas, *Helichrysum* spp. (everlastings) and Kniphofias (red-hot pokers) are among the common flowering plants, while what gives these moorlands their characteristic aspect is the spectacular growth of the giant lobelias and the arborescent *Senecio* spp. Since the representatives of these two genera are for the most part endemic to individual moorlands, they present notable problems of distribution and evolution (see in this connexion especially Hedberg, 1951). These impressions are of course to some extent oversimplified and conditions vary much from one block of moorland to another, though all are physiognomically simple and with few species of vertebrates. However, on some of the moorlands small rodents are very numerous and antelopes occur, notably eland *Hippotragus* on Kilimanjaro and the Crater Highlands. Descriptions of some of the eastern Congo mountains will be found in Chapin (1932) and of East African mountains in Moreau (1936b, 1939, 1944b, 1945), Salt (1954) and Elliott and Fuggles-Couchman (1948).

THE BIRDS OF THE MOORLAND

The presumably resident birds of the most important moorland areas can conveniently be summarized in Table XXI, where subspecific relationships are indicated by the use of letters. Only three species (those marked E) are typically afroalpine. All the others are birds of the lower montane levels and it seems remarkable that they should tolerate the rigorous climate above the timber-line. Besides those listed, other species occur sporadically on the edge of one moorland area or another or forage over moorland; for example, the buzzard *Buteo b. oreophilus*, which is probably dependent on the mountain forest for nesting sites, and the Stonechat *Saxicola torquata*, for the permanent residence of which the vegetation of the lower part of the moorland would seem suitable. The upper limits of the resident species vary between about 12000 and 15000 ft, where in any case the vegetation ceases.

On the showing in Table XXI the moorland of Mt Kenya is richer than any other in resident species of birds. This is probably correct, except that the status of Abyssinia might well be changed if we knew more about it. The richest moorland bird fauna consists, then, of a duck (dependent on open water), a pigmy rail and a snipe (both dependent on boggy ground), a big owl and a buzzard (both dependent on small vertebrates), the lammergeyer (which presumably must range far away from the moorland for much of its food), a francolin (for which the bushy parts of the moorland are suitable), only one group D bird, a swift (which will be very wide-ranging for food), and up to eight passerines. These comprise one miscellaneous scavenger (the raven), two small insect-eaters in bushy growth (*Bradypterus* and *Cisticola*), one ground-feeding insectivorous turdine (*Pinarochroa*), two seed-eaters (the finches *Poliospiza* and *Serinus*), one starling (which feeds partly on molluscs, etc. in the moorland, partly on fruit in the trees of the forest below) and one sunbird the big *Nectarinia* (the exceptionally long beak of which is adapted to reach down into the flowers of the giant lobelias deep within

TABLE XXI

The resident birds of the afroalpine moorlands

	Abyssinia	Ruwenzori	Elgon	Kenya	Kilimanjaro	Crater Highlands
Group A						
Anas sparsa	a	a	a	a	a	—
Gallinago nigripennis	a	—	a	a	a	—
Sarothrura lineata	—	—	—	a	—	—
Group B						
Bubo capensis	a	—	—	b	—	?
Buteo rufofuscus	a	—	a	a	a	a
Gypaetus barbatus	a	—	—	a	a	a
Group C						
Francolinus psilolaemus E	a, b	–	c	d	—	—
Group D						
Apus melba	a	b	—	c	a	—
Apus niansae	—	—	—	—	—	?
Group E						
Bradypterus cinnamomeus	a	b	a	a	c	c
Cisticola brunnescens	a	—	—	—	—	a
Cisticola hunteri	—	a	b	b	b	b
Cornultur albicollis	—	a	a	a	a	a
Nectarinia johnstoni E	—	a	—	b	b	a
Onychognathus tenuirostris	a	b	—	b	—	b
Pinarochroa sordida E	a	—	b	b	c	d
Poliospiza striolata	a	b	a	a	a	a
Serinus canicollis	a	—	a	a	a	a
Number of species	13	8	10	16	11	10

their bracts). On all the moorlands except Mt Kenya several of these species are missing. It is possible to suggest a reason in one or two cases. In particular, the Mt Kenya moorland, which has three water-birds, is much better supplied with watery habitats than Kilimanjaro and the Crater Highlands, which each have one species of water-bird; but no reason can be suggested for several absences—in particular from the francolin niche on half the mountains, from the turdine niche on Ruwenzori, from the sunbird niche on Elgon.

Most of the birds comprising the moorland bird fauna belong to species inhabiting lower montane levels, and show no more than incipient subspeciation. Nowadays such montane species have discontinuous ranges, as discussed in the last chapter, but during the glaciation they presumably did not, except in so far as barriers of forest intervened. *Cisticola hunteri*, a warbler of bushes and heathery scrub, is a particularly interesting case. For one thing, on some of the East African mountains the population varies in colour with altitude in a manner that cannot be recognized taxonomically (Moreau, 1939; White, 1962a, b). Treated as conspecific with these East African birds are those inhabiting mountains mostly between 5000 ft and 9000 ft in (*a*) East Congo, Uganda and West Kenya (*chubbi*), (*b*) Marungu, the massif overlooking the south-west shore of Lake Tanganyika (*marungensis*), (*c*) northern Malawi and south-western Tanganyika (*nigriloris*), (*d*) Mt Cameroon (*discolor*), (*e*) other mountains in the Cameroons (*adametzi*). During the height of the glaciation practically all these different stations would have formed part of the same great montane area, as shown in Fig. 2. It certainly seems at first sight surprising that the narrow corridor of montane conditions connecting the Cameroon mountains with those of eastern Africa which was so effective as a means of communication for forest birds should also have served for a species such as this, which depends on open, though bushy, country.

For the three species peculiar to the moorland, the francolin, the sunbird and the turdine, the problem of distribution is particularly difficult. The distances separating the moorlands are as follows: Abyssinia–Mt Elgon, 400 miles; Mt Elgon–Ruwenzori, 300 miles; Mt Elgon–Aberdares and Mt Kenya, 250 miles; Mt Kenya–Kilimanjaro and Mt Meru, 200 miles; Mt Meru–Crater Highlands, 70 miles. During the glaciation the lower limit of moorland might be expected to be at least as low as 7000 ft. Under those conditions the moorland of Mt Kenya would almost have been connected with that of Mt Elgon, none of the intervening gaps being as much as 30 miles wide; and the distances between the moorlands of Mts Kenya and Kilimanjaro would have been halved. But Mt Elgon would still have been separated from both Ruwenzori and the enlarged Abyssinian moorland by some 300 miles, as can be deduced from the altitudes in Fig. 2.

It is possible that one or more of the three species in question at one time inhabited a lower montane type of environment, which could have given the "necessary" continuous geographical range. The only alternative is active individual dispersal of the type that has peopled oceanic islands. This is least difficult to visualize with the turdine *Pinarochroa*, belonging to a family we are accustomed to think of as strong on the wing; certainly this bird has somehow got across the low and now arid stretches of northern Kenya, between Abyssinia and the Kenya highlands, though there is no evidence that it ever accomplished the equally lengthy journey between Elgon and Ruwenzori. However, in recent times this bird has not been so far adventurous that gene-flow across the short gap between Mt Meru and the Crater Highlands has been enough to prevent subspecific differentiation.

The sunbird *Nectarinia johnstoni* presents a different picture since it has a population outside East Africa on the Nyika plateau of Malawi, over-looking Lake Nyasa from the west and isolated by nearly 500 miles from the next station of this sunbird, the Crater Highlands. The Nyika is an unusual habitat for the bird since it does not much exceed 7000 ft a.s.l. and has no giant lobelias, to the flowers of which the beaks of the *johnstoni* on the afroalpine areas close to the Equator appear to be adapted. The Nyika *johnstoni* seem to be dependent on proteas, as I am informed by C. W. Benson. A sight record of this species at 6500 ft in the Livingstone Mountains, which rise to 10000 ft on the east side of Lake Nyasa (Haldane, 1956), may refer to a straggler from the Nyika or, since these mountains have been so little visited by ornithologists, may belong to a small population resident there. It may be added that the Nyasa birds are smaller and with correspondingly shorter beaks than those which live at so much higher elevations and hence presumably lower temperatures close to the Equator; they could be a case of Bergmann's Rule, according to which, in any species, size tends to be smaller in warmer parts of the range. Moreover, since the Nyasa birds feed on protea blossoms, which have none of the inaccessibility of those of giant lobelias, there is no bar to the beaks being shorter.

Here again there are alternative possibilities to account for the present distribution of this sunbird. If the afroalpine environment has been typical of the species in the past, then even at the height of the glaciation Ruwenzori and the Nyasa station would have been so widely isolated from the East African afroalpines that active dispersal over 300–400 miles would have been necessary. To some extent individual *johnstoni* certainly do wander; whether the Livingstone record is due to straggling from the Nyika is uncertain, but an African collector of mine got a specimen at 6000 ft on the north end of the North Pare Mountains. This bird must have come from the moorland of Kiliman-jaro, some 25 miles away, in easy sight but separated by low hot country and also the forest girdle. However, such a journey is less than one-sixth as extensive as would be needed to disperse *N. johnstoni* from the Kenya afroalpine areas to Kilimanjaro and only one-eighteenth of the gap between the Crater Highlands and the Nyika.

Alternatively, if this bird was originally at home in the conditions of the Nyika, at no more than 7000 ft a.s.l., the equivalent altitude at the height of the glaciation would have been about 4000 ft and the barriers to a continuous range between the lower slopes of all the mountains on which it is now found would have been narrow and relatively unimportant. To this hypothesis a corollary is that at some stage the birds ceased to occupy the lower slopes of the East African mountains and became concentrated on the moorlands. At the lower altitudes they would be in presumptive competition with several other sunbirds, notably *N. kilimensis*, *N. famosa*, *N. reichenowi* and locally also *N. tacazze*, whereas in the afroalpine moorland they have no congener. It must, however, be admitted that these four sunbirds seem to be able to co-exist without obvious ecological segregation and it is not clear why of all of them *N. johnstoni* should have been the one to transfer in East Africa to the restricted areas of moorland.

There remains the case of *Francolinus psilolaemus*, discussed in part by Hall (1963a). Somehow the species got across the great gap between Abyssinia and the mountains of Kenya, yet of all birds a francolin, a creature of short flight and belonging to a family everywhere profoundly non-migratory, seems the most unlikely to have made its way over 300 miles of hostile environment. The most reasonable postulate seems to be that *F. psilolaemus* formerly inhabited lower altitudes, with its lower limit around 1500 m instead of the present 2500 m. Apart from the existence of any forest barriers, this could have given it a continuous range between the highlands of Kenya and Abyssinia in the glaciation. Later, under pressure from other species, presumed to be invading from the south— perhaps *shelleyi* in Kenya and *levaillantoides* in Abyssinia—*psilolaemus* presumably took refuge at the higher altitudes where it is is now confined.

GENERAL COMMENTS

The birds occurring in the non-forest montane areas are not nearly so specialized as those in the montane forest. In the former the seventy-four typically montane species are accompanied by nearly as many (sixty-nine) which are also at home in the lowlands. The corresponding figures for the forest are 120 and thirty-nine. This difference may perhaps be a very recent phenomenon, a result of the almost ubiquitous and incessant disturbance of the unforested part of the montane areas by burning, grazing and cultivation. This tends on the one hand to the extinction of the less adaptable numbers of the original bird fauna and on the other hand, by varying the habitat, to give opportunities for outside species, in this case from the lowland, to insinuate themselves. It is noteworthy, however, that, whereas scores of montane forest species have extremely limited ranges, this applies to only three other montane species (Hall and Moreau, 1962), namely the warbler *Bradypterus graueri*, found only in a few high-altitude swamps on the eastern Congo mountains, another warbler, *Prinia robertsi*, discovered by Benson as recently as in 1946 in the "bracken-briar" tangles on the mountains forming the eastern border of Rhodesia, and the finch *Poliospiza leucoptera*, confined to mountains in an area of about 13 000 sq. miles in the south-western corner of Cape Province. Since very restricted populations of this kind, presumably relict, are so rare among montane non-forest species, it seems that the process of limitation and extinction has not been very active among such birds in recent historical times even though human enterprise in the degradation of habitats has never been so widespread.

The specialized but poor habitats of the afroalpine areas are shown to have a limited bird fauna of which all but three species are derived from the lower slopes of the mountains. Since most of the afroalpine areas must always have been isolated from each other nearly as markedly as they are now, the absence of endemic species on individual mountains is remarkable. It is true that during an interglacial and during the hypsithermal of around 6000 years ago, the increase in temperature would presumably have allowed the timber-line to rise by at least 200 m (Chapter 3); but, the topography of the mountains high enough to carry moorland being what it is, the areas of moorland can hardly have been reduced on any of them to an extent significant for the survival of species. At the same time, there is reason to suppose that in at least two of the three cases the birds at one stage had more continuous ranges at lower altitudes and that their afroalpine association is secondary.

Finally, Abyssinia has a special status, but a very different one, with respect to montane forest and non-forest birds respectively. In the latter it is faunistically predominant, since forty-seven of all the seventy-four species occur there, twenty-two of them nowhere else in Africa, whereas in montane forest birds Abyssinia is extraordinarily poor, as shown in Chapter 11, for no apparent ecological reason. I am unable to suggest why this should be so. It may be added that the status of Abyssinia in this respect owes practically nothing to the fact that it is the nearest of the montane areas to the Palaearctic; only one of the six Palaearctic species that appear on the African montane areas fails to penetrate further south than Abyssinia. The plateau of South Africa, which on geographical and ecological grounds is a likely source of species for the tropical montane areas, seems to have contributed little to the bird fauna. Especially in view of the unbroken montane connexion to South Africa during the glaciation it is not clear why this should be so.

13

Migrants within the Ethiopian Region

In the North Temperate and Arctic Zones of the world the migration of birds is one of the most striking seasonal phenomena. Broadly, it varies in magnitude in direct relation to the extent to which the climate of the breeding grounds varies in the course of the year. From the highest latitudes, where the winter is most severe, nearly all the birds depart in late summer. Also in the eastern part of the North Temperate Zone, where the winter is long and snowy, the great majority of birds depart, while in western European countries with an oceanic climate many birds of species that elsewhere are fully migrant are to be found all the year round. Movements are predominantly southwards in autumn and the most spectacular are those which involve the crossing of the Saharo-Sindian desert belt and the seas to the north of it, to enable the birds involved to winter in tropical and south Africa (see discussion in Chapter 14). But migration behaviour varies greatly, even within the species, according to the population involved. For example, Grey Wagtails *Motacilla cinerea* from eastern Europe travel into eastern Africa and winter from Egypt to northern Tanganyika, while in southern Britain they merely tend to move a few miles from higher altitudes to areas that are lower and more sheltered. The complex pattern possible within a species can be even better illustrated by the Song Thrush *Turdus philomelos* (Ashmole, 1962):

> "Most Song Thrushes from the northern and eastern parts of Europe migrate for the winter, while the western populations contain a high proportion of permanent residents. Most of the migrants move roughly S.W. in autumn, although many birds from the Low Countries, especially Holland, go west to Great Britain. Many Scandinavian birds on their relatively long migration to Spain "leap-frog" those which breed in southern England. The migrants that winter furthest east in Europe tend to come from the more eastern breeding populations, and those furthest west from the more western populations. But birds from a very wide region of Europe may be found together in S.W. France and Iberia in winter. . . ."

No part of the Ethiopian Region is subject to such rigorous conditions as is much of the Temperate Zone, nowhere is there a seasonal blanket of snow and practically nowhere frost that persists all day; but part of South Africa experiences severe night frosts in winter and a large proportion of the entire Ethiopian Region undergoes a prolonged dry season, which over vast areas is unbroken for as much as six months or even more. This means that in many parts of Africa the supply of food, the "carrying capacity", varies greatly in different months of the year, even where the ecological vicissitudes are not complicated by a high unreliability of the rainfall. Grass and tree-foliage come and go. Swamps are formed and then dry up. Fires rage and make wide stretches of country uninhabitable for some species of birds. Rains fail locally and the birds that would normally have bred there cannot do so. All these things occasion an immense amount of coming and going by many species of birds. Even where the climate is far from extreme, in good savanna with bushes, small trees and seasonally long grass, the bird life is remarkably shifting. Good data on this were obtained in the course of a census carried out by E. G. Rowe and subsequently H. F. I. Elliott on a 16-acre plot of typical savanna (with bushes and small trees) at Monduli near Arusha, northern Tanganyika. In the course of 130 counts made at frequent intervals over a period of 37 months a total of 170 species was recorded, including a few Palaearctic migrants. About 40% of all the species were recorded only five times or less, and twenty of the African species only once (H. F. I. Elliott, unpublished). Again, on a plot of 25 ha (62 acres) in Senegal, where censuses were made at intervals over a period 28 months, sixty-six African species were recorded, but only six of them in every month of the year (Morel and Bourlière, 1962).

In these circumstances it would be surprising if the movements of birds within the Ethiopian Region did not show something of the amplitude and complexity of those in Europe, with the additional complication that, besides setting the stage for regular migrations, African conditions on the whole favour nomadism much more than do those of higher latitudes. Over vast areas of the temperate zones trees lose their leaves and herbage withers all together, under the control of declining temperature, with no oases of better conditions, whereas in any section of Africa during a generally unfavourable period of the year, or during a "bad" year when the rains have for the most part failed, relatively attractive conditions may persist in valleys fed by ground water or may be engendered in patches by localized rain-storms. Such spots produce a crop of insects or of seed that can support an influx of birds for a time (whether for breeding or not) and then make it necessary for them to seek food elsewhere. Mostly non-breeding birds are involved in the nomadic existence thus favoured; but one peculiar species, the Wattled Starling (Locust Bird) *Creatophora carunculata* follows locust swarms around and is liable to breed colonially whenever the abundance of young insects ("hoppers") offers enough food.

Movements certainly take place in Africa that are analogous to the "hard-weather movements" of the North Temperate Zone, but induced by drought, not by cold; and some of them are considerable both for the distances involved and for the numbers of birds travelling. For example, in August 1964, following a drought in Bechuanaland, southern areas of Zambia were flooded with finch larks *Eremopterix verticalis* of two sub-species, together with a pipit *Anthus novaeseelandiae bocagei* (observations especially by C. W. Benson and M. P. S. Irwin and myself). None of these birds had been recorded in Zambia before (Benson and White, 1957), and from their known breeding ranges, individual birds had certainly travelled at the least 250 miles and most of them probably much more.

The extent to which bird movements, that could be classed as migrations in the conventional sense, take place within the tropics has been slow to obtain recognition, though Chapin made a pioneer study for Africa (with particular reference to the Congo) in 1932, which still stands alone; and in the latest general work on bird migration Dorst (1962a) has been able to devote to intra-African migration only about half-a-dozen pages, nearly all excerpts from Chapin's study. This comparative lack of recognition is, I think, partly because on average the movements, being between about 20° N. and 35° S., can never be on so spectacular a geographical scale as some of those of Palaearctic birds, and partly because in Africa, especially in the tropics, resident observers, who are best able to detect and define seasonal movements, have been so much more thinly distributed than in the North Temperate Zone. For this reason also, it becomes more difficult in the Ethiopian Region to decide what movements should be ranked as "migration". And it is not surprising that again and again in the African literature one finds some such comment on a species as "extensive movements take place but are imperfectly known". There is also a purely artificial factor, operating in the Ethiopian Region, that tends to hinder a clear and consistent assessment of bird movements. It is that those which cross the lines drawn on the map of Africa by mutual European jealousies in the nineteenth century are much more likely to be recognized and given prominence in the literature on African birds than those movements taking place within the boundaries of a single territory—which may cover nearly 1 million sq. miles. On top of all this, the natural situation is often very complicated, with populations that are morphologically indistinguishable apparently behaving differently. A particularly puzzling example is the parasitic cuckoo *Clamator jacobinus pica*. This appears (and breeds) during part of the year in southern Africa and during another part of the year in the northern tropics, but the position near the Equator is obscure and in the absence of ringing no one can say whether southern birds cross the Equator. In this species there is the peculiar complication that Indian birds leave India after breeding and apparently winter in tropical Africa, mingling indistinguishably with the local birds. (See review of evidence in Friedmann (1964).

For purposes of the present discussion some definition of "migration" is necessary. A prime criterion is that "there is a shift in what may be called the centre of gravity of the population" (Thomson, 1936), which covers partial migrations. A second criterion is seasonal regularity year after year. But partial migration among a local population seems to be a good deal commoner in Africa than it is in the Palaearctic Region; and there may be many examples like that documented by Ward (1965) for the queleas round Lake Chad and described below. However, I shall be able to cite tropical African cases of nocturnal migration by species normally diurnal, which is a most significant characteristic of classic migration; and there is not much doubt that in time there will be African records comparable to those Beebe (1947) obtained of tropical species flying by night through a mountain pass in Venezuela. Several of Beebe's birds were moreover exceedingly fat, another feature of the typical migrant. This does not seem to have been remarked in any African birds except for Chapin's comment that Pennant-winged Nightjars *Macrodipteryx longipennis*, flying across the Congo to their breeding-grounds, were "excessively fat"; and this is a feature that future collectors in Africa should look out for.

Provided there is seasonal regularity I do not regard distance traversed as an important criterion of true migration, especially in face of such an example as that already quoted of Song Thrushes regularly coming from Holland to winter in Britain—a mere 150 miles or so. Particularly in northern tropical Africa, I am prepared to accept, as typical, migration movements of as little as two or three hundred miles, which take birds from one vegetation zone to another, but many movements in fact exceed 1000 miles in each direction. It may be added that a few short-distance migrations that seem to be essentially altitudinal have been reported in southern Africa. For example, the small "robin" *Pogonocichla stellata* to some extent moves from the highland forests on the eastern edge of Rhodesia to the warmer river valleys during the cold season, and similarly from highland Natal to the coastal belt (see Oatley (1966) for other examples). Stonechats *Saxicola torquata* also move downhill in Malawi and from the interior plateau of south-eastern Africa to winter. In Malawi certain forest species, the flycatcher *Muscicapa* (=*Alseonax*) *adusta*, the thrush *Turdus gurneyi* and probably others move altitudinally; and in Rhodesia some sunbirds (*Nectarinia*) do so, *bifasciata* to breed below 4000 ft and *famosa* to winter at low levels. One particularly interesting record of movement in the inner tropics, by birds of montane forest, is given by Serle (1950), who recorded "an invasion" of *Nectarinia preussi* (=*Cinnyris*) *reichenowi* during the "rainy season", i.e. prior to breeding, at Kumba (Fig. 25), about 4° 35′ N. and only 1000 ft a.s.l., "well below the altitudinal range of breeding *C. reichenowi*". The birds must have travelled some 50 miles (from either Cameroon Mountain or Manenguba Mountain) and descended some 3000 ft. So far as I know this record is unique of its kind, but other such movements no doubt remain to be detected.

In his study Chapin (1932) cited tentatively altogether about fifty species, but some of the movements concerned seemed to be rather indefinite, depending (for example) on river levels or, so far as the information went, quite local. Some of the evidence was concerned only with that strip in the north-east of the Congo where Chapin himself worked so long and fruitfully. For purposes of general discussion some of his cases will provisionally be excluded, but on the other hand much more evidence for other parts of Africa has come to hand since he wrote and nearly 100 species qualify as migrants; so that it will now be possible to go rather further than he could in discussing migration within the Ethiopian Region in terms of the areas affected and the species and groups of birds most obviously concerned.

There is no doubt that, apart from the exact results to be expected from ringing, a vast amount remains to be learnt about African migration generally, and before proceeding to the general review I want to give some specific examples that are especially illuminating as showing how much probably remains to be learned. At the risk of giving them too much space and prominence I shall deal with them rather fully, partly because they strike me as being of more than intrinsic importance and partly because I was personally involved and found the stories exciting as they unfolded. We have here cases of thoroughly typical nocturnal migration, apparently long-distance, on the part of species, pittas and quails, for which there was little or no evidence before. I may recall that for a number of years I lived at Amani, at 3000 ft on the seaward face of the East Usambara Mountains, about 100 miles south-west of Mombasa, in a clearing in a luxuriant evergreen forest, where low cloud and often ground-level mist were common. To this extent the situation resembled Beebe's station in Venezuela, already referred to, but it did not occupy a pass in an east-and-west range. On the other hand, it had an unobstructed view to the Indian Ocean 50 miles away and the lighted windows of the research station buildings could be seen for some 20 miles on a clear night.

The pitta story started on the morning of 10 May 1932 when my wife picked up a freshly dead specimen of *Pitta angolensis longipennis* outside our quarters at Amani. It was in worn plumage and its stomach was empty. Neither of the African species of pitta had been regarded as performing migrations though some Asiatic pittas were known to do so. Also no pitta had been recorded within 400 miles of Amani. To the north there were four records in the Nairobi area, which I subsequently ascertained had all been in May and June; at least two of them had entered European buildings and one was described as "starving". The bird's most northerly localities in Africa were in the forests of Uganda, where its status is uncertain to this day. However, Jackson's only personal record there was in May; and I have since found that those Uganda specimens which are dated were taken in May, June and July, a period into which unpublished records of C. R. S. Pitman (*in. litt.*) also fall.

To the south of Amani the nearest record was in extreme southern Tanganyika, where it has since been shown to be a breeding visitor, as in Malawi and Rhodesia. The favourite breeding habitat appears to be dense semi-deciduous thicket at low altitudes—not a widespread habitat, so that the total population of the bird cannot be large.*

On the evening of 26 May 1932, ten days after our first discovery at Amani, a second pitta came fluttering against the lighted windows of our sitting-room about two hours after sunset; and it was another of the thrills of my ornithological life to go outside and put my hands on this gorgeously coloured bird, unhurt as it was. At about the same moment another was caught in the same circumstances half a mile away and there is evidence that two more were around. In subsequent years we had other records at Amani, all of birds at houses and two definitely attracted to light, on 1 May 1935, 15 May 1936 and 7 June 1937. Of these the one picked up dead had an empty stomach. Other records in this north-eastern corner of Tanganyika, which I accept, are of two pittas reported by my most experienced African collector in the Amani forest on a date between 11 and 20 May 1933 (when I was on leave), of one that entered the government school at the neighbouring port of Tanga (date uncertain), and of one shot in the West Usambara Mountains in June 1938.

It will be seen that these records between them point to well-defined, regular and typical nocturnal migration through East Africa, the earliest date recorded being 1 May and the latest 7 June. It is probable that its direction is northward rather than southward, but its goal is uncertain. We never had any indication that pittas "wintered" in the Usambara forests and it seems unlikely that those we saw (or the Nairobi birds either) were making for the Uganda forests which lie far to the north-west. Perhaps the Amani birds were going to the forests of the Kenya coast, but, as J. G. Williams tells me, to this day no pitta has been reported in Kenya except when the odd example turns up in the middle of Nairobi or some such unsuitable place "in April or May". It will be noted that there is so far no sign of a return migration in this part of Tanganyika. But one of these birds was found on 1 December 1931 about 300 miles south-west of Amani in the semi-evergreen Itigi Thicket. (This raises the question that recurs again and again in the present stage of African ornithology: was it on migration or part of a locally breeding population?) There are also records of these pittas evidently on migration far to the west, in the Congo near Lake Kivu, again in April and May. The movement is also discernible much farther south, because one came to lighted windows at Abercorn, which looks down on the southern end of Lake Tanganyika, on 26 December 1956 and five more in December 1959 (Benson, 1957, 1962). It is difficult to

* It is likely that this secretive bird breeds farther south also. There are now odd records all the way to the Cape Province, especially four in Rhodesia November–December and one in April, some of the birds flying against lighted windows. The map of pitta distribution in McLachlan and Liversidge (1957) omits the specimen collected at Tambarara in Portuguese East Africa (Sclater, 1911, p. 405), which, being in worn condition on 18 March, and in the "right" habitat, might well have nested there.

know where the ultimate source of all these birds can be. In any case, if the birds that breed in Rhodesia and Malawi visit Kenya and Uganda for the off-season, they would travel 800–1200 miles.*

It is of interest that the subspecies to which these East African migrants belong, *Pitta angolensis longipennis*, has indeed somewhat longer and also more pointed wings than others of the species, and the second character, if not the first, accords with greater migrational activity. However, the Angolan (nominate) birds are themselves evidently not sedentary, for certain of the specimens known from further north, along the lower Congo, were taken in circumstances suggesting that they were on passage (dates May, June and December).

Up to now long-distance migration has hardly ever been claimed for any East African bird whose habitat is typically evergreen; but the comment of McLachlan and Liversidge that the trogon *Apaloderma narina* "frequently flies into houses at night" is suggestive, and typical nocturnal migration has very recently become evident in another quite unsuspected bird of the lowland evergreen forest, the thrush *Turdus fischeri natalicus*. This had been regarded as a very rare bird, with range from Zululand to Pondoland, and necessarily of very patchy distribution nowadays, owing to destruction of its coastal forest habitat. Yet in 1955 and 1957 six individuals turned up at houses at night, earliest and and latest dates 18 April and 13 May, mostly in the Durban area, on the site of which city this thrush is known formerly to have nested (in November–December). I doubt if we know the full range of the bird today. If we do, it means that typical nocturnal migration is developed for a journey of not more than 400 miles. This example is interesting also because no other African thrush has been suggested as even a partial migrant, except that *Turdus libonyanus* is mainly a visitor to the Gambia during the rains, according to E. M. Cawkell.

The quail story at Amani is noteworthy because although one of the species concerned, the Harlequin *Coturnix delegorguei*, is very well known and is widely regarded as migratory, little had been recorded about any movements in the inner tropics, and the statements about them at the higher latitudes are inclined to be vague, with emphasis on variability. At Amani Harlequins were recorded on 19 May 1934, 19 and 26 November 1935, 9, 10, 15, 17 and 27 December 1936, 5, 8, and 15 January 1941, 15 December 1943, 25 December 1945 and 4 January 1946. Some of the birds were seen to come to lighted windows, others were picked up dazed, exhausted or dead in circumstances that suggested they had hit something. One particularly brightly lit whitewashed wall accounted for several. The quails on each of these dates numbered from one to four—small numbers that would have gone unremarked in a savanna area but of the highest significance as nocturnal migrants in the utterly untypical habitat of montane forest.

* Since this was written an up-to-date documentation of pitta records, embodying unpublished data, has been provided by Benson and Irwin (1964).

In consonance with these occurrences are those on Zanzibar and Pemba Islands—the latter 80 miles from Amani and visible on a clear evening—where the records have been on 14 November 1931, 6–16 December 1936 (a considerable number; gonads inactive) and about the end of April 1943.

Thus there is good evidence for a regular migration of Harlequin Quails through this part of the East African coastal belt around December (earliest date 14 November, last 10 January) with perhaps some return in May. This latter is presumably connected with movements detected inland by H. F. I. Elliott. He tells me that he had Harlequins coming to lighted windows in Nairobi in mid-June one year when he happened to be there; but we are still not certain of the direction of the migration at either season. The same observer tells me that in another savanna locality in Tanganyika about 250 miles north-west of Amani he used to find Harlequins breeding from February to May and then disappearing until December–January. It is not altogether in accordance with this that in Malawi in some years Benson found large numbers of Harlequins passing southwards in October and December, northwards in March and April. North of the Equator the most definite record of movement seems to be Lynes's in Darfur, where Harlequins turned up in July, apparently on their way to breed during the rains on the southern edge of the Sahara. It is doubtful how far the Amani migration fits with any of these other movements, but it may perhaps involve the same population as the Malawi movement and is on the whole in step with Elliott's birds.

An undated incidental observation by Jackson and Sclater (1938, Vol. 1, p. 269) apparently at Entebbe, assimilates the migration of the Harlequin still more to classic Palaearctic examples, for Jackson describes what was evidently typical migratory restlessness that took place on the part of a dozen of these quails on two successive nights after they had been caged for about a month. When released on a subsequent evening at dusk they straightway took wing south over some trees, passing across the observer's front and not away from him. Most unfortunately, Jackson gives no date.

Another quail, the Blue *Coturnix chinensis adansoni* (=*Excalfactoria adansoni*), seems nowhere to have been regarded as a regular migrant. In Malawi indeed it has explicitly been recorded in every month of the year. But, in circumstances like those of the Harlequin, Blue Quail turned up at Amani on 19 November 1935, 10 and 11 December 1936 and on Zanzibar Island on 1 and 9 November 1933, 9 February 1934 and some date in May 1927. It looks therefore as if on a smaller scale this less common species has similar migrations to the Harlequin in the neighbourhood of the East African coast.

Abercorn in Rhodesia has an even better record than Amani for detecting the nocturnal migration of African species, as is described by Benson (1957). One particular house there is situated rather like mine at Amani and high on a ridge, but instead of looking eastwards to the sea it looks down to the south end of Lake Tanganyika. In addition to a Blue Quail on 8 June 1956 (which is near the May date at Amani) and a pitta on 26 December 1956, its lighted windows have attracted specimens of three species of rail, *Crex egregia* and *Sarothrura elegans* on 26 December 1956, *Porzana marginalis* on 31 March 1954. Like all rails, these are rarely seen and only the *Crex* had ever been suspected of migration. Further evidence for the migratory habits of the *Crex* and the *Porzana* has recently been brought together by Benson (1964).

Such occurrences make one wonder how many more species will be added to the list of truly migrant African birds. Meanwhile another type of evidence for partial migration on a large scale is being produced by the ringing activities of the South African Ornithological Society, as reported in successive volumes of their journal *Ostrich*. They already have the following recoveries among others: a night heron *N. nycticorax*, ringed near Cape Town, in Mozambique; Cape Town Buff-backs *Bubulcus ibis* hundreds of miles to the north, as far indeed as Zambia, and Buff-backs from the Rand in Zambia and Kivu; *Anas* (=*Aythya*) *erythrorhyncha* from Cape Town to Zambia and *Anas* (=*Netta*) *erythrophthalma* from the Rand at Lake Naivasha (1800 miles, a year later). With these duck movements may be cited the fact that the Hottentot Teal *Anas punctata* has been recorded several times in Nigeria, though the nearest other occurrences are on the eastern edge of the Congo and no nesting is known nearer than Kenya.

With *Quelea quelea*, the grain-pest to which so much attention has been given of recent years by governments all over the Ethiopian Region, recoveries of ringed birds have suggested that in southern Africa there is, in addition to a great deal of "restlessness" among these birds, a general southerly drift from Malawi and Zambia into the Transvaal and the Orange Free State about October. After breeding, the birds return northwards about the end of March, the distances involved being about 400 miles. The special point about each of these cases is that some members of the species concerned are present in the parts of southern Africa mentioned throughout the year and no populations of the species are recognized as typical long-distance migrants. In Nigeria near Lake Chad, Ward (1965, and *in litt.*) has detected a regular movement of special interest which may well have a counterpart in other seed-eating species. During the dry season the queleas live on the seeds dropped on the ground, but as soon as the rains come (advancing from the sea) such seed as has not been consumed germinates and becomes inedible. The birds have no alternative but to move, and what many of them do is fly south into areas where the rain has already been falling long enough for the new season's crop of seed to form and to have already some nutritive value. Ward estimates that "the centre of their distribution is thus displaced some 300 miles to the SSW" and that they are absent from their home area for only about a month before they are back again to breed in colonies numbering millions. He emphasizes that the timing of the southward migration each year and the distance the birds fly are the result of a delicately adjusted compromise. The short period of absence is noteworthy and many similar movements that are brief but quite regular may well be passing undetected in Africa.

A few illustrations may be given of the difficulty of interpreting, and generalizing about, the movements of tropical birds in the present state of our information. First, the bee-eater *Merops malimbicus* has been found by Elgood (1959) to be a regular immigrant at Ibadan, in southern Nigeria April–September, with, he tells me, the main arrival about June. The breeding status there is still uncertain. Meanwhile, October–May, approximately the other half of the year, in Togo, 400 miles to the west and in much the same latitude and climate, Douaud (1957) found this bee-eater to be a non-breeding visitor. From the relative geographical positions it is difficult to believe that the same individuals are involved, but if not, we have two populations in about the same latitude with very different breeding ecology. Another problem arises over the grass warblers *Cisticola*. Six species have been recorded as breeding in the Gambia, but when I was there in the middle of the dry season I saw only one—*juncidis*. The herbage was at that time nearly all dry and yellow, would become even more desiccated as the season progressed, and would mostly be burnt. The birds presumably migrate, but similar conditions exist for a long distance round the Gambia, so that it is difficult to know where the birds would go to find satisfactory conditions for survival. The same considerations apply in some measure over vast areas of Africa, and the problem of how the grass warblers adapt themselves does not seem to have been discussed. It seems to be generally assumed that in all territories the birds make for the damper spots locally; but if this were the complete answer the over-crowding there would surely be extreme.

With all the foregoing examples in mind it will be clear that a general discussion of migration among Ethiopian birds at the present stage will be imperfectly based. Nevertheless certain categories can be distinguished and generalizations made which are not likely to be upset by further knowledge. To acquire the data I have not attempted a general search of original sources but have relied on statements made in the standard territorial works, with the addition of the cases cited earlier in this chapter. Different parts of Africa of course vary much in the extent to which they have been covered by observers. In South Africa McLachlan and Liversidge have been able to distil the results of generations of resident ornithologists. These are lacking in the northern tropics. There the best data come from Chapin's experience in the north-east corner of the Congo and the critical and meticulous work of Lynes over two seasons in Darfur, the north-west corner of the Sudan. It is fortunate that these two observers worked in nearly the same longitudes, some 800 miles apart, so that often the specific observations of one add to the significance of those made by the other. Similarly, some isolated observations in West Africa acquire more significance when related to Lynes's records. Moreover Douaud (1957) and Elgood (1959) have recently devoted special attention to African migrants in Togoland and round Ibadan in southern Nigeria respectively. Certain of the species they mention as strictly seasonal in their areas, close to the forest belt of Upper Guinea, have not yet been described as migrants in the drier country to the north, though undoubtedly to some extent they must be so. In East Africa generally, notwithstanding the number of people interested in birds, there has been great lack of critical documentation. Data from this area are needed all the more because the natural situation may be complicated by the mingling of off-season birds from higher latitudes with populations of the same species sedentary near the Equator.

As prelude to the discussion that follows, it may be recalled that the ecology and geography of the northern tropics are much simpler than those of the southern (see Figs. 2 and 3). Northwards, apart from some distortion caused by the Abyssinian highlands, latitudinal belts of decreasing total rainfall and increasing length of the dry season run right across from the Atlantic to the Red Sea. The annual flush of vegetation, which comes in summer, gets more and more brief as one goes northward until, about 1000 miles from the Equator, the Sahara supervenes.

On the southern edge of the Sahara the seasonal changes are particularly striking. Gillet (1960) has given a dramatic account of what happens in Ennedi, some 250 miles north-west of Lynes's area and a last outpost of the regular monsoon. He describes how, as the rains travel north, there is a northward movement of tropical birds, as precise and regular as those of the Palaearctic migrants and with the appearance of a true migration. As shallow pools form in what was almost desert they are the scene of "a veritable explosion of life. The water pullulates with innumerable insect larvae, crustaceans and batrachians, all very easy to catch." And he goes on to list the water-birds and others that come to these places. Most of the bird species concerned are mentioned by Lynes or some other observer nearer the Equator as appearing and disappearing in step with the Ennedi observations.

By contrast with the northern tropics, southern Africa extends for some 2000 miles from the Equator and its ecological zonation is altogether more irregular. The picture is greatly complicated by the presence of mountains and high plateaus with cold winter nights and, furthermore, aridity does not increase with latitude right across the continent as it does in the northern tropics; although the south-western part of South Africa is very dry, with highly irregular rains, in the same latitude on the east coast evergreen forest persists. However, one point of resemblance between the two sides of the Equator is that the main rains nearly everywhere fall in the local summer. In South Africa no such explosive seasonal change as that described above for Ennedi has been recorded as a regular occurrence anywhere, but in the arid south-western quarter something like it can happen occasionally after a long drought (as in the Karoo following the unusual rains of 1961, according to Mrs M. K. Rowan); and then the appearance of the birds coming in to take advantage of the good conditions would go down as forming merely a "local movement", which is correct if there is no regularity about it.

From these considerations it might be expected that within the northern tropics the migrations would tend to be more definite, more spectacular and less partial than in the southern tropics, though gener-ally over shorter distances; and on the whole this seems to be what happens. Probably nearly all the African migrants more or less follow the longitudes, travelling backwards and forwards between areas of lower annual rainfall with cooler winter nights and those of higher rainfall with warmer winter nights. Cogent (and complementary) ex-amples of this are provided by the two carmine bee-eaters (Fig. 40). Minor exceptions, in that they are east-and-west movements rather than north-and-south, are those, already mentioned, by some birds travelling to and fro between the interior plateau of South Africa to the lowlands along the Indian Ocean. This would apply also, on a larger scale, to the *Muscicapa adusta* (flycatchers) of the western Cape, which winter near the eastern seaboard (Lawson, 1963). Again, Chapin concluded tentatively that the peculiar swallow *Pseudochelidon eury-stomina* migrates some 500 miles west from breeding grounds in the Congo basin to the coast of Gabon. Other cases of transverse migration, at any rate in part, but on a vast geographical scale, are provided by two storks. *Ciconia* (=*Sphenorhynchus*) *abdimii* breeds during the rainy season in the semi-arid belt south of the Sahara, west all the way to Senegal, but seems to spend the off-season only south of the Equator (Fig. 50). Hence the storks breeding in West Africa (as far as Senegal) presumably gave to make a long journey eastwards (and back) to get round the Gulf of Guinea. The Openbilled Stork *Anastomus lamelligerus* may do the same thing in reverse; its nesting has been established only south of the Equator (in the local dry season) but in the off-season it gets as far west as the interior of Ghana and the Niger above Timbuktu. Both these storks seem to have been noticed very little indeed in the course of the migrations they must make over West Africa west of the Cameroons. A peculiar case is that of the Grasshopper Buzzard *Butastur rufipennis*. It breeds on the summer rains from Senegal to Ethiopia and thereafter migrates into the southern parts of West Africa and in eastern Africa south as far as Tanganyika. In autumn, though not in spring, many of the migrants divagate eastwards all the way to British Somaliland, where Grasshopper Buzzards are "entirely fleeting visitors" (Archer and Godman, 1937-61).

Although it leads to an over-simplification, it is useful to set out, as in Table XXII, the data that can be extracted from published works, especially Lynes's and the standard territorial bird faunas, on species of which at least some populations can be regarded as performing definite migratory movements. Probably no one will agree with every species I have chosen to include as "migrant" or will fail to wonder why some species have been omitted. For example, J. M. Winterbottom tells me that in the western Cape he regards several species as migrants which are not mentioned as such in the standard South African work of McLachlan and Liversidge. Certainly much turns on the factors of observation and interpretation. There may be several species like the Knob-nosed Goose *Sarkidiornis*, which has a definite seasonal migration to and from the edge of the Sahara, but in Zambia and Rhodesia is recorded as having "local movements" and in South Africa movements that are "not understood". However, I think the arrangement in Table XXII is justified, if only to bring into prominence those families in which the migratory habit has been most remarked. On present information no statistical comparisons can reliably be made, but certainly most of the species that are recognized migrants are non-passerines and considerably more non-passerines have been regarded as migrants north of the Equator than south of it. Among them raptors, storks, bee-eaters and the duck family are especially prominent as migrants north of the Equator, and cuckoos everywhere. Probably with increasing knowledge swifts will also take an important place.

It will be seen from Table XXII that in the northern tropics a great variety of birds migrate towards the Sahara to breed, including ducks, herons, the Sacred Ibis, insectivorous kingfishers, nightjars, bee-eaters, cuckoos, swallows, sunbirds, a paradise flycatcher and a warbler (*Acrocephalus*). While these birds all hasten to take advantage of the brief summer abundance to rear their young and thereafter retire south towards the equatorial rainbelt, other species, especially some raptorial birds, storks, another insectivorous kingfisher, a hornbill, a big bustard, another sunbird, a chat *Oenanthe heuglini*, two swallows, and a nightjar, perform the same journeys at the same times of year, but to spend their off-season (and moult) in the sub-Saharan spring after having bred in the dry season nearer the Equator. Of the nightjar in this category, *Macrodipteryx longipennis*, Lynes relates how the east "standards", the greatly modified ninth primary feathers, litter the ground in places. Thus in these latitudes we have a phenomenon unknown in the temperate zones, some species arriving to breed and others, including species of the same families, to moult, alongside one another and simultaneously. We have as yet no certain evidence about how far individuals among these birds move on each journey. Three or four hundred miles would seem to be a frequent distance, with some travelling less and others nearly the whole distance of 1200 miles or so that intervenes between the Equator and the perennially uninhabitable Sahara.

In the southern half of Africa the African migrants present rather a

TABLE XXII

Ethiopian species having typical migrations in some parts of Africa

(1) Each species is entered under the area in which the population concerned breeds. Trinomials are given where the form concerned has often been regarded as a distinct species.
(2) Trans-equatorial migrants are shown in capitals, others in italics.
(3) Those exceptional populations which travel to higher latitudes (on their own side of the Equator) to moult, and not to breed, are marked †. Species of uncertain status in this or other respects are given in parentheses.

| Family | Species breeding in | |
	Northern tropics	Southern tropics
	A. Non-passerines	
Accipitridae	† *Aquila wahlbergi*	
	† *Accipiter badius*	
	Butaster rufipennis	
	† *Buteo auguralis*	
	Melierax metabates	
	† *Milvus migrans*	*Milvus migrans parasitus*
Alcedinidae	(*Ceyx* (= *Ispidina*) *picta*)	*Ceyx picta*
	† *Halcyon leucocephala*	*Halcyon leucocephala*
	Halcyon senegalensis	*Halcyon senegalensis*
		Halcyon senegaloides
Anatidae	(*Anas capensis*)	
	Alopochen aegyptiaca	
	Dendrocygna viduata	
	Sarkidiornis melanota	
Apodidae		*Apus caffer*
		Apus horus
Ardeidae	*Bubulcus ibis*	
	Ardea melanocephala	
Bucerotidae	† *Tockus nasutus*	
Caprimulgidae	*Caprimulgus climacurus*	
	Caprimulgus inornatus	CAPRIMULGUS
		R. RUFIGENA
		COSMETORNIS
	† *Macrodipteryx longipennis*	VEXILLARIUS
Charadriidae	(*Afroxyechus forbesi*)	
	(*Xiphidopterus albiceps*)	
Ciconiidae		ANASTOMUS
		LAMELLIGERUS
	† *Dissoura episcopus*	† *Ibis ibis*
	† *Ibis ibis*	
	† *Leptoptilos crumeniferus*	
	CICONIA ABDIMII	

(continued overleaf)

TABLE XXII (*cont.*)

| Family | Species breeding in | |
	Northern tropics	Southern tropics
Coraciidae	*Coracias abyssinicus*	
	Coracias caudatus	
	(*Coracias naevius*)	(*Coracias naevius*)
	Eurystomus glaucurus (*afer*)	*Eurystomus glaucurus*
Cuculidae	*Chrysococcyx caprius*	*Chrysococcyx caprius*
	(*Chrysococcyx cupreus*)	(*Chrysococcyx cupreus*)
	Chrysococcyx klaas	*Chrysococcyx klaas*
	†*Clamator glandarius*	*Clamator glandarius*
	Clamator levaillantii	*Clamator levaillantii*
	†*Clamator jacobinus*	*Clamator jacobinus*
	Cuculus canorus	*Cuculus canorus*
		Cuculus clamosus
		Cuculus solitarius
Glareolidae		(*Rhinoptilus chalcopterus*)
		(*Cursorius temminckii*)
Indicatoridae	†*Indicator indicator*	
Meropidae	*Aerops albicollis*	
	(*Melittophagus pusillus*)	
	(*Merops malimbicus*)	
	Merops nubicus	*Merops nubicus* (*nubicoides*)
Otididae	(*Choriotis arabs*)	
	†*Neotis denhami*	
Phasianidae		*Coturnix chinensis adansoni*
	Coturnix delegorguei	*Coturnix delegorguei*
Rallidae		*Crex egregia*
		(*Porzana marginalis*)
		(*Sarothrura elegans*)
Threskionithidae	*Threskiornis aethiopicus*	
Upupidae	*Upupa epops* (*senegalensis*)	

B. Passerines

Family	Northern tropics	Southern tropics
Alaudidae		†(*Calandrella cinerea*)
		†*Mirafra nigricans*
Hirundinidae	(*Hirundo abyssinica*)	*Hirundo abyssinica*
		Hirundo albigularis
	(*Hirundo aethiopica*)	*Hirundo atrocaerulea*
	†(*Hirundo daurica*)	*Hirundo cucullata*
		Hirundo dimidiata
	(*Hirundo semirufa*)	*Hirundo semirufa*
		(*Hirundo smithii*)
		Petrochelidon spilodera
	Hirundo senegalensis	(*Psalidoprocne albiceps*)
		Hirundo cincta
	(*Hirundo cincta*)	(*Hirundo paludicola*)
	(*Hirundo paludicola*)	

TABLE XXII (*cont.*)

| Family | Species breeding in | |
	Northern tropics	Southern tropics
Laniidae	*Chlorophoneus sulphureopectus* *Nilaus afer* *Corvinella corvina*	
Motacillidae	†*Anthus leucophrys* †*Anthus novaeseelandiae*	
Muscicapidae	*Terpsiphore viridis*	*Terpsiphone viridis* *Muscicapa adusta*
Nectariniidae	*Nectarinia osea* †*Nectarinia metallica*	
Oriolidae Pittidae Ploceidae	(*Oriolus auratus*)	(*Oriolus auratus*) *Pitta angolensis* *Quelea quelea*
Sturnidae	(*Cinnyricinclus leucogaster*) *Lamprocolius splendidus* *Spreo shelleyi*	CINNYRICINCLUS LEUCOGASTER
Sylviidae	*Acrocephalus baeticatus*	*Acrocephalus baeticatus*
Turdidae	†*Oenanthe heuglini*	†*Oenanthe pileata* *Turdus fischeri*

different picture. Swallows and cuckoos, mostly of the same species as north of the Equator, bulk much larger in the list than any other families. Nearly all of them breed in the more southern areas in relation with the summer rains and then move north. With these may be bracketed, as perhaps the best-established migrants of this type, the same species of kingfishers, of kite (the only raptor and a dry-season breeder), of bee-eater (*Merops nubicus*—see Fig. 40)* and of reed-warbler and of starling as in the northern tropics, together with two swifts, a flycatcher, and a pitta. There are also of course the more elusive species, especially the rails and quails dealt with earlier in this chapter, which probably breed in the southern part of their ranges. A few birds seem to be migrants only in the southernmost parts of South Africa. The African subspecies of the European Quail, *Coturnix c. africana*, is a case in point.

Among the birds breeding south of the Equator there are two interesting exceptions to the foregoing rule. Instead of breeding during the rains and then moving north, the chats of the species *Oenanthe pileata*

---

* The European Bee-eaters *Merops apiaster*, which nest sporadically in South Africa and thereafter disappear (as discussed in Chapter 6), are not included here under intra-African migrants, since their destination is not known.

that breed in a belt across the south-eastern Congo, Zambia, Rhodesia and Malawi do so in the local cool dry season, between April and November, thereafter departing, presumably southwards. This is a parallel to the post-breeding movement of *Oenanthe heuglini* north of the Equator, referred to above; but there is this difference between the species that some populations of *O. pileata*, those breeding in South Africa, seem to be non-migratory. The second southern species that seems to have the same abnormal rhythm as *O. pileata* in Rhodesia is the lark *Calandrella* (= *Tephrocorys*) *cinerea*.

There remain to be considered those species which breed in one half of the continent and at least in part cross the Equator to winter in the other. Such birds so arrange their lives that they never experience the severities of a complete dry season. The most outstanding of these trans-equatorial migrants and the only one of them whose breeding is confined to north of the Equator, is Abdim's Stork *Ciconia abdimii* (Fig. 50). Because of its size, its numbers and its habits, its comings and goings are more remarked than those of any other African migrant. Though some of these storks breed nearer to the Equator, most of them appear with the earliest of the (summer) rains in the semi-arid belt on the southern edge of the Sahara, and there they nest, especially on the huts in the villages, right across from Senegal to the Indian Ocean. After the end of the rains they nearly all flock southward, down through East Africa and over the Congo. Their wintering grounds extend south beyond the Orange River, over 2000 miles from their southern breeding limit (Fig. 50). Two other species, which breed in the north in the rains, to some extent afterwards cross the Equator. The nightjar *Caprimulgus inornatus* in the west does not seem to cross the Equator or travel further than the northern edge of the Congo forest at about 4° N., but in the east it appears in the northern winter all through Tanganyika to about 9° S. By contrast, the Cattle Egrets *Bubulcus ibis* breeding between about 14° and 18° N. in the Sudan are believed by Chapin to cross the Congo forests to winter in the southern part of the basin, while in West Africa the northern breeding birds merely withdraw towards the coast. Again, the Grasshopper Buzzard *Butastur rufipennis*, typically breeding all across the northern semi-arid belt, thereafter penetrates East Africa as far as central Tanganyika, at about 10° S., though others of the species can be seen hanging about at the same season as far north as the Gambia, around 12° N. Much the same applies to the bee-eater *Aerops albicollis*, some of which winter as far south as the central Congo and central Tanganyika.

Three species breeding in southern Africa regularly cross the Equator to spend their off-season. The most remarkable is the nightjar *Caprimulgus r. rufigena* (Fig. 51), which breeds north to southern Angola and Rhodesia during the rains of the southern summer. In the opposite half of the year it is known only in the belt of semi-arid country from northern Nigeria to Darfur, which is then at its ecological best. The migration of the species must be remarkably rapid as well as complete, because it is hardly seen in the huge area between its breeding and its wintering grounds. From the Congo only three specimens are recorded and from the whole of East Africa to this day not one. Another nightjar, the Standard-wing *Cosmetornis vexillarius*, very conspicuous when in breeding

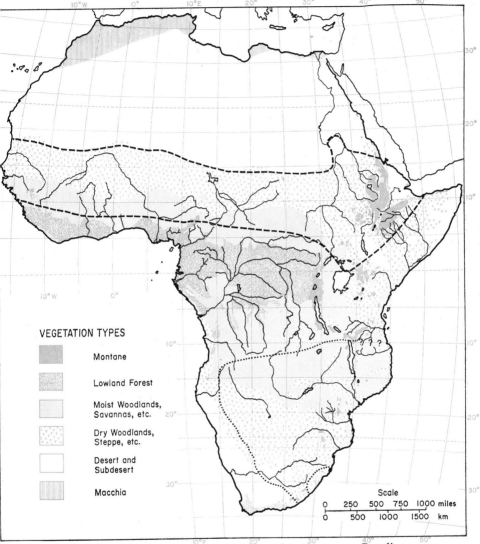

VEGETATION TYPES

- Montane
- Lowland Forest
- Moist Woodlands, Savannas, etc.
- Dry Woodlands, Steppe, etc.
- Desert and Subdesert
- Macchia

Scale

| 0 | 250 | 500 | 750 | 1000 miles |
| 0 | 500 | 1000 | 1500 | km |

FIG. 50. The migration of the stork *Ciconia abdimii.* – – – –, Breeding range; . . . ., off-season range.

plumage, breeds in the southern spring, just before the rains, as far north as the southern Congo and most of Tanganyika. Thereafter the Standard-wing appears north of the Equator, ranging as far as Darfur, and moults there. Some of these nightjars have been reported in East

Africa, close to the Equator, at all seasons and it not known whether these are non-migratory individuals or whether populations from further north and from further south temporarily replace each other with the seasons. Rather the same doubt applies to a third species, the Amethyst Starling *Cinnyricinclus leucogaster*; some have been found in Kenya and

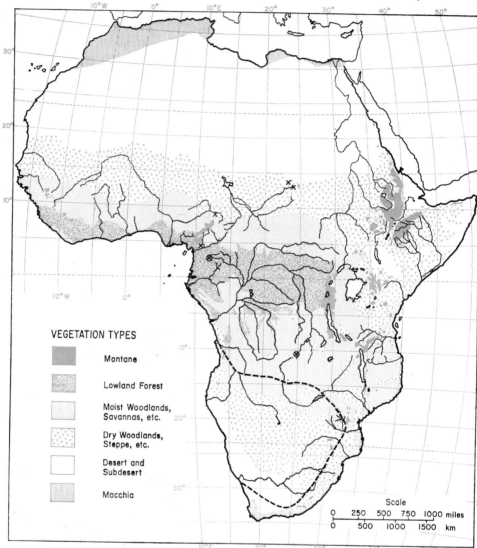

VEGETATION TYPES

Montane

Lowland Forest

Moist Woodlands, Savannas, etc.

Dry Woodlands, Steppe, etc.

Desert and Subdesert

Macchia

Scale

0   250   500   750   1000 miles
0   500   1000   1500 km

FIG. 51. The migration of the nightjar *Caprimulgus rufigena*. ― ― ― ―, Breeding range (September–April); ×, off-season records; ⊗, possible passage migrants.

Tanganyika at all times of the year but the entire population breeding further south in Africa migrates northward after the rains and some birds of this species then reach Abyssinia and the southern edge of the Sudan from breeding areas that only ringing recoveries in the future can define. Amethyst Starlings breeding north of the Equator, a distinguishable subspecies, do not seem to have any comparable migration into the other half of the continent but they move to some extent, so far uncertain, because they disappear from the Gambia after the rains. The swallow *Hirundo atrocaerulea* leaves Malawi and Rhodesia after breeding and then is found in Uganda, but hardly crosses the Equator.

THE MADAGASCAR MIGRANTS

Before proceeding to a discussion of the intra-tropical migrants from a systematic point of view it is interesting to bring into the picture those species which have been reported as migrants between Madagascar and Africa. It may be recalled that this great island has a wide range of habitats, from semi-desert to the richest evergreen forest (see Chapter 17), with a definite rainy season, November–March (as on the opposite mainland), except along the wet eastern rim. It will be seen that the Madagascar species concerned belong to families prominent among the intra-African migrants. Also they fit into the commonest African pattern, namely, of migration nearer to the Equator to spend the off-season. The Madagascar birds, however, of necessity move predominantly north-west rather than north.

Ardeola idae

This Little Bittern, said to breed in Madagascar in the rains (around November), has been collected in Africa, always without sign of breeding activity and nearly always in immature plumage, in Kenya, Tanganyika, Zanzibar and Pemba Islands, Rhodesia and all down the eastern edge of the Congo in the months May–October inclusive. A specimen from Rhodesia was noted as "very fat" on 8 October, a condition presumably preparatory to migration.

Cuculus poliocephalus rochii (Fig. 52)

This cuckoo is present in Madagascar (where it breeds) only from the end of September to April, except that in the wettest areas it may not leave (Milon, 1959). On the mainland it has been recorded in Kenya, Uganda, Tanganyika and the south-eastern Congo as a non-breeding visitor in the months June–September (one recorded by Benson and Benson (1949) from near the head of Lake Nyasa, on 13 September was "very fat"). In fact, on the mainland it alternates with the nominate *C.p. poliocephalus*, distinguished only by smaller size, which breeds from northern India eastwards in the northern spring and subsequently migrates to Africa for the winter.

Eurystomus g. glaucurus

This broad-billed roller, conspecific with others breeding in Africa, breeds in Madagascar from at least October to November and is found there about October to March—Rand's extreme dates are 27 September and 1 April—but whether it leaves entirely is not certain. It has been collected February–November, i.e. mainly in the off-season, in many localities in eastern Africa from Pemba Island and Tanganyika to Rhodesia and Portuguese East Africa, and also over the greater part of

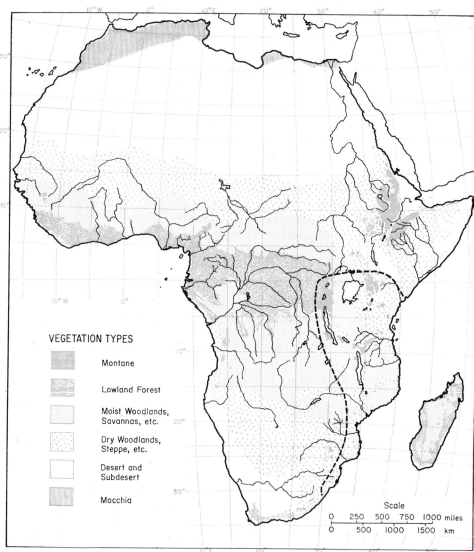

VEGETATION TYPES

Montane

Lowland Forest

Moist Woodlands, Savannas, etc.

Dry Woodlands, Steppe, etc.

Desert and Subdesert

Macchia

Scale

| 0 | 250 | 500 | 750 | 1000 miles |
| 0 | 500 | 1000 | 1500 | km |

FIG. 52. The migration of the Madagascar Little Cuckoo *Cuculus poliocephalus rochii.*

the Congo. The last area is likely to be its main "winter" quarters since in Malawi and Zambia the records are all in October–November and February–April, which suggests that the birds are only on passage through those territories.

Falco concolor

Dorst (1962a) includes this as a species that "also migrates to Africa during the dry season". There is, however, no evidence that this bird breeds in Madagascar; Rand encountered it there in flocks; and there is little doubt that the birds seen in the islands are off-season visitors from their breeding grounds in the latitudes of the Red Sea (see Chapter 4).

Galachrysia nuchalis Madagascar Pratincole

Rand does not give any indication that this Madagascar pratincole leaves the island, but it is quite common locally on the coast of Kenya in August and September (Jackson, 1938; J. G. Williams, *in litt.* 1962) and there are also records in the neighbouring territories of Somalia and Tanganyika. In the latter H. F. I. Elliott tells me he saw the species at Dar es Salaam in August and September of one year; and farther south still on the Zambezi and Lake Nyasa Kirk had October records (Reichenow, 1900–03). Nothing is known of the pratincole in the African continent in any other month of the year and it remains doubtful whether all the above records relate to birds on passage, to or from undiscovered "wintering" grounds, or whether the species is absent from Madagascar for only the three months mentioned, which would be an abnormally short period.

Merops superciliosus

This bee-eater, which may or may not be conspecific with the Palaearctic Bluecheeked *persicus*, has for many years been treated as a migrant between Madagascar and much of Africa, but I regard the evidence as unsatisfactory.

Mackworth-Praed and Grant (1952) accepted that most of the birds seen in eastern Africa arrive from Madagascar in May to breed, and that they spread as far as the Sudan, and Eritrea, but this statement was modified somewhat in the later edition (1957). The true position is probably very different. This bee-eater has been found breeding in Madagascar in September and October but is "common all the year round" there (van Someren, 1947). Indistinguishable birds breed and are common all the year in Somalia, are breeding residents in Pemba Island (confirmed by R. H. W. Pakenham, *in litt.*) have bred at Dar es Salaam on the coast of Tanganyika (Mackworth-Praed and Grant 1957, Vol. 1, p. 799) and have been recorded in Kenya and Uganda in every month. Modern lists for Eritrea and the Sudan do not include the bird at all. Hence in eastern Africa there is no clear evidence for the migration of any population either near the Equator or north of it.

However, in the eastern Congo this bee-eater is a common nonbreeding visitor May–September and twenty skins from Angola are all June–September (Benson, *in litt.*), while the records in Zambia and Malawi are all of birds in transit April–May and August–September.

In Rhodesia the only records are on the northern edge (Zambezi Valley) and again in April and September (M. P. S. Irwin, *in litt.*). These records fall into the off-season of the Madagascar population, but as stated above there is no evidence that birds leave the island. At least some of the *M. superciliosus* seen in Angola and the Congo may well come from another part of the mainland, since Grant saw great numbers of these bee-eaters and found eggs in Portuguese East Africa October–January (Sclater, 1911, p. 695). (It is not clear what is the basis of the statement by McLachlan and Liversidge that this bird is a non-breeding visitor in the Zambezi Valley September–January.)

To sum up, large-scale migration of *Merops superciliosus* from Madagascar remains to be proved.

Phedina borbonica

There is something very odd about this martin. Its breeding season in Madagascar includes October and November. Yet Vaughan reported small numbers in Pemba Island, apparently on many occasions, in the months September–March in years prior to 1930. They have not been reported on the East African coast since then and R. H. W. Pakenham, who made so many contributions to the ornithology of Pemba and Zanzibar in subsequent years, tells me that although constantly on the look-out for these birds he failed to see them. However "hundreds" appeared over Lake Chilwa, Malawi, 28 June to 30 July 1944. It remains to be proved whether this bird is a regular migrant out of Madagascar or not. If it is, it may have the same elusive habits in winter quarters as the House Martin, *Delichon urbica*.

GENERAL DISCUSSION

Reviewing the intra-tropical migrants of Africa and Madagascar as a whole, it will be seen that, apart from raptors and the few water-birds, they are, like the Palaearctic long-distance migrants, nearly all thoroughly insectivorous. But, omitting the water-birds, the Palaearctic migrants include a much higher proportion of passerine species than do the Ethiopian migrants. In fact, among the latter the hirundines appear to be the only important passerine family. In this connexion it is especially interesting that so many swallows move into the extra-tropical latitudes of southern Africa to breed all at the same time not only as each other but also as the three common Palaearctic migrant hirundines are pouring in to winter. On present information the other families of insectivorous passerines, especially the warblers, flycatchers and thrushes, the Palaearctic species of which contribute so massively to the trans-Saharan migration, are little represented among the Ethiopian migrants and the proportion of shrikes that migrate is far lower among the Ethiopian species than among the Palaearctic. These differences presumably indicate that, greatly as the carrying-capacity of some tropical areas of vast extent appears to fluctuate with the seasons, the availability of small insects is not so low as in Palaearctic areas, where the brevity of the potential searching time in a winter day and the necessity for a greater intake to cope with the colder environment are factors also to reckon with. But it must be admitted that these factors do not strike one as likely to be very potent in the Mediterranean part of the Palaearctic, whence there is nevertheless a general exodus.

Among the Ethiopian non-passerines, those families which include an especially high proportion of migrants are the cuckoos, bee-eaters, rollers and nightjars, with the dry-feeding kingfishers as an interesting addition. The first four are all families very poorly represented in the Palaearctic but are there totally migratory. It is noteworthy that the Ethiopian migrant cuckoos are all parasitic species, while the non-parasitic Centropinae are sedentary; and even more parasitic species than are listed may in fact be migratory (especially *Cercococcyx montanus*; see Benson, 1952); but uncertainty is caused by the fact that the species inhabiting dense cover are so likely to be overlooked when not calling. It appears that, except in the more humid parts of the inner tropics, the parasites, having bred at the time of the local flush of vegetation and presumably of larvae, generally find it necessary to move. The rollers, bee-eaters and nightjars, all dependent on large flying prey, presumably find the seasonal fluctuation in food-supply more acute in many areas than do the insectivorous passerines which depend on smaller insects, whether winged or not. The situation is, however, complicated because numerous members of these families find it unnecessary to make long movements.

The five nightjar species excel any other group in the variety of their migratory adaptations, for two species cross the Equator extensively after breeding south of it, two go towards the edge of the Sahara to breed (and some members of one of them thereafter cross the Equator southwards), while the fifth species, confined to the northern tropics, goes towards the edge of the Sahara to moult. Again, it is interesting that prominent migrations have been noted in six raptorial birds north of the Equator, and in only one south of it. Four of the six seek the neighbourhood of the Sahara after having bred in the dry season closer to the Equator; and it must be supposed that with the advent of the rains in this better-watered area the flush of vegetation is great enough to conceal the birds' prey, while with the contemporary lighter rains close to the Sahara it is not.

No doubt food considerations of one sort and another as a rule ultimately determine all the movements discussed, whether directly because the breeding area later becomes ecologically unsuitable or because the wintering area is unsuitable for breeding; but detailed knowledge of each species (and in several species knowledge of different local or sub-regional populations) is needed before discussion can be any better than speculation. A good example is afforded by the quails, which are dependent on seeds at least as much as on insects. It is not difficult to imagine that the long-distance migrations of the Palaearctic *Coturnix* are imposed mainly by the low temperatures of the northern winter. This factor can hardly be adduced in tropical Africa. Temporary destruction of habitat by fires might then be suggested; but this contingency might be expected to be met by local movements rather than by that typical nocturnal migration, which seems to have been evolved in some populations rather than, or more than, in others.

On present information it appears that migration of African species is more obvious and a more important factor in bird ecology north of the Equator than south of it; more species and more families are involved, though the distances traversed are often relatively small because the full Sahara, uninhabitable all the year round, starts only about 1200 miles north of the Equator. The migration in question is essentially one to take advantage of the flush of food consequent on the summer rains, especially in that belt on the southern edge of the desert, where they are limited to only a few weeks. Besides the flood of birds to breed, two other elements swell this migration, both of birds that come to moult where food is abundant, namely some which have nested at lower latitudes in the northern tropics and some which have nested much farther away, in the southern hemisphere. These two elements in the ecological system of the northern tropics of Africa have little counterpart in the southern, and still less in the Palaearctic and Nearctic Regions. Only studies of the ecology of the individual African species involved could explain their peculiar adaptations in this respect.

14

The Immigrant Palaearctic Bird Fauna

Every year birds that have nested in Europe and Asia begin flooding into Africa in August and some of them remain as late as the following May. They occupy the entire breadth of the continent, which extends for some 700 miles west of the westernmost point of Europe, and the entire length, reaching to beyond 30° S. in the Cape Province. Africa south of the Sahara in fact provides a winter home for more than one-quarter of all the species that breed in the Palaearctic Region, especially for nearly all the insectivorous birds of Europe and many from the Near East and Siberia, with some from even as far east as the Behring Straits. Figure 53, on equal-area projection, gives an impression of the geographical situation. In Europe the characteristic summer birds—the swallows, the nightingales, the turtle doves, the cuckoos, and many more that have caught the imagination and enriched the lives of human beings down the ages—could not exist if Africa were not there to provide them with a winter home.

Some years ago, in the endeavour to get some idea of the number of birds leaving western Europe to winter in Africa, I guessed about one bird per 5 acres as the combined density, for Europe generally, of those land-birds which perform this migration (Moreau, 1953). Later this guess received some support from the figures published by Merikallio (1958) for the bird population of Finland—the best of their kind for any country to date. This gave double my hypothetical density, namely one breeding pair to 5 acres. Some parts of Europe have many more species of migrant than Finland, but a much smaller proportion of their surface is occupied by the wooded habitats which harbour so many species of migrant. For this reason, and also because the Finnish figures include a high proportion of a single species, the Willow Warbler *Phylloscopus trochilus*, I thought it reasonable to halve the Finnish density for Europe in general. But since about as many young as adults survive to migrate each autumn the number of migrants entering Africa will be about twice the breeding population, in fact an average of about 250 to the square mile of Europe.

Later (Moreau, 1961), being primarily interested in getting a figure for the number of birds crossing the Mediterranean coast-line into Africa, i.e. travelling between about 6° W. (Gibraltar) and 38° E. (Port Said), I calculated, on the density basis mentioned above, the possible number of migrants coming from the $2\frac{1}{2}$ million sq. miles of Europe west of 38° E., and arrived at about 600 million. If the density

I used was too high, as it may well be, then this estimate is also, but it is based on only a part of the total number of birds to which Africa gives harbourage in the winter. For one thing it includes no waders; for another, as shown in an earlier analysis of the Palaearctic migration system (Moreau, 1952b), nearly as many species as leave Europe west of 38° E. also leave for Africa from what I have called the Central Palaearctic, the area of over 4½ million sq. miles between about 38° and 90° E.

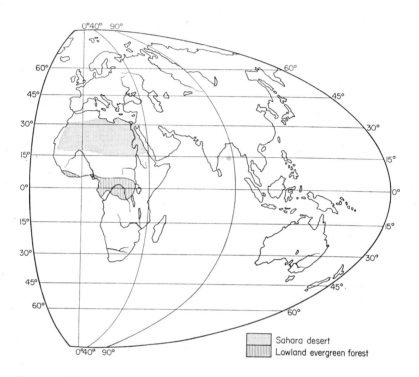

Fig. 53. Africa and Eurasia on equal-area projection. (From Moreau, 1952.)

For passerines the figures are thirty-three species wholly leaving from the Central Palaearctic and twelve partly, compared with forty-three wholly and thirteen partly from Europe. Even east of about 90° E. there are still appreciable numbers making for Africa, including all the Willow Warblers and all the *Oenanthe oenanthe* from as far east as the Behring Straits (180° E.). On the basis of these data and of the 600 million already arrived at for the numbers travelling into tropical Africa from Europe west of 38° E., it can be calculated that something like 1600 million birds, excluding waders, find accommodation each autumn in the 8 million sq. miles south of the Sahara. If they were evenly distributed (which they are not), they would number 200 to the square mile, equivalent to one Palaearctic bird in 3 acres.

This situation appears to be grossly different from that in the New World, for a realization of which I am indebted to discussions with Dr R. A. MacArthur, and I am grateful for being allowed to use his figures. The land surface of America north of 20° N. amounts to about $6\frac{1}{4}$ million sq. miles. On a basis of sample censuses in various types of vegetation and of the areas allocated to these types in standard maps, he has calculated that temperate North America holds about 3500 million males of land-bird species that winter in the tropics. This means that at the time of the autumn migration something like four times as many birds, 14 000 million, are seeking winter quarters. The figure is much the same if 30° is taken as the southern limit instead of 20°; in fact nearly all the birds come from north of 40° N., about one-third from "deciduous woodland" and rather more from "northern coniferous". The American total is thus about nine times as big as that arrived at above for the migrants from Eurasia. Perhaps I have put the latter total much too low, but whether I have or not does not affect a major point of difference between the Nearctic and the Palaearctic migrants in their off-season ecology. It is that only a very small proportion of the American migrants cross the Equator and indeed only a minority seem to reach the continent of South America at all. Now the area of the West Indian Islands is only about 80 000 sq. miles, of tropical Mexico and the rest of Central America 500 000, and of South America north of the Equator 1 million. This gives a total of $1\frac{1}{2}$ million sq. miles, one-fifth the size of Africa south of the Sahara, to accommodate nine times as many birds. On these figures the average density in tropical America would be about fourteen migrants wintering to 1 acre, over forty to 3 acres, compared to one to 3 acres in Africa. Provisional as the estimates of numbers are, they must mean that between the two continents a real difference exists in density unless either (1) the American estimate of population is forty times too high or the Palaearctic forty times too low, or else (2) both are wrong by a factor of 20 in opposite directions. Such margins of error do not seem likely, and we must then accept a probability that in tropical America an area barely one-fifth as large as Africa south of the Sahara suffices as winter quarters for a lot more birds. If this is true, here is an ecological problem on a gigantic scale that demands analysis and investigation.

On present information we can narrow the gap only a little. In tropical and much of southern Africa Palaearctic birds at an even density of anything like one to 3 acres would not readily be noticed, but in fact their density there is very uneven. In the first place evergreen forest, which occupies about 750 000 out of 8 million sq. miles, is hardly frequented by any migrants. In lowland evergreen forest it appears that even the edges are used by no migrant birds except the Honey Buzzard *Pernis apivorus*, the Wood Warbler *Phylloscopus sibilatrix* and locally the Willow Warbler *P. trochilus* (Elgood, *in litt.*). In montane forest I have found in the edges and more open parts Blackcaps *Sylvia atricapilla* and

Garden Warblers *S. borin*; Tree Pipits *Anthus trivialis* and a very occasional Golden Oriole *O. oriolus* have also been reported, but nothing else.*

Here is another important difference from the American situation. Slud (1960) reported a number of North American birds wintering in the wet lowland forests of Costa Rica. In the forests of Colombia around 6500 ft Miller (1963, and *in litt.*) found several species wintering "either in the understory or up at middle level", and he calculated that during the northern winter this involved an increase of 10–15% in the total number of birds inside the forest. There is no evidence for anything comparable in Africa. On the other hand, in other habitats in all parts of Africa south of the Sahara Palaearctic birds not infrequently are extremely prominent. The waders, with so few African competitors, make a great show. Although many species of these to some extent seek the coasts, their concentration in such situations in the northern tropics as the banks of the upper Nile and upper Niger, where the water-level is falling throughout their stay, is one of the ornithological sights of Africa. Even at 5° S., on the shores of Lake Manyara in northern Tanganyika, Palaearctic waders are the most numerous birds for no less than eight months out of the twelve. At the western extremity of the continent, on the lower Senegal River, the Palaearctic waders, in particular the thousands of Ruffs *Philomachus pugnax* and godwits *L. limosa* have the wet places left by the falling flood-waters practically to themselves. Again, in South Africa Dr G. Rudebeck has told me that he found the buzzard of the Eurasiatic steppes *Buteo b. vulpinus* "at least ten times as common" as the local species, *B. rufofuscus*. Moreover, the European Swallow *Hirundo rustica* has repeatedly been noted as out-numbering all the local African hirundines, for example, in the Ituri by Chapin and in the parts of South Africa (Taylor, 1949) where four local species are breeding during the sojourn of *rustica*. Rudebeck (1955) after a careful estimate concluded that European Swallows to the number of 1 million were using a single roost in the Transvaal. Again, native raptorial birds can rarely, even at a locust outbreak, make such a display as the 4000 or 5000 Eastern Red-footed Falcons *Falco amurensis* going to a roost in Malawi (Benson, 1951). (It should, incidentally, be noted that these birds come from furthest Asia, none breeding nearer than Lake Baikal.) In some areas where Palaearctic birds are especially conspicuous, the opportunities for African species to nest are unusually limited. For ex-

---

* I should not be surprised if some time the *Ficedula* flycatchers were found in the canopy of African forest. We know that great numbers, from as far east as Moscow, pour south through Iberia in autumn. Yet they are regarded as "not a very common migrant" in West Africa, "frequenting trees in the vicinity of streams in savanna country" and "more rarely . . . dense bush" (Bannerman, 1953). Travelling by the far west of Europe, as they do, they might be expected especially in the far west of Africa, but they have hardly ever been recorded in Gambia (Cawkell and Moreau, 1963). At the opposite end of West Africa, Serle (1957) knew of only two records in Eastern Nigeria. However, in Western Nigeria there is evidence from several very recent records that Pied Flycatchers "winter in the forest zone, but not necessarily in thick primary growth" (P. Ward, *in litt.*; Elgood *et al.*, 1966).

ample, in a part of the Mbulu highlands of northern Tanganyika unwise land-use and subsequent erosion had reduced the surface to poor
grazing with occasional gullies and very few trees or even bushes. There
I observed that the Yellow Wagtails *Motacilla flava*, Common Rock
Thrushes *Monticola saxatilis*, Red-backed Shrikes *Lanius collurio*, European Swallows *Hirundo rustica* and Pallid Harriers *Circus macrourus* greatly
outnumbered the native birds.

The approximate numbers of Palaearctic species that winter in
different parts of the Ethiopian Region are shown in Table XXIII, I
have derived these figures from specific records in the appropriate
standard works. Inevitably I have had to use my judgment in a proportion of cases, but I have as a rule omitted all records that are doubtful
or that seem to refer in the area concerned solely to passage migrants,
while on the other hand I have included all species that have been
found apparently wintering, however rarely. In most of the territories
there have been additions since the publication of the standard works I
have used, but I have not searched for such additions, since I aim at
comparative figures rather than the unattainable absolutes. To such
figures we have, I think, a satisfactory approximation except for West
Africa, which is discussed separately below.

For the present purpose species that are typically sea-birds in their
breeding areas are included because in Africa most of them to some
extent frequent inland waters and distinction cannot be made. A few
species, especially Ardeidae, cause difficulty because members of them
that cannot be distinguished even in the band without subspecific distinction in both the Palaearctic and Ethiopian Regions, so that a report
of local increase in numbers in places south of the Sahara probably, but
not certainly, is due to Palaearctic migrants. Such birds I have omitted,
so that there are, for example, no herons Ardeidae nor Osprey *Pandion*
in Table XXIII; but ringing recoveries are rapidly providing the evidence that would enable more species to be included (cf. Elgood *et al.*,
1966). It will be seen that the Sudan, with 145 species (eighty-eight
land-birds plus fifty-seven water-birds), has the richest Palaearctic
wintering bird fauna, which is what might have been expected from its
convenient situation on the southern edge of the desert and its wide
frontage on the Palaearctic. Still, the Sudan list of wintering birds does
not include several species, for example, the Sprosser *L. luscinia* and
Roller *Coracias garrulus*, which pass through on both migrations but do
not stay.

Eight hundred miles further from the Palaearctic and with a narrower
frontage on it than the Sudan, Kenya has only 15% fewer wintering
species (seventy plus fifty water-birds). One cause contributing in a
small way to this good showing is that hardly any species press on farther
south without leaving representatives that winter there. Also, a few
species such as the White-throated Robin *Irania gutturalis* come into
Africa too far south to touch the Sudan; subspecies conforming to this

TABLE XXIII

The incidence of Palaearctic migrants in Africa
(by numbers of species)

Family	West Africa	Sudan	Kenya	Rhodesia	Cape Province
Group A					
Anatidae	10	11	9	2	—
Balearicidae	—	2	—	—	—
Charadriidae	4	7	8	2	7
Ciconiidae	2	1	2	1	1
Laridae	10	10	6	1	4
Rallidae	3	4	3	2	1
Scolopacidae	22	22	22	9	16
Total	51	57	50	17	29
Group B					
Aquilidae	5	13	10	12	6
Falconidae	4	4	4	—	4
Strigidae	2	2	1	—	—
Total	11	19	15	12	10
Group C					
Burhinidae	—	—	1	—	—
Phasianidae	1	1	1	—	—
Total	1	1	2	—	—
Group D					
Apodidae*	1	2	1	1	1
Caprimulgidae	1	2	1	1	—
Columbidae	1	1	—	—	—
Coraciidae	1	—	1	1	1
Cuculidae	1	1	2	1	1
Meropidae	1	1	2	2	2
Pitidae	1	1	—	—	—
Upupidae	1	1	1	—	—
Total	8	9	8	6	5
Group E					
Alaudidae	1	2	—	—	—
Emberizidae	—	2	1	—	—
Hiruninidae	3	5	3	3	3
Laniidae	1	5	3	2	1
Motacillidae	5	5	6	2	1
Muscicapidae	2	2	3	2	1
Oriolidae	1	—	1	1	1
Sylviidae	18	22	18	9	3
Turdidae	11	16	10	2	—
Total	42	59	45	21	10
Total groups B–E	62	88	70	39	25

* The Palaearctic Alpine Swift *Apus m. melba* may well winter to some extent in all four of the tropical territories in this table. But it is easily confused with the resident African subspecies and seems to have been collected only in Sudan, Uganda and South West Africa.

pattern are, however, more numerous than species, for example, *Acrocephalus arundinaceus griseldis* from Iraq.

South of Kenya the immigrants thin out; in Rhodesia (which has no coast-line) there are only thirty-nine species plus seven water-birds and in Cape Province twenty-five plus twenty-nine. These latter figures are sufficiently remarkable considering that the Province is at least 5000 miles from the source of any of the birds that reach it. It will be seen that in the progress southwards from the Sudan to Cape Province the passerine total falls off more than the others, from fifty-nine to ten, i.e. to about one-sixth. By contrast, the water-birds and the raptors each fall off by only about half. Once again the differences between the mean geographical ranges of individual species in the several bird groups manifests itself.

In view of its equally good northern position, equally vast size and even wider frontage, the total of sixty-one plus fifty-one water-birds for West Africa is poor compared with the eighty-seven plus fifty-seven for the Sudan; and it is actually lower than the seventy plus fifty for Kenya, a territory lacking all these advantages. Moreover, the ornithological literature of West Africa again and again conveys the impression that observers there have seen few species of Palaearctic migrants and few individuals. Among resident observers, Elgood (1959), in an area around Ibadan in Western Nigeria and some 70 miles inland, had been able in the course of several years to record only twelve species of passerine and four others from the Palaearctic. Moreover, several of these were very infrequent. Rayner's (1962) results at Lagos were even more meagre—clearly not for want of trying—and so were Marchant's (1953) in the south-eastern corner of Nigeria. Serle (1957), bringing together all known data for the whole of Eastern Nigeria, had only four raptor, nineteen passerine and seven other Palaearctic species, nearly half of them represented by only one or two records. The late Père Douaud (1957) in Togoland had much the same totals as Serle's. As a local East African contrast to this, on Kaserazi Island, with an area of no more than 160 acres, close to the southern shore of Lake Victoria, Elliott (1940) recorded sixteen Palaearctic species (apart from water-birds) in the course of a few brief visits.

It has, however, recently become evident that the various results just quoted, which are based on standard works, and are to some extent misleading for West Africa. Now that more attention is being given to the subject there, more species are being added. In Senegal the Grasshopper Warbler *Locustella naevia* and Ortolan *Emberiza hortulana* have been reported by Roux (1959). In Nigeria several species have recently been added, especially with the aid of mist-netting, for example, the Redthroated Pipit *Anthus cervinus* and the Masked Shrike *Lanius nubicus*; so that the number of Palaearctic species now known from Nigeria alone (Elgood *et al.*, 1966) is about equal to that hitherto recorded with certainty for the whole of West Africa in Bannerman (1953). Moreover, ideas about the status of some species are changing. When Bannerman wrote, "very few" Blackcaps *Sylvia atricapilla* were regarded as reaching West Africa and those only from Ghana westwards; they have recently been reported in two localities in Nigeria and seven were actually netted by A. R. Ludlow between 26 February and 27 March 1962 60 miles north-east of Ibadan (P. Ward, *in litt.*).

It is of course unlikely from its geographical situation that the West Africa total will ever equal that of the Sudan or much exceed that of the much smaller Kenya purely because it lies so far west. Several species have their breeding ranges to the east of any part of West Africa, for example *Aquila pomarina*, the buzzard *Buteo buteo vulpinus*, the plovers *Charadrius mongolus* and *leschenaultii* and the turdine *Irania gutturalis*; such species are unlikely to travel west to West Africa regularly to winter unless there are particular ecological inducements for them to do so. Geographically, Nigeria is, of course, better placed to pick up some of these species than is any other part of the sub-continent, as instanced by *Lanius nubicus* mentioned above. The case of the shrikes is, however, a special one. Three species of migrant shrike, the Woodchat *Lanius senator*, Red-backed *L. collurio* and Lesser Grey *L. minor*, breed in Europe directly north of West Africa, but the first species is the only one to winter in that part of the continent, whereas five species of Palaearctic shrike winter in Sudan and three in Kenya. The easterly trend of the Redbacked and Lesser Grey shrikes of western Europe in autumn is in contrast to the south-westerly trend of all the shrikes inhabiting far eastern Europe and Palaearctic Asia; and it is not known whether this exceptional trend is due fundamentally to some ecological inadequacy of West Africa as a wintering ground for shrikes. It is unlikely to be because of exceptionally severe competition in West Africa from other, resident, members of the family, since the Sudan and the East African territories are at least as rich as West Africa in resident Laniidae.

The comparatively meagre lists of migrants quoted above from Douad, Elgood, Rayner, Marchant and Serle all relate to areas having a humid savanna climate, with a mixture of habitats, largely cultivation and regenerating "bush". By what at first sight is a paradox, the drier parts of West Africa seem actually more attractive to Palaearctic migrants than the lusher. In the comparatively harsh climate of Sokoto at 13° N. in Northern Nigeria, Dobbs (1959) recorded from the Palaearctic twenty-two species of passerines (partly transients) and six others. The list of passerines that winter in the Gambia (Cawkell and Moreau, 1963) is actually bigger than this, and the Senegal list bigger again. For Nigeria the new information (Elgood *et al.*, 1966) strongly confirms the previous impression that Palaearctic birds favour the drier north of the country most and has increased the representation of Palaearctic species there. In Senegal, in the course of two winters on a plot of 25 ha (62 acres) of thin Senegalese acacia steppe remote from water, thirty-one species of Palaearctic migrant were recorded, twenty-four of them for all or part of the period December–February, which shows that they were not merely transients (Morel and Bourlière, 1962). There are in fact a number of Palaearctic species which favour only the drier habitats during the winter, as can be checked also for the Sudan in Cave and Macdonald (1955). About ten species, including the lark *Melanocorypha calandra*, bunting *Emberiza caesia*, "Rufous Warbler" *Erythropygia galactotes* and Black Redstart *Phoenicurus ochrurus*, inhabit during their stay nothing better than sub-desert steppe; and another dozen, including Turtle Dove *Streptopelia turtur*, Short-eared Owl *Asio flammea* and *Sylvia rüppellii*, get no farther south than the belt of acacia steppe. The ecological problems involved are discussed later in this chapter.

By contrast with these examples of limitation to comparatively narrow dry belts north of the Equator, most Palaearctic migrants have winter quarters extending over many degrees of latitude. First, there are a few species like the Redthroated Pipit *Anthus cervinus* and the Grey Wagtail *Motacilla cinerea* that in Africa winter both on the Mediterranean coast and also in the tropics. (Indeed, the wagtail, besides wintering in Europe, has a range of winter quarters in Africa from placid ornamental water in Egypt to mountain streams on Kilimanjaro.) Wholly south of the Sahara one finds such winter ranges as those of the Nightingale *Luscinia megarhynchos* from about 15° N. to 15° S. (Malawi). Some ranges are even more extensive: Common Whitethroat from 16° N. (Senegal) to about 25° S. (Transvaal); Willow Warbler, Sand Martin *Hirundo riparia*, Red-backed Shrike *Lanius collurio*, Swallow, Wheatear, from the neighbourhood of the Sahara right to the Cape. It is understandable that birds so numerous in much of their breeding grounds as the Willow Warbler, and so widespread as this species and the Wheatear, breeding across 180° of longitude, should extend widely over Africa and penetrate it to its limits; it is by no means obvious why the Garden Warbler *Sylvia borin* and *Oriolus oriolus* should do so to nearly the same extent. Incidentally, the Willow Warbler has a remarkably wide habitat tolerance in Africa, for I have notes of it in Tanganyika from sea-level to 11 000 ft (on Kilimanjaro) in acacia steppe, cultivation and bushy places of all kinds up to the sub-alpine moorland.

To what extent different breeding populations are segregated in their African winter quarters is not yet known. The winter quarters of different subspecies often overlap more or less completely; for example, the European Nightingale, *Luscinia m. megarhynchos*, the Turkestan *L. m. hafizi* and the Iranian *L. m. africana* (so-called because it was described from a bird collected in 1884 on one of the early expeditions in the Kilimanjaro area) all winter in Kenya. Over much of tropical Africa two or more subspecies of yellow wagtails *Motacilla flava* are found in the closest association (see, for example, Curry-Lindahl, 1958). On the other hand, it may be presumed that some populations which are morphologically indistinguishable are segregated in winter: it is virtually certain, for example, that bee-eaters and whitethroats *Sylvia c.*

communis wintering in Senegambia come from western Europe, while others of these species met with in East Africa come from eastern Europe or the adjacent parts of Asia. In one species, the Swallow *Hirundo rustica*, there is precise evidence from ringing recoveries (Fig. 54), which have given most unexpected results. They have shown among other things that birds from Britain winter in part of the south-east of the Union of South Africa and German birds more than 1000 miles to the north, in an area extending for about 8 degrees on either side of the Equator (Schüz, 1952 and Fig. 3)*. In general climate and in vegetation the wintering areas of the two populations are thus very different, for the German birds stay in the innermost tropics, close to the blocks of evergreen forest and in one of the most equable areas that Africa provides. The British birds actually winter outside the tropics, in an area with a comparatively high annual and daily range and the rainfall mostly in the six summer months. This season is of course when the migrants are there; and during it the most critical element in the temperature, the night minima, average only about 3°C less than in the German birds' area; for example, around 16°C between about 500 and 1000 m a.s.l. compared with around 20°C at 450–475 m a.s.l. (ten stations) within 2 degrees of the Equator in the Congo. However, in this corner of South Africa unseasonable cold snaps do occur and have been recorded as killing numbers of Palaearctic hirundines (Macleod *et al.*, 1953). This hazard diminishes very rapidly with decreasing latitude and is unknown long before the area occupied by the German birds is reached.

It is difficult to imagine how such a degree of isolation between the two populations in their winter quarters has been evolved and why any birds go to the furthest southern end of the continent. The swallows wintering in the wide gap between the British and the German birds are of unknown origin, as indeed are those everywhere in the eastern half of Africa except the extreme south. One noteworthy aspect of this situation is that it must be the product of the last few hundred years. European swallows seem to be so dependent on man-made structures for nesting—more, for example, than House Martins *Delichon urbica*—and on open spaces for feeding, that the population of these birds in Britain and Western Europe generally must have been altogether smaller in

---

* In this connexion, it is of interest that seven European Swallows ringed at Cape Town have been recovered in Russian territory, as far east as the Altai (Hallett and Brown, 1964; Broekhuysen, 1964). Evidently, therefore, many of the swallows wintering in that area belong to the Russian breeding population and the course of their migration must somewhere cross over that of the British birds, most probably somewhere in Southern Africa. Sight records in Europe of Swallows dabbed with colour in Cape Town are difficult to evaluate, but suggest that other breeding populations than Russian may winter in this corner of Africa (*ibid.* and *in litt.*).

Since this was written an up-to-date report on the recoveries of Swallows ringed in Britain and Ireland (Davis, 1965) has to a great extent confirmed the conclusions quoted in the text. Of fifty-eight winter recoveries, one is in Ghana, three in the Congo, and the remaining fifty-five south of 22° S.; of these all but seven are south of 25° S., and east of 24° E. Three others are close to Cape Town.

earlier than in more recent centuries. Indeed it is not easy to visualize any swallow population at all in pre-Roman Britain or contemporary Germany, except in the rare superficial cave or rock-shelter.

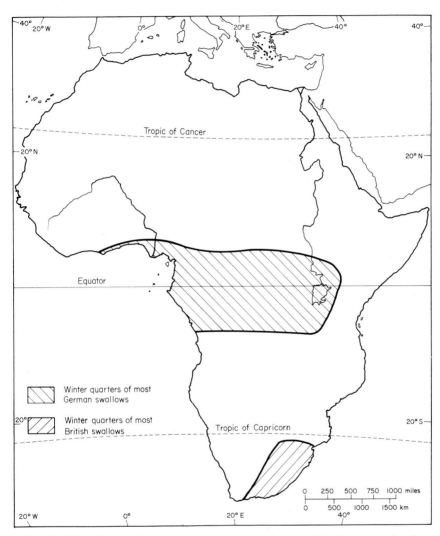

FIG. 54. The African winter quarters of some populations of the European Swallow *Hirundo rustica*. (After Schüz, 1952.)

In general, Palaearctic migrants to Africa seek the same sort of habitat there as they do during the breeding season; for example, the shrikes open ground with suitable perches, the marsh birds marshes, the nightingales dense semi-evergreen thickets (though not evergreen forest), the chats the most open of habitats, with more or less exposed soil, the Pallid Harrier *Circus macrourus* steppe. Only the woodland birds do not accept evergreen forest as an equivalent habitat, but make do with all sorts of tree growth, including acacias (e.g. Icterine Warbler *Hippolais icterina* in Zambia (Benson and White, 1957)) and mangroves (Chiffchaff *Phylloscopus collybita* in Gambia). Conclusive data seem to be lacking, but some species appear to be very closely sedentary and presumably territorial in winter quarters. After a month's interval, a nightingale can be heard singing in the same spot in an area of low dense bushes, as I noticed in Gambia. Other species appear markedly nomadic; in particular, large raptorial birds and storks *C. ciconia* follow locust swarms around. Again, the fact that several Marsh Warblers *Acrocephalus palustris* came to light at my house in northern Tanganyika on nights in November and January suggest that some populations may make intermediate movements within the tropics with all the characters of migration, long before they depart for their breeding grounds.

It is axiomatic that Palaearctic visitors can be accommodated within the Ethiopian Region only when and where food exists for both them and the native birds present for the time being. During their stay the visitors require not only a maintenance ration but also enough food for special needs. For one thing many of them, especially the warblers (Sylviidae), moult in winter quarters, as I am indebted to Ian Newton for reminding me. For another thing, in the northern spring preparatory to their departure from their winter quarters most birds need to put on a great deal of fat, which is essential as a store of energy to sustain them on their northward flight across the desert.

In some parts of Africa the visitors may be taking advantage of a seasonal superabundance of food, in which case other factors than food will determine the immediate local abundances of both Palaearctic and African birds. Where food is not superabundant the possibilities are several. If the native birds are more efficient than the immigrants, then the latter will be restricted to those ecological niches which are not fully occupied by the native birds. If the immigrants are the more efficient, then they could depress the number of native birds, either temporarily by forcing some to move into areas where Palaearctic birds are not superior, or permanently by depressing the numbers of the breeding stock. The first of these two alternatives is a conceivable explanation of something that happens annually in sub-arid West Africa, as discussed later in this chapter; the second alternative has actually been suggested to account for the absence of breeding species of *Anas* ducks in West Africa (see Chapter 7). In the present state of knowledge these ideas must, however, be regarded as speculative.

In those areas where Palaearctic and native African populations react upon each others' numbers, the extent to which they do so will depend on the timing of the Palaearctic birds' sojourn in relation to the local climatic cycle and on the degree to which the carrying capacity of the environment varies seasonally. Other things being equal, one would expect the biggest opportunities for the migrants to exist in the more favourable time of year in those parts of Africa with well-marked wet and dry seasons. Thus, in southern Africa the annual rainy season lasts from October or November, and over a great deal of the country a flush of vegetation, new grass from deep root-stocks, new foliage on deciduous trees, actually precedes the break of the rains and simultaneously most of the local insectivorous birds begin to breed (Moreau, 1950). This presupposes a flush of insect food, and it comes just when the Palaearctic birds are pouring in. They are accommodated at the time when the native birds are making their greatest demands on the ecosystem, but this is at its most lush and to all appearance does not deteriorate seriously until after the end of the rains, around April, when the Palaearctic birds are going or gone.

North of the Equator the situation is reversed. The rains begin in the period April–June and finish September–November, their duration getting shorter and shorter northwards till they diminish to insignificance in the Sahara. This means that the Palaearctic birds arrive each year about the end of the rains and thereafter witness a progressive desiccation of the herbage and some defoliation of the trees. In general, the savanna at first stands high with yellow grass, which is then largely cleared by fire, and above the blackened ground the trees are for the most partly leafless. More hospitable vegetation persists only in pockets and in strips along water-courses. Further north, in the acacia steppe, fire is not so important an agent, but the ephemeral grasses disappear to some extent, through the action of wind, trampling and termites. The few species of trees lose some of their leaves. In such conditions it is difficult for the ordinary observer to imagine how the large number of species, both Palaearctic and Africa, found in this type of vegetation in winter are able to maintain themselves there. Particularly throughout West Africa and the Sudan the Palaearctic immigrants appear to be faced, in both savanna and acacia steppe, with an ecosystem that is continuously deteriorating for most of their stay. Yet, as shown in Table XXIII, the Sudan accommodates fifty-nine species of Palaearctic passerines, West Africa forty-two, compared with only twenty-one in Rhodesia, where the rainy season coincides with the sojourn of the Palaearctic birds, which therefore do not meet with the consequences of desiccation. Furthermore, as must be postulated for all the Palaearctic species and has now been demonstrated for a dozen species (Ward, 1963, and *in litt.*), the feeding conditions in West Africa as late as March and April, after six months presumptive deterioration of the eco-system, are good enough for the migrants to accumulate the fat necessary to carry them north across the Sahara in spring with little or no chance of refuelling for 1000 miles. Also at least some of the birds put on this fat in a very short period. Finally, as shown by Elgood *et al.* (1966), in Nigeria the number of Palaearctic species accommodated in winter actually increases from the humid south to the arid north, so that more than twice as many live in the area which has started a dry season of seven months when they arrive than in the area that has a total dry season of only four months.

Our ignorance of the eco-system is such that we do not know how these apparently paradoxical results are achieved. There are three possible explanations. The impression of ecological deterioration, which seems so obvious to the ordinary observer, as noted above, may be fallacious. The populations, of both African and Palaearctic species, may waste away at least as fast as the local economy deteriorates. And there may be a temporary withdrawal of the African birds which reduces the total demand on the food supply as a whole and so gives the visitors a bigger share in what is available. All three of these suggestions could have some validity and could contribute to a solution of the problem, but the last factor could do so only locally. Since practically the whole of the northern tropics undergoes a dry season centred round the winter solstice, the presumed deterioration in the economy takes place at the same time throughout and any relief of pressure on the food supply brought about by withdrawal of native birds can be only local.

An important pioneer contribution to the problem has been made by the monthly censuses on the 25-ha plot in the "Acacia steppe" of Senegal (Morel and Bourlière, 1962), already mentioned (Figs. 18 and 19). The area is at 16° 55′ N., with only about 300 mm (12 in.) of rain per year, all falling within the period July–October, and with no permanent water. The sparse trees are only up to about 12 ft high, there are very few bushes and the ground cover is so poor that its dry weight varies only between about 592 kg per ha in the late rains and 188 kg in the late dry season (if it escapes burning). However, the arboreal vegetation during the dry season is not so unfavourable as might be expected, because several species of tree are then in fruit or flower and the acacias come into new leaf some time before the rains break. This, together with a flush of grass on burnt-over ground, is of course a feature of dry Africa generally, as discussed in Chapter 2. In so far as the new growth stimulates a flush of insect food, the migrants could be benefited during the last few weeks before their departure for the north.

The figures Morel obtained in the 28 months of census vary widely, even in the same month in successive years. It does, however, definitely emerge that the greatest density of African birds (up to 8 per ha) is attained about the end of the rains, September–October, which is the season when the Palaearctic birds, newly arrived, also are most numerous (around 2 or 3 per ha). With the progress of desiccation both types of birds become scarcer, the African species decreasing to less than 2 per ha by the latter part of the dry season, when in April the last migrants leave. The process by which the number of the African birds is reduced is not known. It is true that six species, a nightjar, three cuckoos, a bee-eater and a courser, are listed as disappearing altogether, presumably by emigration, but none of these occur in the study area in such numbers as could much affect the statistics of density. I think that apart from natural wastage there may be unobtrusive withdrawal southwards by a number of species. Such a movement would be in accordance with what has been noticed 200 miles farther south, in the Gambia (Cawkell and Moreau, 1963), most obviously with several cisticola species and *Turdus libonyanus*, and is suspected to occur to some extent with many more. Indeed, it is probable that both partial and complete withdrawal southwards in the dry season is very widespread in the northern tropics, as indicated in Chapter 13; but, because the geographical range of most (sub-)species is so extensive that birds which have moved a few hundred miles would not be individually distinguishable, the magnitude of the movement would be ascertained only as a result of ringing and of replicated census work. If indeed withdrawal southwards takes place on a large scale every dry season, it would tend to leave room for Palaearctic birds in the more northern zone, and at the same time would raise the density of African birds farther south, banked up against the barrier of the Guinea coast (and of the Congo forest farther east). The resultant accumulation might mean that the Palaearctic birds find more difficulty in accommodating themselves in the more humid zones nearer the Equator than in the drier nearer the Sahara. This is only a tentative suggestion and it leaves open the problem of why the adjustment has taken its present form instead of the African birds remaining relatively undisturbed in the more northern zones and the Palaearctic birds concentrating in the more southern.

At the opposite end of the ecological scale from the dry habitats, which have just been discussed, we have evergreen forest, which, as already noted, is utilized by very few migrants indeed, even those classed as woodland species in Eurasia. The ecology of this vegetation type changes little in the course of the year and its breeding bird population is thoroughly sedentary. If this population is limited by food, then there will not at any time be a surplus on which visitors from the Palaearctic could rely. Another factor in discouraging their appearance inside evergreen forest could perhaps be its extremely low light-intensity; this does not apply to the canopy, but as Christopher Perrins has reminded me, most of the Palaearctic migrant species, even those breeding in woodlands, do not feed in their canopy as do, for example, the American Parulidae.

Virtually nothing is known about the process by which the individual Palaearctic birds establish themselves in their tropical winter quarters or to what extent interspecific hostility is shown between the migrants and the resident African birds with which they share a habitat. One whole large group, the waders, seems in fact exempt from competition, since Africa has no closely related resident species; but amongst the insectivorous land-birds, which make up the vast majority of the immigrants, the prospects of competition would seem to be extensive, as, for example, between *Phylloscopus* spp. and *Eremomela* spp., which are all warblers foraging in trees. If any interspecific hostility is indeed shown, then it is likely to be most marked where birds that have breeding territories are concerned. In this matter there would be a marked difference on the two sides of the Equator. During the weeks when the immigrants are seeking to establish themselves in Africa, in the northern tropics the dry season is beginning, and the main breeding season is ending. In southern Africa these conditions are reversed; the rains are breaking and the local birds are nesting. Hence on behavioural factors alone the visiting birds might find it more difficult to establish themselves in southern Africa than in northern; but the effect might well be insignificant compared with the apparently far more favourable ecological conditions in the south than in the north, as discussed above. Moreover, in the north since the breeding season is ending the number of native birds will be at its highest point in the year, while in the south it will be at its lowest until the new crop of young is produced.

The manner in which the Palaearctic birds travel to and from their winter quarters in tropical Africa and beyond is a problem that has fascinated me ever since I went to live in Egypt in 1920 and got an inkling of its magnitude. We now know that birds from Eurasia converge on Africa in autumn throughout an immense quadrant. A few travel nearly due west across the northern end of the Indian Ocean (Moreau, 1938), great numbers more or less south-west across Iraq and Arabia (see especially Marchant, 1963) and the remainder round and across the Mediterranean. These last then have to cross the Sahara, a belt about 1000 miles wide, in which the heat and aridity are so great that they can obtain no food or drink except in the small and widely separated oases or along the Nile Valley, a mere strip 10 miles wide or less, in an east-and-west extension of 3000 miles of desert. In spring, of course, the same journey must be accomplished in reverse. Certainly, except so far as the Nile or the oases are utilized, the journey is among the most arduous regularly performed by migrants anywhere in the world and is undertaken by a wider variety of species. Ornithologists were long in facing these realities. When I went to Egypt in 1920 the literature on the subject still presented hypothetical lines along the coast of the Red Sea, the Nile Valley and the Atlantic coast as the "routes" by which the multitudinous Palaearctic migrants travelled to and from the tropics. No attempt had been made to synthesize such scanty indications to the contrary as already existed and no one had bothered to remark that the Red Sea coast was as inhospitable to land-birds as any part of the Sahara. When I discovered for myself that wherever I went in Egyptian territory, however far from the Nile, during the migration seasons, I saw migrants, I thought I was a pioneer (Moreau, 1927). In fact I had been anticipated so far as the spring migration was concerned by the more spectacular results of Geyr von Schweppenburg (1917) on his great journey through the western Sahara. In the last forty years evidence bearing on the subject has been accumulating from many sources, and in 1961 I made a general review of the available data and discussed it from the physiological and ecological point of view. I concluded that the evidence (from its nature more conclusive for the spring than from the autumn) points to a broad-front migration at both seasons, in fact that birds cross the Mediterranean and the desert in all longitudes, not concentrating on the oases or on the Nile.* This is not to say that all species cross in all longitudes. In fact we know they do not. Particularly notable exceptions are the big soaring birds, the eagles and storks, which go to great lengths to avoid crossing wide stretches of water and cross from Europe almost exclusively by way of Gibraltar and the Bosphorus. Such species consequently enter Africa only at the north-western and north-eastern extremities but they comprise only a small minority among the migrants and it does seem that great numbers of birds fly over all parts of the Sahara, at both migration seasons.

The conditions the migrants encounter differ markedly between autumn and spring. On the southern edge of the Mediterranean the vegetation is in a poor state after the unbroken drought of the summer months and in any case over most of the stretch of nearly 1500 miles from Tripoli to the Delta of the Nile, desert comes close to the sea. Hence in that broad belt birds that come to earth directly they have crossed the Mediterranean, though they may find a little shade here and there to rest in, are unlikely to find much in the way of food and almost certainly no water. On the other hand, by the time they have crossed the Mediterranean most of them need be only 500 miles or less from their last meal; and evidence is accumulating that after they have completed the sea crossing they still have plenty of the fat, which we now know it is the rule for long-distance migrants to put on before they start their journeys. Also, in autumn northerly and north-easterly winds predominate at all altitudes and as their average speed throughout the 24 hours is some 10 m.p.h. they can powerfully aid the migrants on their way, since the air-speed of small passerines is little more than 20 m.p.h. Finally, when these migrants have crossed the desert they find the steppe on its southern edge in the best condition of the whole year, around the end of its brief rainy season.

In spring, conditions for the migrants are not nearly so good as they are in autumn. In the first place, for most species the distance they have to cover in practically unbroken flights is if anything longer than in autumn. The steppe on the southern edge of the Sahara, which they found green when they arrived in the autumn, has been rainless ever since and hence continuously deteriorating. Therefore, in order to accumulate the necessary fat for the desert crossing the birds may have to take off further south than the latitude in which they land. Moreover, in spring the winds are predominantly against birds travelling north-wards unless they are moving east of north at altitudes of over 2000 m. On the basis of wind-speeds and the air-speeds of birds, I have calculated

* A subsequent study (Moreau, 1967), shows that these conclusions apply also in the main to water-birds.

that most passerines, for the successful crossing of the Sahara that is performed by so many millions, would need to be able to fly continuously for 50–60 hours without refuelling. The work of Nisbet (1963) on some American migrants suggests that such a duration should be well within the powers of the Palaearctic species concerned. Also, when the birds reach the Mediterranean coast in spring they find it in a better state than in autumn; though over much of the distance between Tripoli and the Delta there is no vegetation better than a strip of low scrub a few miles wide. At present we have no accurate knowledge of how much fat the Palaearctic migrants in general accumulate before the journey, but it can surely not be less than the percentages of body weight—up to over 40%—which has been found in birds crossing the Gulf of Mexico (Odum, 1960). Examination of a small sample of birds taken near Lake Chad in March has in fact given a wide variety of figures, but the highest have been 33% fat in a wheatear *O. oenanthe*, 39% in a Yellow Wagtail *Motacilla flava* and 40% in a Subalpine Warbler *Sylvia cantillans* (Ward, 1963).

It is important to realize that the circumstances of Palaearctic migration are constantly changing and those we see today have never occurred before. There have been three main variables: (1) the ecological fluctuations in Eurasia due to the climatic vicissitudes, especially the violent alternation of glacial and inter-glacial periods; (2) the ecological fluctuations in Africa, in part, but only in part, directly related to the glaciations, as discussed in Chapter 3; (3) the effects of man's activity. The first group of fluctuations have had recurrent effects on the bird population throughout the Pleistocene, with an enormous reduction in the total number of birds breeding in European and northern Asia whenever the ice-caps and the periglacial zones were extensive (Moreau, 1954). Indeed, as shown in Fig. 55, at a glacial maximum, with the ice south to the Severn estuary and Berlin, there was practically nothing better than tundra north of the Loire, the Alps and the Carpathians, and no deciduous broad-leaved woodland between the Caspian and the Atlantic except in the most favourable parts of Iberia, Italy, the Balkans and the fringe of Anatolia. Hence during such a severe climatic phase the number of birds, especially insectivorous birds, in need of winter quarters in Africa would have been only a small fraction of the numbers during an interglacial or even during the hypsithermal around 6000 years ago, when the climate was warmer and woodland could grow further north than it does today. The last minimum in the Eurasian population of potential migrants was no more than 18000 years ago, the last maximum only about 6000.

In Africa the changes were complex, as described in Chapter 3. They were recurrent in so far as montane and lowland conditions alternated with the glacial changes in predominance over most of tropical Africa; but apart from these, important changes affected wide areas, such as the desiccation that extended Kalahari conditions north and east around 12000 years ago. None of these sub-continental changes were more important than those affecting the Sahara, which lies across the path of all the European migrants. On the occasions when its amelioration coincided with a glacial maximum and a minimal European bird population, it seems that only a part of the species which now penetrate tropical Africa would have to go so far south to find winter quarters as they do now; and this effect, coupled with the smaller total of the migrant population, would greatly reduce the demands made by Palaearctic migrants on the eco-system of the Ethiopian Region.

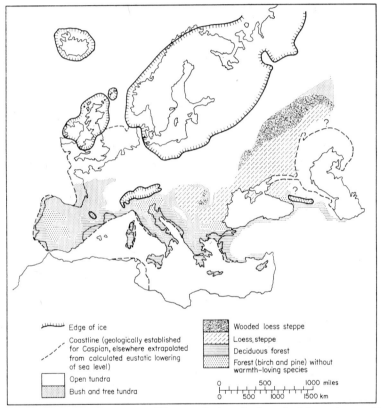

Fig. 55. The vegetation zones of Europe during the last glacial maximum. (After Büdel, 1951.)

A special set of conditions ruled in early Neolithic times, some 6000 years ago, when the world temperature was higher than now and at the same time the Sahara was more humid. On the one hand woodland extended farther north in Eurasia as it does now and man had yet to make serious depredations upon it, so that the woodland birds, which provide so large a proportion of immigrants to Africa, would have reached their highest numbers at that time. So in all probability did the duck and wader population, since the whole of Europe was free of ice in summer and nothing had been done to drain the swamps. However, the Mediterranean basin, with shores undrained and "unimproved", would have provided wintering grounds for these species on a scale far greater than in historical times. In Africa the western Sahara seems to have been under macchia or open woodland and the hills east of the Nile less inhospitable, so that where now the desert is too severe for any migrant but *Oenanthe deserti* to find a lodgment altogether some 2 million more square miles of habitable land existed. In the following centuries the deterioration of the Sahara was catastrophic. Hence the last 5000 years or so have probably witnessed an increase in the length of many migratory journeys and certainly a very great increase in the physiological demands they make. This may mean that the present high level of fat accumulation, and indeed the whole behaviour pattern of trans-Saharan migration as we know it, are to a considerable extent the result of evolution during the same period.*

Finally, we have to consider the impact of the latest influence of all, widespread human exploitation of the land. In the last 2000 years the Eurasian areas of woodland, of scrub and of swamp, which between them provide nearly all the Palaearctic migrants, have been drastically reduced, to be replaced by farmland. It cannot be doubted that the numbers of most migrant species have diminished correspondingly. Indeed, in its most intensive state, when copse and hedge are gone, agricultural land supports hardly any trans-Saharan migrants. Similarly in tropical Africa, where really intensive, large-scale agriculture has been developed, conditions suit but few migrants, except that water-birds may benefit from irrigation. So far, however, much agriculture is still small-scale and so primitive that areas cleared of forest or savanna woodland remain for much of their time rank with herbage, regenerating bushes or coppice growth from stumps: such vegetation seems more attractive to most migrant species than the climax communities that it has been replacing, and inasmuch as this is so the face of sub-Saharan Africa is more welcoming to Palaearctic birds than ever before.

On the whole, it would seem that, with a combination of reduced numbers of migrants and enhanced opportunities for them to winter in Africa, pressure upon winter accommodation may at the moment be less than it has been in the recent past. This state of affairs will alter again as intensive, "clean", agriculture gains ground in Africa and large-scale drainage schemes rob water-birds of their habitat. Owing to the nature and the scale of human interference in both the Palaearctic and the Ethiopian parts of the migration system, the present situation is certainly unique.

* Since migrants have recurred in successive years at identical localities in Tunisia after crossing the desert (Moreau, 1963) and since much of this journey must be accomplished at night, it is necessary to postulate that the star pattern provides an aid to their navigation. If this is accepted, then it appears, for astronomical reasons, that adaptation in this respect must have taken place at much the same rate as in other features. "The stars that [are postulated to] guide the migratory warbler from Europe to Africa today are quite different from the ones that guided its ancestors 6000 years ago. In a relatively small number of generations reliance has gradually and imperceptibly been transferred to a new set of stars" (Agron, 1962).

15

The Bird Faunas of some African Vegetation Types

In this chapter outlines are given of the bird faunas associated with four different vegetation types of tropical Africa, namely, to start with the driest first, (*a*) "Wooded steppe with abundant *Acacia* and *Commiphora*", (*b*) "Woodlands with abundant *Colophospermum mopane*", (*c*) "Woodlands, south-eastern areas, with abundant *Brachystegia* and *Julbernardia*" (i.e. Brachystegia Woodland=*miombo*); (*d*) "Moist forest at low and medium altitudes" (i.e. lowland forest and intermediate forest), (*e*) "Montane evergreen forest" (Keay, 1959). The method I have adopted is to list by families, and the number of species in each, the birds found in two or more sample areas of each vegetation type (Tables XXIV and XXV). Such a list approximates to the "community-stand" of Kendeigh (1961), but it is not possible to designate communities in the sense of this author, since for his purpose "it is important to have quantitative information on the size of the populations to evaluate the importance of each species before community classification is attempted". It is, I think, true that practically nowhere in Africa do we have satisfactory statistics of population for the present purpose; as it happens, those censuses which have been carried out mostly deal with disturbed habitats (e.g. Winterbottom, 1936) and none provide results than can be utilized here. Actually, the problems of censusing birds in evergreen forest have not yet been solved and in drier areas, where the seasonal fluctuation in numbers is great, a satisfactory technique for evaluating census results has not been elaborated. For the climax types of vegetation mentioned above it is, however, possible to compile from various sources sample lists of species known or presumed to breed on the spot. Since the samples are widely separated, there is between them much geographical replacement of species, but for each vegetation type the composition by families shows a good deal of agreement. Further, although we are far from being able to define the "niche" of most species with any precision, enough is known about the kind of food they take—e.g. seeds, fruit (more or less fleshy), insects or combinations of these—for it to be possible to calculate in each bird fauna what proportion of the species are drawing on what type of food. As will be seen from Table XXVI, in terms of food-utilization on these lines the agreement between different samples of each vegetation type is good, and the differences between the types are probably genuine. The results will, I hope, stimulate research into the problems of the production and utilization of bird food in terms of biomass and of change in the course of the year,

Table XXIV

Bird communities of Acacia Steppe, Brachystegia Woodland and Mopane Woodland

	Acacia Steppe				Brachystegia Woodland		Mopane Woodland	
	Senegal	Tanganyika A	Tanganyika B	Kalahari	Zambia	Rhodesia	Zambia	Rhodesia
Group C								
Burhinidae	1	–	1	1	–	–	1	–
Charadriidae	1	1	1	1	–	–	–	–
Glareolidae	2	2	2	2	1	–	1	1
Otididae	2	–	2	3	–	–	2	1
Phasianidae	2	2	7	1	2	3	2	2
Pteroclidae	1	1	1	2	–	–	1	1
Struthionidae	–	–	1	1	–	–	–	–
Turnicidae	1	–	1	1	–	–	–	–
	10	6	16	12	3	3	5	4
Group D								
Alcedinidae	1	1	2	–	3	2	1	–
Apodidae	–	–	–	–	–	–	–	–
Bucerotidae	2	4	3	2	2	–	2	1
Capitonidae	1	3	3	1	3	3	1	–
Caprimulgidae	1	2	1	1	3	3	1	1
Coliidae	1	1	1	1	–	–	–	–
Columbidae	4	3	4	3	1	2	2	2
Coraciidae	2	1	1	2	2	1	2	1
Cuculidae	3	1	3	?	4	3	2	–
Indicatoridae	–	–	1	–	3	3	1	–
Meropidae	1	1	1	1	1	–	1	1
Musophagidae	–	1	2	–	–	–	1	–
Phoeniculidae	2	1	2	1	2	2	2	2
Picidae	2	3	2	3	4	3	3	3
Psittacidae	2	1	2	–	1	1	2	2
Upupidae	1	1	1	1	1	1	–	–
	23	24	29	16	30	24	21	14

Group E

Family								
Alaudidae	2	2	2	10	1	1	1	2
Campephagidae	–	–	–	–	2	2	–	–
Corvidae	–	–	–	–	–	–	–	1
Dicruridae	–	1	1	1	1	1	1	1
Emberizidae	–	2	2	1	2	2	1	–
Fringillidae	–	–	1	–	3	3	1	–
Hirundinidae	–	–	–	–	–	–	–	–
Laniidae	2	6	5	4	6	4	4	5
Motacillidae	–	1	–	1	2	1	–	–
Muscicapidae	2	2	3	3	9	6	2	2
Nectariniidae	1	2	1	–	5	4	1	2
Oriolidae	1	–	1	–	2	1	2	–
Paridae	–	2	1	2	3	3	1	1
Ploceidae								
Estrildinae	3	9	5	3	–	–	1	1
Others	7	7	7	4	2	2	5	3
Pycnonotidae	1	1	–	1	–	–	–	1
Sittidae	–	–	–	–	1	1	–	–
Sturnidae	3	4	5	1	2	3	2	2
Sylviidae								
Cisticolae	1	–	1	2	3	2	–	1
Others	5	6	4	5	5	3	1	2
Timaliidae	–	–	–	1	–	–	–	–
Turdidae	2	2	1	4	6	3	1	1
Zosteropidae	–	1	1	–	1	1	–	1
	30	48	41	43	56	43	24	26
Total groups C–E	63	78	86	71	89	70	50	44

but I wish to emphasize that I regard this chapter as offering no more than the roughest preliminary sketch.

Ideally the sample bird faunas should come from homogeneous vegetation occupying island sites of equal area and well documented botanically. In practice this is unattainable, the nearest thing being provided by some isolated montane forests. However, in each sample I have tried to achieve homogeneity so far as possible by omitting those species whose presence is not dependent on the actual vegetation type under consideration, for example those dependent in dry country on the lush vegetation associated with incidental drainage channels and flood-plains. Explicitly the samples leave untouched that great and increasing proportion of the surface of Africa which is under cultivation or "induced" vegetation in various stages of recovery from clearing and utilization. Such areas involve so complicated a mosaic of poorly definable sub-types that they present special, and most difficult, problems of their own from the ecological point of view. This has the effect that some of the most numerous birds in Africa, for example the weavers dependent on crops and flood-plain vegetation, do not come into any of the statistics in this chapter.

Since the avifaunal comparisons that follow are concerned with dry-land habitats, group A birds are omitted, though occasionally a plover or, among forest birds, *Lampribis*, could be counted a member of the community. The raptors and scavengers are also omitted because the records of them are of doubtful validity for the present purpose, since the bigger species have such great individual ranges that any circumscribed area is unlikely to be able to claim one as a breeding bird. Thus, in the first Tanganyika area, discussed below, some 3 sq. miles of acacia steppe, the only strictly resident birds of prey were in my experience *Falco biarmicus* and *Melierax poliopterus*, but species of vulture and eagle were often seen passing overhead. In the comparisons that follow it is therefore logical to omit also group B, and they are confined to groups C, D and E.

ACACIA STEPPE

SENEGAL

The first example, and the only one in which the observations have been systematically directed to enumeration, is provided by the census work in Senegal (Figs. 18 and 19) at about 16° 25′ N., on a plot of 25 ha (62 acres) little above sea level in an extensive acacia steppe that is devoid of permanent water (Morel and Bourlière, 1962). The rainfall averages only about 300 mm (12 in.), with much variation between years but always confined within the months July to October, so that more than eight months of complete drought is the rule, with very high temperatures during the latter half of that period. The sparse trees, none more than 12–15 ft high, consist mainly of *Acacia* species. Since they are leguminous trees, producing in their pods seeds like small peas, it might be supposed that they would be an important source of food for granivorous birds, especially doves and game-birds. Actually, however, Morel informs me that the only bird for which he has positive evidence that it consumes *Acacia* seeds is a small local parrot *Psittacula krameri*, which also eats *Acacia* flowers. The dozen other species of trees, which are present in small numbers, are some of them valuable sources of small fruit, especially *Grewia mollis*, *Salvadora persica*, *Balanites aegyptiaca* and *Zizyphus jujuba*. Furthermore, most of the fruit is borne in the dry season, together with some flowers, which help to maintain insect life, and the defoliation does not at any time approach completeness. The ground cover consists mainly of the grasses *Cenchrus biflorus*, *Chloris prieurii* and *Aristida* sp. (Bourlière, *in litt.*), growing only some 8 in. tall in the drier places (though over 2 ft in depressions). With the advance of the dry season the grass becomes yellow and brittle, the sandy soil shows between the tufts and, even if not burnt, the dry weight per hectare drops from about 592 kg at the end of the rains to 188 kg (about 2–3 oz to the sq. yd) later. This is, however, not infrequently cleared altogether in patches by fire. Nowadays wild mammals other than rodents are insignificant in this area, but cattle and small domestic stock frequent it.

On the 62-acre plot in this environment Morel recorded sixty-six African species (and in the winter also Palaearctic migrants which are discussed in Chapter 14) in the course of two-day monthly censuses maintained over 28 months. It is true that of these sixty-three only six were so sedentary or common that they were met with in the course of every single count; but, on the other hand, of the sixty-six, forty-nine species were proved to breed in the plot or in the vicinity, and only six to eight (most of which breed) are known to be seasonal visitors. Thus it seems that the list of sixty-six species can be accepted as a reasonable approximation to the bird community of this area of acacia steppe. It may be noted incidentally that this is more than twice as rich as the communities recorded in areas of slightly lower rainfall (9 in. per year) in Arizona—twenty-eight and twenty-nine species (Hensley, 1954). (See Chapter 8 for another comparison between an African area and Arizona.) Moreover, in addition to the African species, twenty-eight Palaearctic species have been recorded by Morel in his plot, twenty-one of them in the period December–March, which shows that they were not merely transient migrants. Hence they were presumably making demands on the local resources in the middle of the dry season.

TANGANYIKA A

These data come from Mkomasi (Fig. 16), 4° 40′ S., which is at about 1500 ft a.s.l. in a rain-shadow between the mountains of Usambara and Pare, rising on each side to over 7000 ft. The Mkomasi rainfall, with great annual variation, averages about 10 in. per year, mostly falling about April–May, but with occasional rain around November.

The gap between the mountains forms a corridor through which the Somali fauna characteristic of the arid lowlands of eastern Kenya reaches the acacia steppe of Tanganyika Masailand. Such characteristic Somali animals as the Gerenuk *Lithocranus walleri* and the Vulturine Guinea-fowl *Asryllium vulturinum* are present and nearly all the species are distinct from those found in the savanna only a few miles away. The dominant trees are the spindly *Acacia drepanolobium*, often growing more densely than in the Senegal area just quoted, while *Commiphora* spp., *Grewia* spp., *Zizyphus* spp. and *Salvadora persica*, all berry-bearing, appear here and there. Trees higher than 12 ft are rare—a small Baobab (*Adansonia digitata*) and *Delonix* at long intervals. General defoliation seems more severe in the latter part of the dry season than in the Senegal steppe just discussed. The herbaceous cover is all short and ephemeral, and shady bush cover is confined to a few low clumps. Fire is not an important factor here, but every dry season the light ground-cover is almost completely removed by desiccation, wind and harvester-termites. Wild ungulates are not so rare at Mkomasi as in the Senegal area, but there is (or was) practically no grazing of domestic stock and the big pachyderms are absent.

Information about the birds in the acacia steppe of Mkomasi was obtained by my wife and myself in the course of a number of brief visits during several years, which greatly amplified the preliminary list published in Moreau (1935a). From the present point of view the observations were unsystematic, being directed mainly to breeding habits, but the list of species used in compiling Table XXIV can be regarded as giving a good idea of the bird fauna found in an area of some 3 sq. miles. For one thing, nests of most of the species concerned were found by us in this area, while on the other hand, care has been taken to exclude any species we observed whose presence is suspected of being dependent on the river valley bounding the area on the south.

It will be seen that with seventy-eight species the Mkomasi community is richer than the Senegal with sixty-three (there are only fifteen species in common). Some such difference might have been expected from the fact that West Africa as a whole is poorer in species than sections of eastern Africa (Chapter 7). In this case the balance is determined by the passerines; in particular, Mkomasi has six shrikes compared to two in the Senegal plot, nine estrildine weavers to three and also a bunting and two finches, neither of which families appears in the Senegal list. The latter is, however, actually the richer in group C birds, because no bustard, button-quail or stone-curlew has been recorded in the Mkomasi area though they occur not far away. In group D, where Senegal and Mkomasi score almost the same, 23:24, the most interesting feature is the accumulation at Mkomasi of hornbills—three *Tockus* spp. and *Bucorvus*—and three barbets. Two of the latter are *Trachyphonus* spp., a genus living locally in termite mounds and not entering West Africa at all.

TANGANYIKA B

The third block of data for acacia steppe comes from the Tarangire Game Reserve (Fig. 17), some 600 sq. miles around 36° E., 4° S., about 140 miles WNW. of Mkomasi. The list has been compiled from the records of Dr Hugh Lamprey, who lived in the Reserve, and has kindly given me all the details in this section (see also Lamprey, 1963). Although classified in the "Vegetation Map" as acacia steppe, the area lies at a higher altitude than the two already dealt with, around 3700 ft a.s.l., has a much higher rainfall (averaging 24 in., which falls in the period late October–May) and, conformably, it has a richer vegetation. For the present purpose the species on which the statistics are based are confined to those known or believed to breed in the acacia-commiphora steppe and associated grassland.

The general aspect of the area is of parkland, most of the trees being 25–30 ft in height. *Acacia tortilis* is the dominant species, with other *Acacia* spp., *Delonix*, *Kigelia* and a few *Adansonia* (baobab), *Salvadora persica*, *Commiphora* spp. and the giant *Euphorbia bilocularis*. The last three species provide much fruit, which is also contributed by bushy growth up to about 10 ft high, mainly of *Grewia* spp. and *Maerua*. Except on the barren stretches of hard-pan the grasses mostly belong to perennial *Cenchris*, *Digitaria* and *Cynodon* spp., growing 2–3 ft high, but ground cover practically disappears towards the end of the dry season, except along the drainage lines. Big game is very plentiful.

As might be expected from the nature of the Tarangire area, its bird list is rather longer than those of the other two areas just discussed, with eighty-six species in groups C–E against seventy-eight and sixty-three. This preponderance is due to the greater prevalence of ground-birds at Tarangire; in passerines it appears actually to be poorer than Mkomasi, but this may be due in part to the precautions taken to exclude from the Tarangire list species believed not to be wholly dependent on acacia-commiphora steppe.

KALAHARI

I am indebted to Mrs B. P. Hall for the information in this section, obtained by her during one month's dry-season study in the south-western Kalahari around 24° S., 22° E. The local rainfall, extremely variable and sporadic, averages barely 10 in. per year, falling mostly in the period November–March. There is no dependable surface water, no permanent river for more than 200 miles, and no growth of tall trees or thick vegetation along drainage channels. The tree vegetation is an open stand, very heavily dominated by a number of species of *Acacia* with some *Terminalia* sp., few of the trees exceeding about 15 ft in height. Several fruit-bearing species of tree are present in small numbers, especially *Grewia*, *Diospyros*, *Ziziphys*, *Fadogia*, *Ficus* and *Commiphora* (G. L. Guy, *in litt.*). There is a fairly ample herbaceous ground cover, in the main consisting of perennial tuft grasses (Rattray, 1960) and the wooded areas are interspersed with stretches of pure grassland. A limited amount of burning takes place. Defoliation of the trees was not very advanced in June and is probably never very severe, so that in this respect the area resembles that cited above for Senegal. Big game is plentiful.

Mrs Hall tells me that her list, analysed in Table XXIV, can be regarded as representing the bird fauna that could be found within a 2-mile radius, that is, within about 12 sq. miles, which included areas of open grassland. It will be noted that the list includes no sunbird and no cuckoo. In view of the migratory habits of the cuckoos, especially in South Africa (see Chapter 13), their absence from Mrs Hall's list is surely a seasonal effect, and so may be the absence of sunbirds. It will be seen that, amounting as it does to seventy-one C–E species, it is within the size-range already worked out for the three acacia steppe bird faunas further north, but the Kalahari total owes a good deal to the extraordinary proliferation of larks—ten species compared with only two in each of the other areas.

BRACHYSTEGIA (MIOMBO) WOODLAND

I have had only a little personal acquaintance of brachystegia wood-land (Figs. 12 and 13) and I am indebted to C. W. Benson for providing me with a botanical description by Fanshawe (1959) and with a demon-stration in Zambia. I have also benefitted by discussion with N. R. Fuggles-Couchman, with M. P. Stuart Irwin and with W. R. Bain-bridge at Livingstone.

Brachystegia woodland dominates southern Africa between about 5° and 17° S., an area with a single well-defined rainy season about October–April. This woodland is typically a closed formation of trees up to about 50 ft high, which viewed from the air appears for most of the year an all-embracing green blanket. Some fourteen mostly legumin-ous species out of about nearly fifty canopy species are widespread dominants, the commonest belonging to the genera *Brachystegia* (5 spp.) and *Julbernardia* (2 spp.). It might be supposed that the *Brachystegia* seeds would be important food for the larger granivorous species but Benson tells me that a bustard *Eupodotis ruficrista* is the only species he has proved to eat these seeds.

The great preponderance of light foliage means that the shade is never at all dense. Moreover, the great majority of the trees are decidu-ous, but not all at the same time, so that the woodland is never entirely leafless. The new foliage, assumed considerably before the rains break, is variously coloured, yellow to red, more suggestive to a European of early autumn than of "spring". Some ninety species of smaller trees have been recorded in brachystegia woodland, nearly all of them semi-deciduous. About twenty species in some fifteen genera are widely distributed, including three *Strychnos* and three *Uapaca* spp., with edible fruits. Undergrowth of shrubs is patchy and often absent over considerable areas, but the species of these and of the ground flora are numerous. About forty different grasses occur, a dozen of them abundantly, but the most important food grasses, such as support the large number of bishops (*Euplectes* spp.) on the flood plains, are generally absent. Although the total number of plant species recorded in brachystegia woodland is about 700, "individual stands . . . rarely carry more than seventy woody species and say eighty herbaceous".

Fires sweep through a great deal of the brachystegia woodland in the dry season and vast areas then for a few weeks present a barren appearance. The ground is blackened, the trees are partly denuded of leaves, and their grey stems are reddened towards the base by the accumulation of termite runs. However, there is no continuous deterioration of conditions throughout the dry season; on the contrary, as early as the end of June, some trees are beginning to flower, with corresponding encouragement of some types of insects, while the trees start their leaf-flush, and insect food increases further, some weeks before the rains break late in October. At the worst stage of the vegetation there may, however, be some local movement of the birds to the more favourable areas. These may be along drainage channels and on "deeper, sandier, better drained soils", where the woodland includes some semi-deciduous or even evergreen thicket. But this is very unevenly distributed and even when there are patches "covering many square yards" (Fanshawe) they could contribute little towards harbouring most of the species or the numbers of birds that populate brachystegia woodland.

For being able to include details of the bird communities of brachystegia and also of mopane woodland (see below) I have to thank the kindness of C. W. Benson in Zambia and M.P. Stuart Irwin in Rhodesia, who are residents with prolonged field experience. Both the observers quoted have taken precautions to exclude species whose presence within the vegetation types concerned seems to be due even at the best time of year to the lusher vegetation found especially along drainage lines; but in such an investigation as this there will always be some proportion of uncertainty depending on assessments by the individual observer.

The basis of the Zambian community, as summarized in Table XXIV, is Benson's list of species "typical of a couple of square miles" in particularly rich brachystegia woodland near Kasama; the basis of the Rhodesian community is Irwin's records over an area of about 4 sq. miles in Mashonaland. They agree in showing that the brachystegia bird community is rich, for the Zambian total is eighty-nine species, that for Rhodesia seventy. Two factors may contribute to the difference: one is that the total Rhodesian bird fauna is poorer than the Zambian, as shown in Chapter 7; the other is that the brachystegia woodland in Mashonaland is a somewhat poorer type than that at Kasama. This latter factor may be acting also within Rhodesia, for Irwin tells me that

in brachystegia near the limit of its range in the dry west of the territory
he recorded only forty-eight species of birds.

MOPANE WOODLAND

Colophospermum mopane, also belonging to the Leguminosae, dominates
the deciduous woodland characteristic of the hotter and dryer low-lying
parts of Zambia and Rhodesia (Figs. 14 and 15) and the northern
Transvaal. It is especially well developed in the Luangwa and Zambezi
Valleys. Like brachystegia woodland, it occurs in areas with an
unbroken dry season of at least five months. For discussing this vegetation type with me I am indebted especially to R. G. Attwell and W. R.
Bainbridge. Mopane may form dense thicket, open parkland or a
fine stand of trees 40–50 ft high under which few grasses survive. The
seeds of *mopane* are so large that they are probably of no use as bird
food. Several of the numerous associated trees provide berries.

The figures for the mopane community in Table XXIV are derived
from Benson's unpublished list of fifty species breeding regularly in
Zambian mopane woodland (as well exemplified in the Luangwa
Valley) and from Irwin's list of forty-four species compiled in about a
dozen square miles of "stony rough ground", near Kariba, on the
Rhodesian side of the Zambezi well away from the river. The mopane
bird communities are clearly much poorer than those of brachystegia,
except of the poorest type. Personally, on a brief visit to mopane areas
of the Luangwa Valley in the middle of the dry season I was struck by
the evident paucity of birds. Possible reasons for the more exiguous bird
communities in mopane, as suggested to me by W. R. Bainbridge are
(1) that this species is a much heavier dominant in its vegetation type
than are the *Brachystegia* spp., so that the mopane woodland is floristically more homogeneous; (2) that mopane woodland, occupying hotter
and drier country than brachystegia woodland, has a longer and more
severe period of defoliation and a briefer period of flowering. Both these
last two features would contribute to greater seasonal fluctuation in the
amount of food available to birds.

In composition, the bird lists for the northern and the southern
mopane samples show important differences. Although geographically
so near, only two-thirds of the forty-four species on the Rhodesian list
are also on the Zambian. The differences most calling for comment are
that the Zambian list includes one (dry-feeding) kingfisher, two cuckoos
and one honeyguide, while there are no birds of these families on the
Rhodesian list. For the last three species, which are parasitic, potential
hosts are by no means lacking in the Rhodesian community, so there is
no obvious reason for their absence; Irwin tells me, however, that at
the time he made the original survey on which his list is based he was
struck by the fact that as soon as he entered the mopane woodlands of
the valley he seemed to leave behind the numerous cuckoos of the surrounding brachystegia.

EVERGREEN FOREST

A general account of lowland evergreen forest has been given in Chapter 2. It may be recalled that the main features are the tall smooth trunks, devoid of side-branches except near the top, the rich lianas and epiphytes, the deep shade below, and the scanty undergrowth, mostly of seedling trees and saplings. Grasses and herbs are almost confined to spots below openings in the canopy. Several authors have been impressed by the stratification of life in tropical forest. Beebe (1917, pp. 81 and 87) in Guiana divided the habitat into "the Floor of the Jungle", the "Lower Jungle" up to 20 ft, "the Mid-Jungle" up to 70 ft and above this "the Tree-tops". For ornithological purposes in Africa "the Floor' is not a necessary sub-division. Chapin (1932) and Moreau (1935a) independently in their different areas divided the bird species into (a) those feeding on the ground or in the lowest few feet of the undergrowth, but nearly all nesting in the undergrowth, not on the ground, (b) those nesting in the undergrowth but foraging upwards by way of the saplings and lianas, often into the lower edge of the canopy, (c) those passing their lives in the tree-tops, rarely or never coming near the ground. The relevant divisions of the forest have been called the Ground Stratum, the Mid Stratum and the Tree-tops. Harrison (1962) working in Malaya and Australia with special attention to feeding habits, has used four zones—Ground, Middle, Canopy and Upper Air —the last of which would, so far as African forest birds are concerned, be needed only for swifts.

Most of the birds of the ground stratum are passerines, especially babblers, bulbuls, thrushes and flycatchers, with one or two species of game-birds, doves and centropine cuckoos. As Chapin has pointed out, some of the passerines of the ground stratum live much in association with driver-ants, eating not so much them as the creatures they flush. Characteristic birds of the canopy are hornbills, barbets, turacos, shrikes and starlings. In the mid-stratum, which extends up to the lower side of the canopy, the mixed foraging parties, which include bulbuls, drongos, warblers, flycatchers, cuckoo-shrikes, insectivorous weavers, are a typical phenomenon and often provide, at long intervals, the only obvious signs of life in the gloom of the forest. (For a recent discussion of foraging-parties, see Winterbottom, 1949.) In the montane forests these three strata also exist, but the trees are not so tall, lianas are much reduced or absent, the canopy is not so continuous and the gloom close to the ground not so marked. The floristic resemblances between the montane forests cited below are strong.

In the pages that follow a number of local bird faunas will be adduced for lowland forest, intermediate forest and montane forest localities, and analysed by families in Table XXV. Except for special reasons stated, I have avoided citing bird faunas that I think may be impoverished by reason of extreme geographical isolation or of habitat attrition.

TABLE XXV

African forest bird faunas

	Tropical lowland			Tropical intermediate				Tropical montane					Temperate
	Primary		Secondary										
	Bwamba	Budongo	Gambari	Amani	Cholo Mt	Soche Mt	West Usam-bara	Kiliman-jaro	Mt Kenya	Ruwen-zori	Kabobo	Cameroon Mt	S. Africa Knysna
Group C													
Phasianidae	1	2	—	—	—	—	—	—	1	1	1	1	—
		—		—	—	—	—	—	—	—	—	—	—
Group D													
Alcedinidae	4	3	1	—	—	—	—	—	—	—	—	—	—
Apodidae	1	3	—	1	—	—	—	—	—	—	—	—	1
Bucerotidae	8	4	8	1	—	—	1	1	1	—	—	—	—
Capitonidae	5	7	6	3	3	2	2	1	1	2	—	2	1
Caprimulgidae	1	—	—	—	—	—	—	—	—	—	—	—	—
Columbidae	6	6	3	3	4	3	3	3	3	2	1	3	4
Coraciidae	1	1	—	—	—	—	—	—	—	—	—	—	—
Cuculidae	5	2	4	3	—	—	—	—	—	—	—	—	2
Indicatoridae	2	1	2	2	—	—	1	—	—	—	—	—	1
Meropidae	1	—	—	—	—	—	—	—	—	—	—	—	—
Musophagidae	3	2	1	1	1	1	1	1	1	1	1	1	1
Phoeniculidae	2	—	2	1	—	—	—	—	—	—	—	—	—
Picidae	4	4	2	—	—	—	1	1	1	1	2	2	2
Psittacidae	2	1	1	1	—	—	1	1	1	1	—	—	—
Trogonidae	1	1	—	1	1	—	1	1	1	1	1	—	1
	46	35	30	17	9	6	11	9	9	8	5	8	13

Group E

Campephagidae	–	–	1	1	1	–	1	1	2	1	1	1	1
Dicruridae	1	1	1	1	1	1	–	–	–	–	–	–	–
Eurylaemidae	1	1	1	1	1	–	–	–	–	–	–	–	–
Fringillidae	–	–	–	–	–	–	1	2	1	1	1	2	1
Hirundinidae	1	–	2	1	3	3	1	1	1	1	1	–	–
Laniidae	5	–	8	1	2	2	2	1	3	3	3	3	2
Muscicapidae	14	14	7	4	2	1	3	3	3	5	3	2	4
Nectariniidae	8	6	2	3	1	1	1	2	2	4	2	3	2
Oriolidae	1	1	–	1	–	–	1	–	1	1	–	1	–
Paridae	–	1	–	–	–	–	–	–	–	1	–	–	–
Ploceidae													
Estrildinae	4	6	4	3	1	2	1	2	3	4	3	2	2
Others	6	4	5	2	1	1	–	1	1	2	1	3	–
Pycnonotidae	15	12	14	9	5	4	5	3	3	2	3	4	2
Sturnidae	1	2	3	4	–	–	3	3	3	2	1	1	1
Sylviidae	13	9	4	6	5	4	4	3	4	6	8	6	4
Timaliidae	3	5	3	1	–	–	2	1	1	3	4	1	–
Turdidae	7	6	–	5	4	5	6	4	4	5	6	4	3
Zosteropidae	1	–	–	1	–	–	1	1	1	1	1	1	1
	81	68	55	44	27	24	32	28	33	42	38	34	23
Grand total	128	105	85	61	36	30	43	37	43	51	44	43	36

Also, so far as possible I have used data from localities which I believe to have been adequately explored, but even so, except where otherwise stated, it is unwise to attach importance to apparent gaps in a bird fauna. The emphasis in discussion will be on the positive records.

LOWLAND EVERGREEN FOREST—PRIMARY

I know of no circumscribed patch of primary lowland forest in West Africa or the Congo (Fig. 5) for which we have an exhaustive list of birds. Some idea of the magnitude of a local fauna of this kind can be obtained from the fact that the whole Upper Guinea forest of some 70 000 sq. miles holds 182 species of birds, as noted in Chapter 9, and the bird fauna of any particular locality in it will clearly be a good deal smaller. The most useful minimum figures I can adduce come from two forests in Uganda, the Bwamba and the Budongo (see Fig. 33 for location). The Bwamba is the ultimate eastern salient of the great Congo (Lower Guinea) forest. The van Somerens' (1949) list for the area includes the birds of the clearings and also those of the montane area that slopes up from the Semliki Valley to the flank of Ruwenzori. It is however possible to select from the list those species which belong strictly to forest and are typically lowland. This list has been supplemented by Williams(1951) and by unpublished date kindly supplied by Stuart Keith. An analysis of the lowland forest bird fauna thus obtained appears in Table XXV; and by inference from the van Somerens' map and text it can be estimated that the records come from traverses in an area of about 20 sq. miles. It will be seen that group D species number forty-six, the passerines eighty-one and with the addition of the local guinea-fowl *Guttera edouardi* the total becomes 128 and is probably approaching completeness. Thus it appears that an area of about 20 sq. miles is inhabited by nearly half as many species of birds as the entire Congo (266, as already quoted) and nearly two-thirds as many as the 70 000 sq. miles of the whole Upper Guinea forest.

The Budongo forest (Fig. 56), which has been described botanically by Eggeling (1947),* is isolated at about 3600 ft a.s.l. on the east side of Lake Albert. It has an area of about 136 sq. miles and lies about 120 miles from Bwamba. Since the whole area south of the lake would have been well above the montane limit during the glaciation, the present bird fauna, which is typically lowland, must have established itself subsequently, presumably round the ends of Lake Albert, and most probably from Bwamba. Published information about the birds is confined to incidental references to about sixty species, but this total has been supplemented by three lists with which I have kindly been supplied, one of sight records by Miss P. Allen, one by Dr H. Friedmann, of species collected in 1963 by the Knudson-Machris Expedition, and one

---

* The "feel" of this particular forest has been admirably conveyed by Reynolds (1965), who spent some months studying the chimpanzees in it.

by Stuart Keith of species collected and species seen by himself and Robert Smart. Combined, these sources give for the Budongo 106 forest species, all but one typically lowland, two game-birds, thirty-five group D non-passerines and sixty-nine passerines. This list, is, however, presumably not complete; for one thing a neighbouring forest, the Bugoma,

FIG. 56. The Budongo forest, Uganda. Photo: by courtesy of the Ministry of Information Broadcasting and Tourism, Uganda.

of about 116 sq. miles, also isolated east of Lake Albert (Fig. 33), includes among incidental published records of about sixty forest species a few not recorded for the Budongo but likely to occur there. Hence the bird fauna of this latter forest most likely amounts to about 110 species, a few less than the Bwamba forest, which is not similarly isolated.

LOWLAND EVERGREEN FOREST—SECONDARY

For the whole of the data under this head I am indebted to John Elgood, who has generously given me the use of it. The list is derived from a number of unsystematized traverses through the Gambari Forest Reserve, an isolated patch of high old secondary forest occupying about 80 sq. miles some 15 miles south of Ibadan, in Western Nigeria. It is botanically rich; on a traverse 400 yd long seventy-five species of tree were enumerated in about 10 yd on either side, i.e. in a total area of about 2 acres (less than 1 ha).* Within the Gambari as a whole, and excluding birds that appear to be dependent on clearings rather than on actual forest, Elgood has recorded thirty species belonging to group D and fifty-five to group E (passerines), besides seven raptors. This means that the 80 sq. miles of the Gambari secondary forest houses more than half as many species of birds as all the Upper Guinea forests put together, totalling 70 000 sq. miles. Moreover, the list of the eighty-five D plus E species in the Gambari (of which Elgood classes only fifteen as "rare") is evidently not quite complete, since it includes none of the Turdidae and only four of the Sylviidae (compared with thirteen in Bwamba). On the other hand, Gambari counts six forest barbets against three in Bwamba, which arouses doubts about the completeness of the Bwamba list. However, if for comparative purposes the totals for the two forest areas are accepted at their face value, they show that the Gambari secondary forest accommodates some three-quarters as many species of birds as the Bwamba primary forest. Although not fully comparable, data from Owerri in southern Nigeria reported by Marchant (1953) are worth recalling. Amplifying and discussing what he wrote, he tells me that all his forest records in that locality came from a strip about 200×600 yd, that is, a total of about 25 acres. In that area he found in the course of numerous visits a total of seventy-three species which are known or likely to nest there. Parenthetically, since one tends to think of the weaver-birds as graminivorous, it should be mentioned that all the nine species in the Gambari are wholly or mainly insectivorous.

---

* Although they are not all fully comparable, it is useful to reproduce some statistics of tree-variety elsewhere. Richards (1952, pp. 8, and 50) records seventy species on about 1 ha in an unspecified locality in southern Nigeria (the general area in which the Gambari Reserve is located) and seventy-four species on 1·4 ha in Côte d'Ivoire, figures close to those of the Gambari, while later (p. 380) he suggests 300 species for the evergreen forest of the Côte d'Ivoire as a whole. In Amazonia on two different plots of 1 ha each Dobzhansky (1959) cites sixty and eighty-seven species, the average of which is close to the West African figures just given; in British Guiana, Richards (p. 8) records 166 species within a radius of 5 miles (say 80 sq. miles), and, by far the richest of all, 227 species have been reported among 559 trees per hectare in Malaya (Wyatt-Smith, 1952).

INTERMEDIATE EVERGREEN FOREST—TANGANYIKA

Here I can draw on long and intimate personal experience of the forest at Amani (Fig. 57), 5° 30′ S., at around 3000 ft in the East Usambara Mountains of northern Tanganyika, 50 miles inland from the

FIG. 57. Forest at Amani, Tanganyika. Photo: P. J. Greenway.

Indian Ocean, the rainfall averaging nearly 2000 mm (80 in.) per year, mostly falling in two rainy seasons. The result is an evergreen forest of exceptional luxuriance for East Africa; it is botanically very rich, with a number of endemic species and a wealth of epiphytes, but no estimate of the number of tree species is available. As shown in Fig. 7, the temperature/altitude gradient is abnormal in the Usambara Mountains and the consequence is that there at 3000 ft it is no hotter than it is at 5000 ft in Kenya; and no doubt partly as a result of this a number of typically montane species are present, while at the same time other typically lowland species, occurring in the lowland forest towards the seaward foot of the East Usambaras, find their upper limit at about this altitude.

Something like 100 sq. miles of this type of evergreen forest survives on the much-dissected plateau of the East Usambara Mountains. The sixty species, which are analysed by families in Table XXV, are all to be encountered within not more than a couple of square miles in the vicinity of Amani. It will be seen that most families are less amply represented than in the Bwamba or the Gambari forests, and the Muscicapidae, with only four species compared with thirteen and eight respectively, are exceptionally poor. On the other hand, the thrushes and the starlings are actually more numerous than in the other forests.

INTERMEDIATE EVERGREEN FOREST—MALAWI

A number of small isolated mountains within 25 miles of Blantyre, in Malawi, which reach 4500–5000 a.s.l., retain areas of evergreen forest that have been well worked by Benson (1948). The bird species are predominantly montane, but one-third of them are lowland and it seems reasonable to classify these forests as "intermediate" for the present purpose. The forests are very small; the biggest, that on Cholo Mountain, is given by Benson as only 3 sq. miles while others are no more than ¼ sq. mile. The Soche forest, which seems to be of special interest, is recorded as only ½ sq. mile. The number of tree species is remarkably limited; J. D. Chapman, the local Forestry Officer, tells me that the Cholo forest counts only thirteen species of canopy trees, eighteen of understorey, four woody climbers and twelve "small trees and shrubs" —a total of little over thirty species.

Because these forests are so exiguous and because Malawi is geographically marginal for the montane species of birds, it comes as a surprise that Benson found thirty-six and thirty species, respectively, in the Cholo and Soche forests, as shown in Table XXV.* Moreover, these

---

* By contrast, we have the extraordinary case of Endau Mt in Kenya, referred to in Chapter 11. This reaches about the same altitude as Cholo and Soche Mts, and has a much more extensive forest, over 10 sq. miles, dominated by montane trees, which, however, holds no more than seventeen species of groups D and E birds, together with the eagle *Stephanoaetus*. It includes no woodpecker, turaco, barbet, drongo, nor ploceine weaver, while seven of the eighteen species, including the only hornbill (*Tockus*), both flycatchers and both bulbuls, are not typically forest birds at all. Thus the Endau bird fauna must be regarded as anomalous, defective and apparently impoverished, for reasons that are not clear (see discussion in Tennent, 1964).

bird faunas may have been larger very recently, since a thrush that was breeding in the Soche forest in 1924 could no longer be found there by Benson.

MONTANE EVERGREEN FORESTS—EAST AFRICA

Only a few miles inland from the first (Tanganyika) area of intermediate evergreen forest described above, but effectively isolated by a trench-like valley nearly down to sea-level, the West Usambara massif rises to 7500 ft and carries at least 150 sq. miles of montane forest. Incidentally, the western, inland edge of this massif falls almost sheer, with in one area what is virtually a cliff a mile high, to the acacia steppe locality dealt with above under the heading Tanganyika A (pp. 279–280). A map and some details are given by Moreau (1935a), while a botanical description of part of the area will be found in Pitt-Schenkel (1938). He listed sixty-three species of tree and tells me that he thinks the total number occurring above about 1500 m would be about ninety. Trees bearing edible fruit are numerous. As shown in Table XXV, the bird species in these montane forests total forty-three. The forests vary, however, a good deal in composition from place to place and it may well be that all forty-three species would not be found in any single square mile.

Kilimanjaro (nearly 20000 ft) and Mt Kenya (17000 ft), the biggest mountains of East Africa, at 3° S. and on the Equator respectively, have also had their bird faunas well explored. They have been listed and compared by Moreau (1936a, 1944b, 1945). With a few minor changes, which subsequent experience has shown to be necessary, these sources supply the information for Table XXV. On both mountains the forest takes the form of a belt between the cultivation of the lower slopes and the moorland of the upper. The forest rings are broadest and richest in the south and east facing the monsoons. On the opposite, rain-shadow, side they become narrower and on Mt Kenya almost disappear. On neither mountain is the forest in contact with a block of lowland forest, though a few lowland species of birds occur in the lower part of the forest belts. In the 54000 acres of the South Kilimanjaro Forest Reserve, which has its upper limit at 8600 ft, ninety-eight species of trees have been recorded, as J. A. Fraser, the local Forestry Officer, informs me. He thinks that the total number in the entire Kilimanjaro forest would not exceed 150. In places on both mountains tree-growth continues to over 10000 ft, the total montane forest areas being perhaps 250 sq. miles on Kilimanjaro and rather more on Mt Kenya. This latter mountain possesses a conspicuous belt of bamboo *Arundinaria alpina*, which is practically absent from Kilimanjaro, but it does not seem to be of ornithological importance.

At the highest forest levels, where the tree species are virtually reduced to *Hagenia abyssinica* and *Erica arborea*, many of the montane species of birds no longer appear, but for lack of precise altitude data it is necessary to treat each forest without subdivision. There is also some uncertainty about the extent to which some typically lowland forest species, found in the lower edges of the forest belt, penetrate upwards and so qualify for inclusion in the montane forest community. This factor must be regarded as potentially increasing somewhat the figures given in Table XXV, perhaps by ten to twelve species. In any case, Mt Kenya (forty-three species) has undoubtedly a slightly larger forest bird fauna than Kilimanjaro (thirty-seven) but for this no ecological reason can be suggested. In particular it has two very similar *Geokichla* spp. (the ecological relations of which are not known), while Kilimanjaro has only one,* and the latter mountain also unaccountably lacks the flycatcher *Trochocercus albonotatus* otherwise met with very generally on East African mountains to both north and south. Over thirty species are common to both forests.

MONTANE EVERGREEN FORESTS—EAST CONGO

Ruwenzori, the third greatest mountain of Africa, carries perhaps 200 sq. miles of montane forest. This differs from those hitherto considered in that it is in contact at its lower edge with a great mass of lowland forest, for it forms part of the eastern rim of the Congo basin, with the richest part of the Congo forest at its foot. A list of the montane species can be extracted from Chapin's volumes (that which he gives on p. 252 of his first volume under the heading "Birds typical of the mountain forest zone of the eastern Congo" cannot be used directly because it includes some species, such as "*Cisticola chubbi*", which are not forest birds, and some forest species which have not been found on Ruwenzori). There seems to be no doubt that the Ruwenzori forests have a somewhat more ample passerine bird fauna than those of Kilimanjaro and Mt Kenya, with markedly increased representation of several families, as shown in Table XXV. Some 400 miles to the south of Ruwenzori, another good sample bird fauna is that of the Kabobo massif, which has been explored ornithologically under the direction of Prigogine (1960). Rising from the north-western shore of Lake Tanganyika, it carries perhaps 150 sq. miles of forest above 5000 ft. The list given by Prigogine is enhanced by excellent altitude data, so that the actual montane forest species relevant for the present purpose can easily be extracted. It will be seen that with forty-four species the Kabobo montane forests also have a bigger bird fauna, at least so far as passerine families are concerned, than Kilimanjaro or Mt Kenya. From one point of view this is made possible by the fact that, as shown in Chapter 11, the mountains of the eastern rim of the Congo basin as a whole form the richest area in Africa for montane forest birds. The problem of how the greater number of species is accommodated on individual mountains in

---

* Mackworth-Praed and Grant (1955), copied by White (1962b), credit Kilimanjaro with both *Geokichla gurneyi raineyi* and *G. piaggiae kilimensis*, but this is an error; only the second of these birds has been found on this mountain.

TABLE XXVI

Food exploitation by birds of groups C–E in different sample communities

Vegetation	No. of species in whole community	No of species dependent on					
		Invertebrates		Fruit		Seeds and roots	
		Equivalent total*	Proportion of attention	Equivalent total*	Proportion of attention	Equivalent total*	Proportion of attention
Acacia steppe							
Kalahari	71	46	0·65	3·5	0·05	21·5	0·30
Senegal	63	34·5	0·56	8	0·13	18·5	0·30
Tanganyika A	78	41	0·54	10·5	0·13	24·5	0·32
Tanganyika B	86	48	0·56	11	0·13	26	0·30
Mopane woodland							
Zambia	50	33	0·66	6	0·11	10·5	0·20
Rhodesia	44	28	0·64	4·5	0·10	10·5	0·24
Brachystegia woodland							
Zambia	89	71	0·77	9	0·10	8	0·09
Rhodesia	70	50·5	0·72	7	0·10	8	0·11
Evergreen forest							
Lowland primary							
Bwamba	127	96·5	0·76	24·5	0·19	4	0·03
Budongo	103	69·5	0·67	24·5	0·24	4·5	0·04
Lowland secondary	85	58	0·68	21·5	0·25	2	0·02
Intermediate	61	39	0·64	17·5	0·28	3	0·05
Montane							
Cameroon	44	29·5	0·67	8·5	0·19	4·5	0·10
Ruwenzori	51	34·5	0·67	11	0·21	5·5	0·11
Kabobo	43	32	0·74	5·5	0·13	4·5	0·10
Mt. Kenya	43	23	0·53	14·5	0·34	4·5	0·10
Kilimanjaro	37	19·5	0·53	13·5	0·36	3	0·08
West Usambara	43	28	0·65	12·5	0·29	2	0·05
Subtropical	36	24·5	0·68	7·5	0·21	3	0·08

* Note: The equivalent total is arrived at by counting "mainly dependent" on a given type of food as one unit of exploitation, "partly" (but to an important extent) as half and summing the results. The proportion of attention is calculated on the equivalent total divided by the number of species in the community. It is to be emphasized that the figures have no relation to density of individuals of particular species or to biomass.

a type of forest apparently so similar to that of Kilimanjaro and Mt Kenya, remains to be investigated.

MONTANE EVERGREEN FOREST—MT CAMEROON

Remote from the montane forests so far considered, this forest, which is exceptionally wet in parts but ceases as low as 7000 ft (Chapter 11), has been discussed by Keay (1955), but no list of trees is available. It appears from Serle (1950) and Eisentraut (1963) that thirty-seven montane forest species of birds are found on the mountain. In addition, six species that are typically lowland ascend into the montane forest on this mountain, which makes the total of forty-three species included in Table XXV.

EXTRA-TROPICAL EVERGREEN FOREST—SOUTH AFRICA

Finally it is interesting to consider the bird fauna of the Knysna forest, close to the southern shore of the Cape Province and some 12° outside the tropics (about the same distance as northern Morocco). It has been fully described by Phillips (1931), who listed fifty-six tree species under "Climax high forests". Its bird fauna, which I am indebted to C. J. Skead for verifying for me, is a hotch-potch. Of its thirty-six group D plus group E species, five are South African endemics, eight are typically montane forest species in the tropics, sixteen are known in the tropics as forest species either lowland or adaptable, and seven are not regarded in the tropics as typical forest species at all. However, even with these auxiliaries the variety of D plus E species is smaller than in any of the forests discussed above except for the little forest of Cholo Mountain (equal) and the vestigial Soche.

DISCUSSION

It is useful to summarize as in Table XXVII the available data on the forests mentioned.

It will be seen from the totals in Table XXVII that the samples of lowland evergreen forest quoted have much more varied bird faunas than those of the higher-altitude and extra-tropical forests. In variety of birds the Amani intermediate forest exceeds, and the little relict Malawi intermediate forests nearly equal, the much larger montane forests. Thus, prima facie, a broad correlation exists in evergreen forest between warmth and the species diversity of the birds. What, then, are the ecological factors immediately responsible? When in Chapter 8 we compared certain tropical African bird faunas with bird faunas in the North Temperate Zone and found the former richer, it became clear that part—though only part—of the difference was due to the prevalence in the African areas of species dependent on fruit and on large insects, both of which could be had all the year round in the African habitats but not in those of the Temperate Zone. It is not clear that there are significant differences in this respect between the lowland and montane forests of tropical Africa, so that the search for other factors becomes all the more necessary.

TABLE XXVII

African evergreen forests and numbers of bird species

Forest	Estimated area*	Estimated no. of tree species	Recorded no. of bird species
Lowland primary			
Bwamba, Uganda	(20 sq. miles)	?	128
Budongo, Uganda	(116 sq. miles)	?	106
Lowland secondary			
Gambari, Nigeria	80 sq. miles	75+	85
Owerri, Nigeria	25 acres	?	73
Intermediate			
Amani, Tanganyika	(2 sq. miles)	?	60
Cholo, Malawi	3 sq. miles	30	36
Soche, Malawi	300 acres	?	30
Montane			
West Usambara, Tanganyika	150 sq. miles	90	43
Kilimanjaro, Tanganyika	250 sq. miles	150	37
Mt Kenya	250+ sq. miles	?	43
Ruwenzori, E. Congo	200 sq. miles	?	57
Kabobo, E. Congo	150 sq. miles	?	44
Mt Cameroon	100 sq. miles	?	43
Extra-tropical			
Knysna, Cape Prov.	?	56	36

* The areas in parentheses are not ecologically segregated, but are samples of larger forests.

In the deciduous forests of North America MacArthur and MacArthur (1961) concluded on the basis of numerous censuses and subsequent statistical treatment, that "bird species diversity can be predicted in terms of the height profile of the foliage density. Plant species diversity, except by influencing the profile, has nothing to do with bird species diversity." Subsequently, however, bird species diversity was found to be "proportional to foliage height diversity" (MacArthur *et al.*, 1962). I should like to see this tested in the African forests, montane and other. On purely subjective impressions of profiles of which I have experience, I doubt whether the conclusions would be found applicable.

An alternative, and superficially more attractive, hypothesis is that the species diversity of the birds varies with the species diversity of the plants, which in evergreen forest practically means that of the trees. It would not be impossible on the botanical side to test this hypothesis because much basic information is contained in the unpublished records of forestry departments. I have given such scraps of information as I have been able to collect in the foregoing pages, not because anything can be built upon them as they stand, but as an indication of magnitudes. Thus, both the West Usambara and the Kilimanjaro montane forests, as a whole, housing thirty-seven and forty-three bird species respectively, are respectively composed of "perhaps ninety" and between 100 and 150 species of trees. Yet the Cholo Mountain forest, with some thirty species of trees, and the Knysna with fifty-six each house thirty-six species of birds. The figures are not such as to yield statistically significant results, but they do not point to close correlation. The Gambari forest, with seventy-five tree species in 2 acres, probably has a more complex botanical environment than the montane forests, and with this its more ample bird fauna, eighty-six species, might perhaps be correlated, but certainty is lacking.

It remains to compare the figures for the African forest bird faunas with those of some temperate broad-leaf woodlands (which are, however, deciduous). In Europe it is difficult to decide what species should be regarded as originally woodland birds, since the habitats have been so disturbed and broken up, with multiplication of edges. Also, several species of birds seem to use the woodlands only incidentally or marginally. Even for English woodlands (with less than a score of tree species), which have recently been discussed by Yapp (1962), some doubts remain, but the number of bird species certainly does not exceed thirty. This falls within the North American range, for according to Kendeigh (1961), in numerous censuses the total of breeding species was from nineteen to thirty-three, usually twenty-three to thirty. These North-Temperate totals are only about one-third of those worked out above for lowland forest in Africa, and also slightly less than those of the African montane forests, or that of the remote extra-tropical Knysna forest of the Cape Province, or even, if any allowance at all is made for area, that of the relict 300 acres on Mt Soche in Malawi. The make-up of the northern communities is of course very different from those of the tropics, for they include no purely fruit-eating species but, in England, as many as six tits (Paridae), compared with none, or at most one, in African forests.

GENERAL DISCUSSION

On the basis of the bird faunas listed in Tables XXIV and XXV, re-summarized in Table XXVII, it appears that lowland evergreen forest communities are the richest in species, with around 120, while montane evergreen forests have only about one-third. Brachystegia woodland, with up to eighty-nine species, is next richest to lowland evergreen and acacia steppe ranks nearly as high, with sixty-three to eighty-six, while mopane woodland is much poorer, about the same as montane evergreen forest, with up to about fifty species. The most noteworthy feature about these comparisons is that there is no consistent relation between the number of bird species and the biomass of the vegetation nor probably between number of bird species and number of plant species, since acacia steppe has about three-quarters as many bird species as either lowland evergreen forest or brachystegia, although it is a simpler environment, with sparser and far more puny trees belonging to a very limited number of species. Furthermore, the herbage in the acacia steppe is probably less in mass and variety than in the other vegetation types except forest.

A noteworthy feature, for which no explanation is readily forthcoming, is the variation in proportion of passerine birds in the total C–E bird faunas of the different communities. It is highest in the six montane forests, 0·74–0·86, and the three intermediate forests, 0·72–0·80, while for the lowland forests the figures are 0·63–0·65, much the same as the brachystegia, 0·61–0·63. The acacia steppe, 0·48–0·62 and the mopane, 0·48 and 0·59, are lower still.

Striking differences between the various communities listed are found when we compare the representation of individual families. Most of them do not call for separate mention because they are governed by two principles. The first is that, as shown in Table V, a few families such as the Pycnonotidae (bulbuls), are everywhere characteristic of evergreen forest and so could not be expected to have strong representation in any deciduous community, while others, such as the larks, are similarly not to be expected in forest. Second, since lowland forest and brachystegia have the richest bird communities, most families will be more abundantly represented in them than in any others. Within this class a direct interrelationship worth mentioning is that the two parasitic groups, the cuculines and the honey-guides, are most richly represented in brachystegia and next in lowland forest, the two communities which provide the biggest variety of passerine hosts for the cuckoos and of barbet and woodpecker hosts for the honeyguides.

Leaving aside the cases governed by the two general factors mentioned above, the following points are worth noticing, as calling for examination in the field.

GROUP C

The acacia steppe is on the whole much richer than any other community in these birds, while no evergreen forest has any representative of group C but Phasianidae, and often only a single species. Brachystegia, with only one other family represented, is nearly as limited.

GROUP D

Bucerotidae (hornbills) are best represented in lowland forest, less in acacia steppe and least in montane forest and brachystegia, although in the last two vegetation types the trees are bigger and presumably provide more and larger holes for nesting-sites than does acacia steppe. This suggests that food, rather than nesting-sites, is the

factor governing the number of species of hornbills; and this is not expected in view of the fact that both insects and fruit contribute to the
food of hornbills in general. Caprimulgidae (nightjars) have three
species in brachystegia, compared with one in most other deciduous
communities and none in most forests. Picidae (woodpeckers) conform
to the second generalization mentioned above, in that there are most
species in brachystegia and lowland forest. It is, however, difficult to
understand why the montane forest should be so much poorer in woodpeckers, poorer too than all the deciduous woodlands, with only one
species, or exceptionally two, to get their food from a dense stand of
big trees.

GROUP E

Larks (Alaudidae) are represented by only one or two species in each
of the deciduous communities named (none of course in forest), except
that there are ten in the Kalahari sample. This is one aspect of the
wealth of lark species in south-western Africa (and the ecologically
rather similar Horn of Africa), as noted in Chapter 10; but this does
nothing to elucidate the problem of how such a number of allied species
manage to co-exist in a limited area. Shrikes (Laniidae) are unusual in
that their species are fewest in montane forest and the Senegal acacia,
even mopane woodland being superior in this particular family. Tits
(Paridae) are best represented in brachystegia and only one species
appears in any of the forests. The weaver species (Ploceidae) are most
numerous in lowland forest, where they are nearly all insectivorous,
whereas those in the other vegetation types listed are seed-eaters. Of the
latter a notable feature is that the acacia steppe is richer in both
estrildine and other (ploceine) weavers, especially the former, than
either brachystegia or mopane.

It must be presumed that differences of the nature noted above for
individual families of birds are related in part to the diversity in the food
resources of the relevant type in the communities concerned and in part
to the extent of the competition encountered by the birds of any given
family in any given community. Since we know nothing precise about
the food resources of any habitat and have only a broad general idea of
the feeding habits of the birds species concerned it is impossible at present
to carry this analysis further. The information does, however, suffice for
us to obtain some idea of the extent to which in each vegetation type
the bird community depends on each main category of food, (a) invertebrates, (b) fruit, (c) seeds and roots.* For this purpose I have on
the basis of literature and personal knowledge made a snap assessment
of each species and allocated each as "mainly" (or wholly) taking for
example insects (species in point being any of the flycatchers), or
"partly" insects and "partly" seeds (e.g. a lark). This enables us to say

---

* Nectar, which is admittedly a distinct category, is here omitted; it is taken in tropical
Africa only by sunbirds (Nectariniidae) and Zosteropidae and is apparently always additional to insects, etc.

in each community how many species are dependent "mainly" and how many "partly" on each broad category of food. Further, by reckoning each "mainly" as one point and each "partly" as half, it is possible to calculate for each bird community what proportion of feeding attention is devoted to each main category of food and hence to compare the importance of each category in the resources (from birds' point of view) of each habitat. The results are shown in Table XXVI. It must be emphasized that, for this purpose, in each community every bird species is treated as if it were represented by the same biomass—a convention rendered necessary by the absence of census data. For similar reasons seasonal changes in the ecology are perforce ignored. Since these factors tend to operate similarly within each vegetation type but differently between types, they do not invalidate comparisons between communities of the same type but they may to some extent do so between types. With these caveats in mind the following comments may be made.

(1) In all communities more than half the feeding attention is devoted to invertebrates. On the limited number of samples the proportion is highest in brachystegia and lowland forest, lowest in acacia steppe; montane forest shows wide variation, which I cannot explain, particularly the low figures for the East African mountains of Kilimanjaro and Mt Kenya. Some enlightenment might be forthcoming if the feeding on vertebrates could be analysed in terms of foraging behaviour, including preferences defined for feeding strata, but I feel that knowledge is insufficient for this at present.

(2) More attention is given to fruit in evergreen forest than in the other communities, but again there is rather high variation. On the other hand, acacia steppe, brachystegia and mopane all rank very low, under 0·15.

(3) Attention given to seed is decisively highest in acacia steppe, and is consistent at about 0·3. Next comes mopane, with all the forest and brachystegia communities very low. The large number of seed-eating species accommodated in the acacia steppe is particularly noteworthy.

The foregoing comparisons and their admittedly unsatisfactory nature suggest the need for basic research into the food resources of each vegetation type and into the exact niches of the bird species present. This would be least difficult in the deciduous habitats and nowhere more likely to give good results than in the acacia steppe amongst the seed-eaters. Here many of the bird species concerned are common and easy to collect, while the plant species are relatively few. On the other hand, to the ordinary observer clear-cut ecological segregation between many of the seed-eating species, at any rate of approximately the same size, cannot be envisaged and patient work at all seasons of the year will be needed to define the feeding range of each species and the precise niche peculiar to it.

The Bird Faunas of the West-coast Islands

General Introduction

The resident land-bird faunas of the islands off Africa, including Madagascar, make a valuable comparative study. They are summarized by families in Table XXVIII, working first down the west coast, Canaries, Cape Verdes, Gulf of Guinea islands, and then up the east coast, Madagascar, Comoros, Zanzibar with Pemba and Mafia, and finally Socotra. These are all the important islands with resident land-birds wholly or mainly of African stock. The Mascarenes, Mauritius, Réunion and Rodriguez, are omitted as having presumably been populated from Madagascar; the Seychelles are not likely to have been populated direct from Africa; St Helena possesses only one indigenous land-bird, a *Charadrius*, though it evidently had others a few centuries ago (Ashmole, 1963a); Ascension Island perhaps never had a land-bird other than a rail (? *Rallus* sp.) that became flightless there (Ashmole, 1963b).

Any species, the occurrence of which is believed to be due to human introduction, is omitted from Table XXVIII. Group A and group B birds are included for the sake of completeness, but they are of minor interest for the present purpose. Most water-birds are of little zoogeographical significance and neither they nor the group B birds are of value for statistical purposes, the first because the presence of a single pond or swamp on an island can have so much effect on its water-bird species, the second because raptors and owls often go unnoticed or uncollected.

From Table XXVIII the following families found breeding on the African continent, but on none of its islands, are omitted: Balaenicipitidae, Balearicidae, Pelecanidae, Sagittariidae, Struthionidae, Coliidae and Pittidae. It will be seen that this table includes only one group D family, the colies, and one passerine, the sub-oscine philepittas. The Laridae and the Phalacrocoracidae are retained in Table XXVIII solely for those species which typically breed on fresh water.

The islands dealt with in this and succeeding chapters are of two types, those which are on the continental shelf and those which appear never to have been connected with the mainland. Of the continental islands Fernando Poo, Zanzibar and Mafia are typical examples, and can be accepted as having been cut off from the mainland only in Late Pleistocene times as a result of the post-glacial rise of the ocean level. Such islands are too young for evolution on their surface to have proceeded far and as a rule they are sufficiently close inshore for birds to fly across not rarely and hence for gene-flow from the continent to hamper the divergence of the insular population. Such islands moreover start their independent existence with the full complement of species forming the continental biomes occupying their surface at the time of their separation. Subsequent change is likely to be in the direction of impoverishment: isolated populations are particularly liable to extinction for obscure reasons, even if it is not brought about by that human interference from which all the African islands have suffered with varying degrees of severity.

Two islands on the list, Pemba and Sokotra, are of intermediate status. The geological history of Pemba is uncertain, but it is probably, like Sokotra, an "old" continental island dating from Plio-Pleistocene times, so that opportunity for evolution on its surface has been considerable. The remaining African groups seem never to have been joined to the mainland, or, in the case of Madagascar, to have been separated very early in the history of the Class Aves. Hence these "oceanic" islands have been colonized by chance dispersal from the mainland. The factors determining the birds such an island receives are the prevailing winds, the nature of the adjacent mainland, the individual and specific mobility of the birds breeding there and, in some of the African islands, the passage of Palaearctic migrants. Successful establishment of a species on an island depends partly on the arrival of male and female within each other's sexual lifetime, so that in this respect gregarious species will have an advantage;* partly on the island's possessing a habitat sufficiently like that to which the new arrival is native; and partly on no species of too closely similar ecology being already established, as referred to again below.

So far as the actual process of colonization is concerned, while of course the manner in which the various oceanic islands were originally populated can at best only be inferred, we have before us one most remarkable modern example of the colonization of oceanic islands. Within thirty or forty years of their introduction to New Zealand a number of European species of birds, which as it happens are all typically non-migratory, had dispersed in all directions (hence not in close dependence on prevalent winds) and had established themselves on islands up to 550 miles away (Williams, 1953). In these cases the actual establishment on the several islands may have been facilitated by the presence there of man-made habitats; but, apart from this, unless tropical birds in general are less liable to disperse than the Temperate-Zone species concerned, this antipodean example gives some indication of the frequency with which potential colonists may make their landfall on all the African islands.

The smaller the island the less the opportunity for diversity of habitat and the smaller the bird fauna the island can support. The earliest birds to establish themselves on an oceanic island may, however, be able to

* So have mammals over birds, since a single pregnant female can suffice.

Table XXVIII

The resident land-birds of the African islands

	Canaries	Cape Verdes	Gulf of Guinea				Madagascar	Comoros				East Afr. Isl.			Sokotra
			F. Poo	Príncipe	S. Tomé	Annobon		G. Comoro	Mohéli	Anjouan	Mayotte	Mafia	Zanzibar	Pemba	
Group A															
Anatidae	—	1	—	—	—	—	10	—	—	—	—	1	3	3	—
Anhingidae	—	—	—	—	—	—	—	—	—	—	—	—	—	—	—
Ardeidae	—	2	2	2	3	1	12	1	2	4	1	4	10	9	—
Charadriidae	3	1	—	—	—	—	4	—	—	—	—	—	—	—	—
Ciconiidae	—	—	—	—	—	—	2	—	—	—	—	—	—	1	—
Jacanidae	—	—	—	—	—	—	1	—	—	—	—	1	1	1	—
Laridae	—	—	—	—	—	—	2	—	—	—	—	—	—	—	—
Pandionidae	1	—	—	—	1	—	—	—	—	—	—	—	—	—	—
Phalacrocoracidae	—	—	—	—	—	—	1	—	—	1	—	—	1	1	—
Phoenicopteridae	—	1	—	—	—	—	—	—	—	—	—	—	—	—	—
Podicipitidae	—	—	1	—	—	—	3	—	1	—	—	1	—	—	—
Rallidae	—	—	3	—	3	—	11	—	—	—	—	2	3	5	—
Recurvirostridae	—	—	—	—	—	—	1	—	—	1	1	—	—	—	—
Rostratulidae	—	—	—	—	—	—	1	—	—	—	—	—	—	—	—
Scolopacidae	1	—	—	—	—	—	1	—	—	—	—	1	1	1	—
Scopidae	—	—	—	—	—	—	1	—	—	—	—	—	—	—	—
Threskiornithidae	—	—	1	1	1	—	4	—	—	—	—	1	—	1	—
	5	5	7	3	8	1	54	1	3	6	2	11	19	22	—

Group B

Aegypiidae	1	—	—	—	—	—	—	—	—	—	—	—	1	1	3
Aquilidae	1	7	6	3	2	3	2	3	11	—	1	—	2	3	3
Falconidae	2	1	—	—	1	1	1	1	3	—	—	—	—	1	3
Strigidae and Tytonidae	1	1	2	1	2	2	1	2	6	1	2	—	3	1	2
	5	9	8	4	5	6	4	6	20	1	3	—	6	6	9

Group C

Burhinidae	—	1	1	1	—	—	—	—	—	—	—	—	—	—	1
Glareolidae	—	—	—	—	—	—	—	—	1	—	—	—	—	1	1
Mesitornithidae	—	—	—	—	—	—	—	—	3	—	—	—	—	—	—
Otididae	—	2	4	1	—	—	—	—	—	—	—	—	—	1	1
Phasianidae	—	—	—	—	1	1	1	1	4	—	1	—	—	—	3
Pteroclidae	1	1	1	—	—	—	—	—	1	—	—	—	—	—	1
Turnicidae	—	—	—	—	—	—	—	—	1	—	—	—	—	—	—
	1	4	6	2	1	1	1	1	10	—	1	—	—	2	7

(continued overleaf)

TABLE XXVIII (cont.)

Group D	Canaries	Cape Verdes	Gulf of Guinea				Madagascar	Comoros				East Afr. Isl.			Sokotra
			F. Poo	Príncipe	S. Tomé	Annobon		G. Comoro	Mohéli	Anjouan	Mayotte	Mafia	Zanzibar	Pemba	
Alcedinidae	—	1	6	2	1	—	2	1	1	1	1	4	4	4	1
Apodidae	2	1	6	2	2	—	4	2	1	2	2	2	3	3	—
Bucerotidae	—	—	1	—	—	—	—	—	—	—	—	—	1	1	—
Capitonidae	—	—	3	—	—	—	—	—	—	—	—	1	2	1	—
Caprimulgidae	—	—	1	—	—	—	2	—	—	—	—	1	1	1	1
Columbidae	4	1	5	3	5	2	4	5	6	5	5	4	6	4	3
Coraciidae*	—	—	1	1	1	1	7	1	1	1	1	2	2	1	—
Cuculidae	—	—	5	1	1	1	13	—	—	—	—	3	3	2	1
Indicatoridae	—	—	1	—	—	—	—	—	—	—	—	—	—	—	—
Meropidae	—	—	—	—	—	—	1	—	—	—	1	1	1	1	—
Musophagidae	—	—	3	—	—	—	—	—	1	—	—	—	—	—	—
Phoeniculidae	—	—	—	—	—	—	—	—	—	—	—	1	1	—	—
Picidae	1	—	3	—	—	—	—	—	—	—	—	—	2	—	—
Psittacidae	—	—	2	2	1	—	3	2	1	2	—	1	1	1	—
Trogonidae	—	—	1	—	—	—	—	—	—	—	—	1	—	1	—
Upupidae	1	—	—	—	—	—	1	—	—	—	—	—	—	—	—
	8	3	39	10	10	3	37	11	11	11	10	20	27	18	6

Group E															
Alaudidae	1	4	–	–	–	–	1	–	–	–	–	–	–	–	1
Campephagidae	–	–	1	–	–	–	1	1	1	–	–	1	1	–	–
Corvidae	2	1	1	–	–	–	1	1	1	1	1	1	1	1	1
Dicruridae	–	–	1	1	–	–	1	1	–	1	1	–	1	–	–
Emberizidae	1	–	–	–	–	–	–	–	–	–	–	–	–	–	2
Eurylaemidae	–	–	1	–	–	–	–	–	–	–	–	–	–	–	–
Fringillidae	6	–	1	1	1	–	–	–	–	–	–	–	–	–	1
Hirundinidae	–	–	1	–	–	–	2	–	–	–	–	2	2	2	1
Laniidae	1	–	1	–	1	–	–	–	–	–	–	2	3	–	1
Motacillidae	2	–	–	–	–	–	1	–	–	–	–	1	1	1	1
Muscicapidae	–	–	9	1	1	1	6	2	1	1	1	3	3	1	–
Nectariniidae	–	–	13	2	2	–	2	2	2	1	1	4	5	2	1
Oriolidae	–	–	1	–	1	–	–	–	–	–	–	–	–	–	–
Paridae	1	–	–	–	–	–	–	–	–	–	–	–	–	–	–
Philepittidae	–	–	–	–	–	–	4	–	–	–	–	–	–	–	–
Ploceidae															
Estrildinae	–	–	12	2	3	–	1	1	1	1	1	6	9	5	–
Others	2	2	10	1	3	–	4	2	2	2	2	4	6	3	1
Pycnonotidae	–	–	10	–	–	–	2	1	2	1	1	3	3	–	–
Salpornithidae	–	–	–	–	–	–	1	–	–	–	–	–	–	–	–
Sturnidae	–	–	4	2	1	–	1	–	–	–	–	1	1	1	2
Sylviidae															
Cisticolae	–	–	–	–	–	–	1	–	–	–	–	1	1	1	1
Others†	5	3	14	–	2	–	10	1	2	1	–	2	4	2	1
Timaliidae	–	–	4	–	–	–	7	–	–	–	–	–	1	–	–
Turdidae	3	–	9	1	1	–	4	2	1	1	–	2	3	–	–
Vangidae	–	–	–	–	–	–	12	–	1	–	–	–	–	–	–
Zosteropidae	–	–	2	2	2	1	1	2	1	1	1	–	–	1	1
	24	10	95	13	18	2	63	16	15	11	9	33	45	20	15
Grand total	53	26	147	26	40	7	184	35	34	35	27	70	105	73	27

* Including (in Madagascar and Comoros) Leptosomatinae and Brachypteraciinae. † Including *Regulus*.

command a wider range of resources, i.e. occupy a more extensive niche, than they could in the community from which they came. Successive arrivals will be able to establish themselves if a suitable unexploited food is available, but if they have similar or overlapping food requirements to the birds already there, they will be able to establish themselves only if from the start they show greater efficiency in securing some part of the food supply which the original species has hitherto utilized. Destruction of the natural habitat by man may, if only partial, increase the diversity of the insular bird fauna by providing an additional set of conditions, so that, for example, a seed-eating bird of open country can get a foothold on an island originally covered with forest. In practice, human interference has nearly always been to the bad, resulting in the impoverishment of insular bird faunas to a degree that is exasperatingly uncertain when such a bird fauna is being discussed. It is a frequent feature of insular biology that a community may appear unbalanced; for example, the woodpecker niche is unoccupied on Pemba, an island with plenty of trees only 30 miles off the mainland. We are left guessing whether, human interference apart, this is the direct result of the chances of oversea colonization or whether for reasons we cannot discern the insular environment is less favourable than it appears to us.

On a small oceanic island evolution may be rapid: on the one hand the small number of individuals facilitates the establishment of any mutation that is not unfavourable; on the other hand-birds that reach such islands have freed themselves from the web of ecological checks and balances in which they were enmeshed in their closely-knit continental community. For example, on the islands there are no predators until raptorial birds appear, so that loss of flight can take place with impunity, and changes in average individual size very often take place. On the whole the tendency appears to be towards gigantism rather than dwarfing, and this preponderance is very marked in the islands of the Gulf of Guinea. The reasons for this are obscure and no doubt multiple, since the tendency does not seem in such cases to have any relation to Bergmann's Rule—to the effect that body size tends to increase in cooler climates—or to the presence of a closely related species from which ecological segregation is necessary (see discussion in Amadon, 1953); and in the present state of knowledge any discussion of size-modification in relation to food supply can only be speculative.

Another feature of insular birds is that in the absence of close relatives, ready differentiation from which conduces to successful reproduction, a population is liable to lose its specific recognition characters, of voice or plumage, so that birds on oceanic islands are liable to be duller and to have less developed songs than their nearest relatives elsewhere. Again, adaptation to different food is likely to facilitate changes in the beak. All such changes may contribute to reproductive isolation, so that if after an interval new individuals from the parent stock reach the island, interbreeding does not take place. Conceivably the niche occupied by the island stock will be sufficiently different from that of the newcomers from the mainland for old colonists and new to co-exist. Repeatedly we see that this situation, commonly referred to as a case of "double invasion", is indicated by the presence of two birds, clearly close relatives, on the same island. Within an archipelago double invasions can occur between closely related (congeneric) birds of different islands, as in the classic case of the Galápagos finches (Lack, 1947) and in the *Zosterops* of the Mascarene Islands (Moreau, 1957).

Since the circumstances of each African island or archipelago differ a good deal, the discussions on them in these pages will not follow uniform lines. Two groups, the Canaries and the Cape Verdes, are probably exceptional in the extent to which the vegetation has been degraded and changed, so that the biological significance of the individual island bird faunas has been correspondingly diminished. Partly on this account, partly because the components of these two archipelagos are so numerous, each of these bird faunas is listed in Table XXVIII as a whole, but all the other islands are listed singly. In my discussions I have throughout concentrated attention on the positive evidence. The negative is of very uncertain significance; and I have tried to avoid the common error of remarking that such and such a species "has not reached" the island in question. In face of the massive evidence for extinction over vast continental areas (often through the effects, primary or secondary, of human intervention), it seems wrong in most cases to assume that if any particular species has not been found by an ornithologist, i.e. virtually in the last 150 years, on a given island that seems ecologically and geographically likely for it, then that species never has been established there.

THE CANARY ISLANDS

The most recent general accounts of the Canary Islands bird fauna are those of Volsøe (1951, 1955), who gives also a full discussion, Knecht (1960) and Bannerman (1963).

The group consists of seven main islands (Fig. 58), all of volcanic origin and probably never connected with the mainland. They are all close to 28° N., from 60 to 300 miles west of the coast of Sahara Español, and they show much ecological diversity. The two inshore islands, Fuerteventura (680 sq. miles) and Lanzarote (340 sq. miles) are almost desert except for the cultivated patches and are nowhere higher than 800 m a.s.l. The other five main islands are more mountainous, rising in Tenerife to 3707 m (over 12 000 ft) and the rainfall, all coming in the winter months, is more adequate on the northern and western slopes than elsewhere. The lowest levels are mostly arid, but above about 1000 m the mountains were originally forested, on the lower slopes with a peculiar broad-leaved forest in which the laurel family (Lauraceae) predominate, and above that with an indigenous pine. Habitat destruction has been severe, perhaps for the most part in the five centuries of

European occupation. The now almost desert coastal zones are believed to have carried at one time *Juniperus phoenicea*, a widespread element of the Mediterranean macchia, and the mountain forests have been diminished gravely (see Volsøe's maps). Hence bird extinction may have been severe, but we have no means of knowing the extent. A pointer is the fact that the chough *P. pyrrhocorax* is known only from one of the outermost islands, La Palma, which it is not likely to have reached unless it at one time inhabited others in the group.

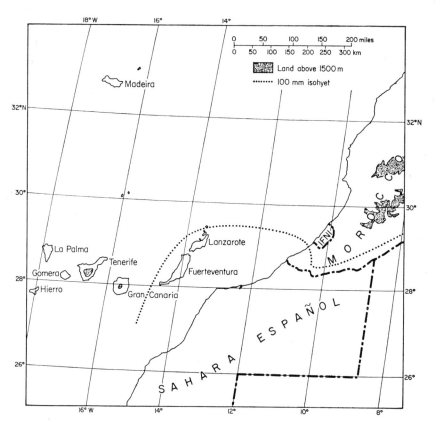

FIG. 58. The Canary Islands in relation to the limit of the Sahara.

The land-bird fauna of the archipelago as a whole is as shown by families in Table XXVIII. In addition to this total of fifty-three species, a few others have been reported as breeding, most of them in the middle of the last century, but none are of significance. From the total of fifty-three I exclude for the purpose of the present discussion the three shore-bird Charadrii and the Osprey *Pandion*. This leaves forty-nine species, by no means all of which occur on all the islands. In fact the richest island, Tenerife, counts only forty and, if the two low arid islands in the east, which are ecologically in a separate category, are kept apart, the number of species diminishes with, though not in proportion to, size of island, as in Table XXIX.

TABLE XXIX

	Area (sq. miles)	No. of breeding species	Additional doubtful species
Tenerife	790	41	1
Gran Canaria	590	38	1
Palma	280	34	—
Gomera	150	32	3
Hierro	110	24	5
Arid islands			
Fuerteventura	675	26*	1
Lanzarote	340	26*	2

* Omitting species breeding only on outlying islets.

Except for the few in the Canaries that have been regarded as endemic species (see below), all the insular birds also occur in Morocco. There is no typically Ethiopian species among them; three, limited to the dry eastern islands, are typically Saharan; the rest are Palaearctic. Of these, a number breed on the opposite mainland as far or nearly as far south as the Canaries. A few others, the Woodcock *Scolopax rusticola*, Chiff-chaff *Phylloscopus collybita* and Grey Wagtail *Motacilla cinerea*, appear in Morocco only in winter, the first only sparsely. Some others are high-mountain birds in Morocco, typically sedentary except perhaps for very local movements in cold weather: the Chough *P. pyrrhocorax*, Robin *Erithacus rubecula*, Firecrest *Regulus igneicapillus* and Rock Sparrow *Petronia petronia*. These are therefore separated from the Canaries by a barrier of lowland as well as the barrier of sea. Evidently the southward and downward shift of the biomes during a glaciation would have made it easier for several of the species now inhabiting the Canaries to invade them. On the other hand, Ethiopian species may never have had much opportunity for colonizing the Canaries from the mainland immediately opposite the islands. It is true that, as shown in Table III (p. 72), nineteen typically Ethiopian species exist in Morocco today, but it is not known whether an extensive West African bird fauna ever reached so far north. In any case, under the present ruinous ecological conditions of the Canaries at low altitudes, the chances of survival by any of the Ethiopian, as distinct from Saharan, species that might have established themselves there are not good.

Endemism in the Canary Islands at the specific level is high in the plants, but not in the birds. Although three-quarters of the birds are distinguishable subspecifically from continental populations, only two can unquestioningly be accepted as full endemic species, the Blue

Chaffinch *Fringilla teydea* and the pigeon *Columba junoniae*. Taxonomically, another pigeon, *Columba bollii*, the pipit *Anthus berthelotii*, the Stonechat *Saxicola dacotiae* and Canary *Serinus canaria* are borderline cases, to be regarded certainly as forming superspecies, if not conspecific, with *C. trocaz* of Madeira (and the Wood Pigeon *C. palumbus*; cf. Goodwin, 1959), *Anthus campestris*, *Saxicola torquata* (following Volsøe, 1951) and *S. serinus* respectively. Of all these six species *C. junoniae* is the most peculiar and probably the oldest of the island forms. The Blue Chaffinch presumably represents the first of two waves of Chaffinch stock. The second of these has indeed been in the archipelago long enough to produce three endemic subspecies, but in view of the evidence from elsewhere that "good" subspecies can be produced in much less than post-glacial time there is no reason to postulate that the present *F. coelebs* stock reached the islands any earlier than the last glaciation. The same applies to all the other birds that have one or more endemic sub-species (four in *Parus caeruleus*) in the islands. This is not to deny that the original colonization that gave rise to the existing populations of these birds may have been older than the last glaciation; but it is cer-tainly surprising that for only two species, *Fringilla teydea* and more especially *Columba junoniae*, does there seem strong reason to think further back in time.

On the whole the bird fauna of the Canary Islands strikes one as having a much more modern aspect than might have been expected. Only *Columba junoniae* and *bollii* are so strongly different from birds present today in north-western Africa and/or Iberia that there is much room for discussion about their affinities. These birds might either have been evolved in the islands or be colonists from the mainland where the parent stock has ceased to exist. In either case these two pigeons can hardly have stood alone. They must have formed part of a contemporary bird fauna and there is nothing to suggest what the associated species were. If these have disappeared without descendants, it could be the result of habitat destruction or of direct extermination by human beings; alternatively, and more probably, the earlier island birds could have found it impossible to compete with newcomers whose coloniza-tions were facilitated by the conditions induced on the mainland by a glaciation. Yet this did not happen with the two *Columba* spp. of the laurel forest, which prima facie constitute a double invasion by allied stocks. Nothing in Canarian bird ecology is more extraordinary than that the staple food of the two birds should, by all accounts, have been the fruit of the same species of tree, but if a critical investigation of their ecology could have been made the story might have been different. The information we have may perhaps have been based on observations made during a season of superabundance.

The few really dry-country birds, especially courser, bustard and Desert Bullfinch, are likely all to be post-glacial newcomers in the archi-pelago, even though they are subspecifically distinguishable from the

Saharan populations. If they reached the islands during an interglacial the improved vegetation during the ensuing glaciation, not at that time exposed to human destructiveness, is likely to have wiped them out again; and during the glaciation, moreover, there would have been no desert to sustain a parent stock on the neighbouring mainland coast.

While there is little endemism at the undoubted specific level, sub-speciation in the islands has been so brisk that only about one-quarter of their fifty birds are indistinguishable from continental populations. Moreover, twelve of the species have developed more than one sub-species within the archipelago. This shows that in some species isolation is more effective than might have been expected, in view of the fact that the water-gaps are only about 15–50 miles between neighbouring islands and that the western five are high enough to be in sight of each other. By contrast, the Spanish Sparrow has spread from the easternmost islands to the westernmost in the course of about a century.

As summarized by Volsøe, a main tendency is towards a darkening of plumage, not only as compared with the continent but also from east to west within the archipelago. This is in accord with Gloger's Rule, since the outer islands are the more humid. Also, in four-fifths of the insular subspecies the wings are shorter than in continental birds, but in the few birds whose body-sizes have been tested they are not commensurately smaller. Furthermore, in half the insular subspecies the beaks are comparatively big and in only one is the beak smaller than on the continent. At present the reasons for the prevalence of short wings and big beaks remain uncertain.

The question of the habitat range of individual species in peculiar insular conditions has been examined with more attention in the Canary Islands than in probably any other group (Lack and Southern, 1949; Cullen *et al.*, 1952). In the absence of congeners several species are considered to occupy a wider habitat range than they do on the continent; the most noteworthy examples are the Blue Tit *Parus caeruleus* and the Chiffchaff *Phylloscopus collybita*. The Blue Tits are the only members of their genus on the Canary Islands; they are found in a wide variety of habitat, including pine forest, which is not typical for this species. In the Canaries they evidently fill all the food niches that in Western Europe are divided between *P. caeruleus*, Great Tit *P. major*, Coal Tit *P. ater* and Crested Tit *P. cristatus*,* The Chiffchaffs of the Canaries breed not only in woods, but in low scrub without trees, of the type which in Great Britain would normally be occupied by the Willow

---

* The fact that the beaks of the Canarian Blue Tits (four subspecies) are abnormally long may be connected with their wide range of food; but Snow (1954) has a different interpretation. He points out that these birds also have long tarsi and for him these elongated appendages of the body, both bill and tarsi, are to be explained by Allen's Rule, according to which appendages shorten in cooler climates and lengthen in warmer. He regards the long bills in the Canarian tits as representing "the end-terms of a climatically-related trend" which he has shown to be well marked through Europe and North Africa.

Warbler *P. trochilus*. To such cases there is one insignificant contrast: the two *Fringilla* spp. divide the woodland habitats between them; the coniferous forest is occupied exclusively by the presumed earlier colonist *F. teydea*, so that on the islands *F. coelebs* has a more restricted habitat than is typical for the species.

CAPE VERDE ISLANDS

The Cape Verdes (Fig. 59) are an oceanic group, comprising about a dozen main islands, within an area of about 200 miles each way, centred round 16° N., 24° W. and between 350 and 550 miles west of Cape Verde. Opportunities for their colonization are thus poorer than

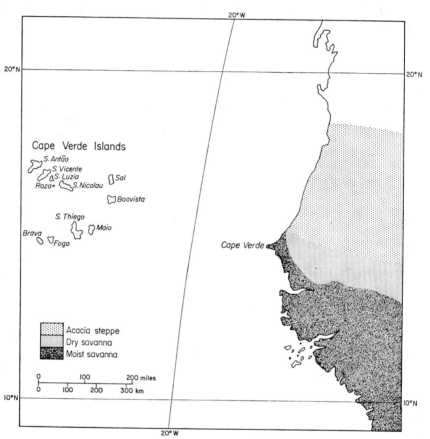

FIG. 59. The Cape Verde Islands in relation to the continental vegetation belts.

for the Canary Islands, both because of the greater distance that birds resident on the African mainland would have to cross, and because of the diminished chances of picking up stray Palaearctic migrants. The nearest part of Africa, Senegal, is an area where two thoroughly Ethiopian types of vegetation, acacia steppe and savanna, meet. But the flora of the islands cannot be definitely allocated, being "on the whole more tropical than Mediterranean"; while different groups of invertebrates vary in the extent to which they are dominated by tropical or temperate-zone elements (Monod, 1957). The dozen main islands of the archipelago vary from flat near-desert to rugged mountain masses, the highest of which, Fogo, is an active volcano. The ornithology and the environments were discussed by Bourne (1955), and to the data he gave there seems nothing to add, except for one or two taxonomic reassessments.

At the time of the islands' discovery about 1460 they were wooded and the mountain slopes are believed to have carried broad-leaved forests of non-Ethiopian type, something resembling the laurel forests of the Canaries 900 miles to the north-east. But subsequent habitat destruction has been extremely severe, "the last remnants of the indigenous cover are now restricted to inaccessible cliff ledges"; extinction of the original bird fauna has presumably been almost commensurate, especially since monkeys as well as cats and rats have been introduced, and the existing bird fauna is hence a great deal less rewarding as a basis for discussion than it would otherwise be.

Taking the archipelago as a whole, there seem to be twenty-eight breeding species, allocated between families as in Table XXVIII, but they are by no means all on any one island. Of the twenty-eight, two, the guinea fowl and the waxbill *Estrilda astrild*, can be dismissed as introductions. Two are endemic species. The more remarkable is a large-billed lark, originally described as *Spizocorys razae*, confined to the single arid islet of Raza (3 sq. miles) and certainly one of the most restricted species in the world. It seems to be a modified skylark (Hall, 1963b) with an enlarged beak, and was presumably derived from the same stock as *Alauda arvensis*, which does not today extend south of Morocco as either a breeding bird or a winter visitor. There is no means of knowing whether the Raza larks are relics of a population formerly widespread in the islands or on the mainland as well. The possibility cannot be excluded that the specific characters have been evolved *in situ*, with a rapidity facilitated by the restricted population. The peculiar digging habits reported of *razae* show that it would not be in competition with the two other more "ordinary" larks of the insular bird fauna, the *Eremopterix* and the *Ammomanes*, but the Bifasciated Lark *Alaemon alaudipes* also digs.

The second endemic species *Acrocephalus* (=*Calamoecetor*) *brevipennis*, while superficially a typical reed warbler, is also of uncertain ancestry, all the more so because the specific characters in the genus *Acrocephalus* are subtle. *A. brevipennis* closely resembles two sibling Ethiopian species *A. rufescens* and *A. gracilirostris*. Hence it may be an example of loss of specific characters (usually manifested in dullness of plumage) which is a not uncommon character of birds of small islands. Until recently neither *A. rufescens* nor *A. gracilirostris* was known from nearer than

Nigeria, some 2000 miles from the Cape Verdes, but now the former has been discovered in Senegal (Morel and Roux, 1962), on the continent opposite the islands, and so has another Ethiopian reed-warbler, *A. baeticatus*, which could also be closely related. Hence the immediate ancestry of the insular population is very doubtful indeed. It is noteworthy that *A. brevipennis* has most successfully adapted itself to the artificial habitats of the islands, for it frequents a variety of cover not related to a watery habitat and in behaviour it resembles a small turdine (a group absent from the Cape Verdes). This adaptation is paralleled by that of *A. baeticatus* in similar circumstances on Pemba Island on the east coast of Africa (see Chapter 18) and has occurred also in acrocephaline warblers in the Pacific (see, for example, Baker, 1951).

Of the remaining species in the islands, four, the courser, the *Alaemon* and *Ammomanes* larks and the raven (*C. ruficollis*), can be accepted as immigrants from the Sahara. Three others are definitely Ethiopian, the dry-feeding kingfisher *Halcyon leucocephala*, a sparrow *Passer iagoensis* and a lark *Eremopterix nigriceps*; eleven, all non-passerines, are distributed over much of both Europe and the Ethiopian Region; six are typically Palaearctic, so that this element bulks proportionally larger than anywhere on the continent within the tropics.

Individually, these Palaearctic elements are worth listing. (1) The duck *Anas angustirostris* breeds from Morocco to Transcaspia. (2) The local kite is an endemic subspecies, not, as might be expected, of the Black Kite *Milvus migrans* which is so widespread in the Old World, both temperate and tropical, but of the Red *Milvus milvus*, otherwise not breeding south of the Canaries and Morocco and not known to winter anywhere further south. (3) The local swift, which Lack (1956) regards as an endemic subspecies of *A. apus*, is likewise the most southerly breeding population of its species, but migrants from Europe penetrate all Africa. (4) The Blackcap *Sylvia atricapilla* of the Cape Verde Islands is indistinguishable from those which breed in Europe, Morocco and the Canaries, and winter in Senegal, on the mainland behind Cape Verde itself. (5) *Sylvia conspicillata*, a bird of the Mediterranean and the Maghreb, has breeding populations in Madeira and the Canaries as well as in the Cape Verde Islands. The populations of these three island-groups are regarded by Vaurie (1959) as forming a single subspecies, though he notes that there is a slight colour cline, with the palest in the south. (6) *Passer hispaniolensis*, the insular birds of which are indistinguishable from those of the Maghreb, breed no nearer than in the Canaries and at 28° N. on the continent, in the Sahara Español some 700 miles to the north. It may be added that the only pigeon in the Cape Verdes is a form of Rock Dove *Columba livia*, a species which we are accustomed to think of as a purely Palaearctic species, but which in fact lives also in Senegal and further east in tropical West Africa.

It will be seen that, because their present continental ranges lie so far to the north at all seasons, it is necessary to postulate colonization during the conditions of a glaciation for two of the six species, the duck and the kite, and to these may be added the parental stock of the Raza lark. The situation of *Sylvia conspicillata* is difficult to interpret but perhaps the same postulate can be made here also. The population of the Spanish Sparrow, a sedentary bird, could be another glacial derivative; but the recent history of the bird in the islands raises a suspicion that it has not been there long and it could be a human introduction. The Swift and the Blackcap could well have reached the islands as drifted migrants; the fact that the Blackcaps are not differentiated may be due to very recent arrival or to enough intermittent gene-infusion to maintain the island population unchanged. A third possibility, that the local conditions do not encourage divergence, cannot be excluded, though this seems most unlikely.

It is interesting that of the species shared with both Europe and tropical Africa two that are widespread in the latter area breed there only at high altitudes connoting cool climates, namely the quail *C. coturnix* and the buzzard *B. buteo*; so that these also are most likely to have established themselves in the Cape Verdes during a glaciation. One of the Ethiopian elements on the islands, an endemic form of *Passer iagoensis*, is vastly isolated from its continental relations, for the species is absent from the whole of West Africa, without any obvious ecological reason, and is first found 3000 miles to the east, in the Sudan. This is an even more extensive isolation from its relatives than in the case of *Acrocephalus brevipennis*, noted above, and with less differentiation.

Besides two endemic species, subspecific endemics are numerous in the archipelago. If we omit from the calculation the "water-birds", and also the Osprey *Pandion* and the *Neophron* vulture, neither of which shows geographical variation anywhere in their world range, two-thirds of the insular birds are different from their relatives on the mainland. They do not, however, show the marked subspeciation within the archipelago which has occurred in the Canaries, a difference for which no explanation can be offered. The only generally accepted instance of such subspeciation in the Cape Verdes is that the kestrels *Falco tinnunculus* of the south-eastern islands are distinguishable from those of the north-western —an unexpected case of this phenomenon because the individuals of this species are so individually mobile. As Professor A. J. Cain has pointed out to me, the result suggests that there has been selection for sedentariness among the insular kestrels. The most general trend among the island endemics as a whole is, as Bourne has noted, shortness of wing and leg, and many are also darker than their mainland relatives; but the relation of wing-length to body-size does not seem to be known.

A large proportion of the species in the islands show differences in their ecology from the mainland birds. These are in part correlated with the absence of closely allied species (so that, for example, *M. milvus* has to some extent the feeding habits of *M. migrans* elsewhere) and in part correlated with the peculiarities of the island environment (so that *Ardea purpurea*, which is normally a swamp bird, feeds on dry ground in the Cape Verdes and takes different prey accordingly). The two *Sylvia* spp., the Blackcap and Spectacled Warbler, which are secured from competing with each other by their difference in size, both occur widely in great numbers, feeding not only in trees and shrubs, but in towns, around houses, along rock faces and on the ground.

Bourne has designated nine habitats, including five that are purely man-made, and has tabulated the species and their relative abundance in those habitats on three of the islands. Each of these habitats has six or seven species. One habitat, "ravines", has, for example, the (dry-feeding) kingfisher, Blackcap, Spectacled Warbler, *Estrilda* and two species of sparrow, as well as the introduced guinea-fowl. The extent to which *Passer iagoensis* and the more recently arrived *P. hispaniolensis* overlap ecologically seems to be considerable and it will be interesting to see how this situation develops.

THE GULF OF GUINEA ISLANDS

These four islands (Fig. 60) form a most interesting series, the ornithology of which has been discussed by Amadon (1953), especially from the taxonomic and evolutionary point of view, with a complete bibliography.* They lie on a straight line, diverging somewhat from the main north-and-south direction of the African coast: Fernando Poo, some 2000 sq. km (800 sq. miles), 32 km (20 miles) offshore and only 50 km from the top of Cameroon Mountain; Principe, 126 sq. km, 220 km to the SW. and 220 km from the mainland; São Tomé, 100 sq. km, 146 km SW. of Principe and 280 km from the mainland; Annobon, 15 sq. km, 180 km SW. of São Tomé and 340 km from the mainland. All four islands are volcanic mountain masses and probably post-Miocene, built up, that is to say, in the course of the Pliocene and Pleistocene. Fernando Poo is on the continental shelf; the other three islands have, with little doubt, never been connected with Africa or with each other. All four, especially São Tomé and Principe, have been afflicted with plantation agriculture for centuries but they remain heavily vegetated.

The islands have been somewhat neglected by modern ornithologists but can probably be regarded as adequately known in the faunistic sense. Biologically they are not, apart from tiny Annobon (see especially Fry, 1961), the main modern contributions being those of Basilio (1963) for Fernando Poo and of Snow (1950) for São Tomé and Principe on the basis of brief visits. It is difficult to assess the amount of extinction on the islands. Some of the endemic species have certainly adapted themselves to plantation conditions; but there is one species of São Tomé,

---

* For a sophisticated mathematical discussion of the number of bird species on these islands, see Hamilton and Armstrong (1965); but it may be doubted whether the different conditions of the various islands have been given due weight.

Neospiza concolor, a monotypic genus somewhat doubtfully placed in the Ploceine Weavers, that has not been recorded since 1888.

In compiling the insular lists one encounters some uncertainty as to whether some of the species found there occur naturally and permanently on the islands or not; and this applies to about one-eighth of all the species recorded for Principe and São Tomé. Most of the uncertainty relates to weavers, estrildine and other, a canary and perhaps the

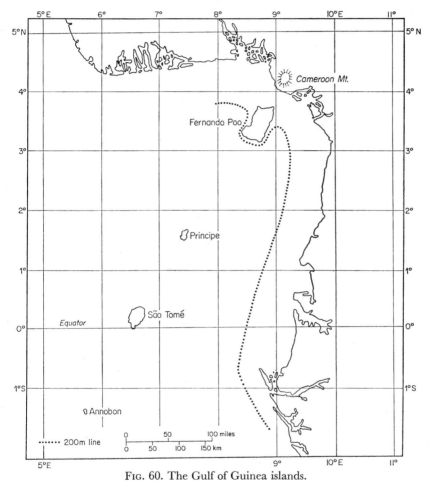

Fig. 60. The Gulf of Guinea islands.

parrots; members of these groups are especially liable to be carried about as pets, individuals taken on the islands show no significant divergence from mainland populations, and moreover, as Snow has pointed out, some of the island populations of these birds have been transitory. One ploceine, *Euplectes aureus,* a very distinct "bishop", is particularly difficult to assess. It was common on São Tomé in 1928–29, but Snow did not see it there in 1952. Otherwise it is known only from the coast of

Angola (Traylor, 1963), far across the sea to the south-east. On the whole I think it more likely that it was introduced from the continent and briefly flourished on the island (as some other mainland birds have done), rather than the reverse.

Fernando Poo, much the largest of the islands, is also the highest, reaching 2850 m (over 9000 ft). It is separated from the shore of the Cameroons by a strait only 60 m deep. Evidently it was formed by the post-glacial rise in sea-level and, from the data quoted in Chapter 3, it can be calculated that isolation by this means took place some 11 000 years ago. In this volcanic area there may, however, have been changes in the sea-bottom which would greatly modify this date; at the same time, since the strait is flanked by the mountainous mass of the island at one side and the bulk of Cameroon mountain at the other, its width of about 20 miles cannot have been much altered. The climate of Fernando Poo is very wet, the vegetation rich. Most probably the entire island, except for the uppermost 500 m or so of Clarence Peak, was under heavy forest, much of which survives. As expected from the size, height and proximity of the island to the mainland, the bird fauna of over 140 species (Table XXVIII) is quite rich, especially seeing that there was probably no natural open country.

A complete annotated list of the birds has recently been provided by Basilio (1963). As will be seen from Table XXVIII the bird fauna appears to be a well-balanced one, with no important gap in family representation except for the absence of Phasianidae. True, some others are poorly represented: it is odd that only a single hornbill is present, though there are half a dozen on the opposite mainland coast, and no barbets except three of the smallest size (*Pogoniulus* spp.), compared with a dozen on the mainland. In general, however, Fernando Poo has a very close faunistic resemblance to Cameroon Mountain and the immediately surrounding lowlands. Three-quarters of the island species are among those I have classified in Chapter 5 as typically forest birds; some others are everywhere associated with more open humid country while a few, such as the crow *Corvus albus* and the quelea, are typically associates of man and his cultivation and have probably arrived in recent colonial times following clearing of forest.

Fernando Poo has only one endemic species of bird, the aberrant zosteropid, *Speirops brunnea*, but subspecific endemism is considerable. Amadon accepted thirty-seven insular subspecies and queried several more. This means that fully one-quarter of the insular birds show some divergence from the mainland populations, a proportion which compares closely with the proportion of montane species on Cameroon Mountain that show divergence from the other montane populations in the Cameroons (see Chapter 11). It may be inferred that the existence of a water-gap up to about 20 miles wide for something like 11 000 years has had much the same effect on the rate of subspecific evolution as about the same width of lowland forest has had with regard to the montane birds of this area over an indefinitely longer period. However, when we come to consider the islands on the continental shelf of the East African coast, cut off by about the same width of water for about the same period of time (Zanzibar and Mafia) or for much longer (Pemba), we shall find that endemism at the subspecific level is far lower than the one-third in Fernando Poo. I have no doubt that the significant factor in this result is the far higher proportion of forest species, about three-quarters, in the Fernando Poo bird fauna, compared with less than one-tenth in any of the East African islands; for species typically confined to forests are, it seems, more sedentary than those of other vegetation types and hence gene-flow between isolated populations of forest birds is especially limited.

It is convenient to consider the next two islands, Principe and São Tomé together, although the specific differences are great and are the more striking to the observer because certain endemics peculiar to each island are (happily) among the birds most prominent on them (Snow, 1950). The smaller island, Principe, only 50 sq. miles and rising to 3200 ft (barely 1000 m), would seem to have been naturally covered with forest, but "development" had already proceeded far by 1909 (Alexander in Bannerman, 1914) and Snow found that it had been nearly all replaced with plantations except at the highest elevations. If we exclude those species in Amadon's list which he queries and also Quelea, which I find difficult to believe is a genuine colonist, the bird fauna is limited to twenty-six species, including two herons.

São Tomé, eight times the area of Principe, rises much higher, to 2024 m. It too is thickly vegetated, but mostly with plantations, and the original forest remains only on the steeper mountain sides. There is some doubt about the original state of the island. The rainfall has been quoted as 1000 mm (=40 in.), which is normally too low to support evergreen forest; but on such a mountainous island the results obtained with a rain gauge would depend greatly on its aspect and elevation. It has been claimed that "the presence of such well differentiated endemics [species] among grassland or savanna birds as *Prinia molleri*, *Lanius newtoni* and *Ploceus grandis* indicates that there has always been some open country on this island" (Amadon); but the evidence from these species is not good. It is true that especially the first two of the species named would, from the ecology of their nearest relatives on the neighbouring continent, not be expected to be typically forest birds, but the information from the island itself is inconclusive or conflicting. Concerning *Lanius newtoni*, "no naturalists have made any notes" and concerning *Ploceus grandis* we know only that Boyd Alexander reported it "in the vicinity of [not "in"] the cocoa-plantations on the lower ground" (Bannerman, 1915, 1953). Snow failed to see the shrike or the weaver at all and found the *Prinia* abundant in every habitat, "in the mountain forests" as well as in the streets of the town of São Tomé and in all sorts of plantations. At first sight, better evidence for the pristine state of the island is provided by the fact that the São Tomé quail *C. coturnix*, unquestionably everywhere a species of open country, is accepted as an insular subspecies averaging "darker above than those from the mainland"; but the fact that the difference is merely on average makes it unimpressive and does not rule out recent colonization since the clearing of forest for agriculture, especially in view of the evidence for subspeciation in House Sparrow in 30–100 years, quoted in Chapter 1.

The species admitted as resident on São Tomé total forty against twenty-six on Principe but, as will be seen from Table XXVIII, most of the difference is in the group A and group B birds. No bird of prey or owl of any sort is known to inhabit Principe, an apparent gap in its bird fauna which is hardly likely to be genuine. However, although São Tomé is eight times as big and much more mountainous, it has only the same number of group D species (ten) as Principe, in exactly the same five families, though eight of the species are different. (This compares with thirty-three species in thirteen families on the "continental" island of Fernando Poo.) Among other non-passerines, Principe has a dry-feeding (forest) kingfisher and a parrot (the Grey), which São Tomé lacks. In fact, with two fruit-eating pigeons, the Grey Parrot and two starlings, whole-time fruit-eaters are strong on Principe and it would be extremely interesting to know their diet. The starlings, the endemic species *Lamprotornis ornatus* and an endemic subspecies of *Lamprotornis splendidus*, being evidently derived from the same mainland stock, indicate a double invasion and it is surprising that in the face of such putative competition, for nest-holes if not for food, the second invader should have been able to establish itself. Moreover the birds do not differ in size. The only other Principe birds treated by Amadon as belonging to the same genus are the sunbirds *Cyanomitra olivacea* and *C. hartlaubi*. They do not, however, seem to be closely allied as are the two starlings. *Cyanomitra hartlaubi* is probably related to *C. reichenbachii* of West Africa; *C. olivacea* is widespread in the forests of tropical Africa generally. The Principe Olive Sunbirds are of the same subspecies as those of Fernando Poo and, since the species has elsewhere much capacity for geographical variation, it may be inferred that the Principe birds are derived from comparatively recent arrivals from Fernando Poo.

Among the passerines the absence of shrikes, orioles and perhaps warblers in Principe is noteworthy. This island is also badly off for weavers, for even of the two estrildines I have admitted the status of one is uncertain. Ecologically, the absence of a shrike may in part be compensated by the presence of the mainland drongo, which is, however, not of the forest species (and does not appear on São Tomé). I say "perhaps" for the warblers because the bird counted as a flycatcher (following Amadon) forms a monotypic genus, *Horizorhinus dohrni*, "considered a babbler by some, a flycatcher by others; the possibility that it is a thrush also deserves consideration" (Amadon); while Snow, with experience of it in the field, recorded that "in size and movements it corresponds to a Wood Warbler *Phylloscopus sibilatrix*". Thus it is clearly one of the intermediate passerine types, of which other examples occur especially among the "rare birds" of the African mainland (Hall and Moreau, 1962). *Horizorhinus* might be a relict representative of an old undifferentiated oscine stock now no longer recognizable on the mainland or it might be a descendent of some colonist of a modern family which had lost its characters in isolation. Although counted for the present purpose as a flycatcher, *Horizorhinus* certainly does not sound from Snow's description as if it would be in any way a competitor of a hawking flycatcher. It is then all the more surprising that there should be no Paradise Flycatcher *Terpsiphone* on Principe, for birds of this genus occupy all the other three islands, even tiny and remote Annobon, and are moreover widely dispersed in distant islands of the Indian Ocean.

While Principe has its one enigmatic endemic genus, São Tomé has two. One is the dingy heavy-beaked *Neospiza concolor*, which has been tentatively placed in the Ploceinae and also in the Fringillidae. Since it has not been seen since the 1880's it may be that we shall never know more about it than can be gleaned from the existing skins, one in London and one in Lisbon. If it is allied to the *Poliospiza* finches, as some have thought, then it is the earlier member of a double invasion of this stock; the later, *Poliospiza rufobrunnea*, has itself been in the islands long enough to develop into an endemic species, with distinguishable populations on São Tomé and Principe.

The other "difficult" endemic in São Tomé, *Amaurocichla bocagei*, was for a long time placed amongst the babblers, but is currently regarded as an aberrant warbler. However, its affinities among African birds are profoundly uncertain and Amadon even considers the possibility that the bird may be descended from some "stranded" Palaearctic migrant. São Tomé also resembles Principe in having three congeneric pairs of species, in *Columba*, *Cyanomitra* and *Ploceus*. However, the two pigeons, whether regarded as endemic insular species, as in the past, or not (Amadon), belong to different superspecies, though they probably eat much the same fruits. The two *Ploceus* spp. are also very unlike. One, *grandis*, is a giant relative of the widespread mainland *P. cucullatus*, which is almost entirely a seed-eater. The other, *P. st. thomae*, is highly differentiated, among other characters being devoid of yellow and having a strong musky smell like a Fulmar *Fulmarus glacialis*. It seeks its food by climbing like a nuthatch *Sitta* on trunks and twigs and belongs to the thin-billed group of insectivorous weavers, which includes *insignis* and *angolensis*. There is indeed no likelihood of competition between *st. thomae* and *grandis*. It may be added that *cucullatus* also has been collected on São Tomé, but the specimen is indistinguishable from mainland birds and the species is not regarded as native to the island.

The third congeneric pair of species on São Tomé, the sunbirds *Cyanomitra newtoni* and *C. thomensis*, both endemic species, probably derive from two colonizations by the same stock, which has also given rise to the *Cyanomitra hartlaubi* of Principe and, as noted above under that species, to the mainland *C. reichenbachii*. On São Tomé, where two representatives of this stock meet, one has become the smallest in the species group, the other, *thomensis*, not only the biggest in the group, but bigger than any sunbird in Africa. Moreover, it has developed quite abnormal habits, "digging its beak into the bark and using it as a probe; the whole manner of feeding more like a creeper than a sunbird" (Snow, 1950). Clearly the retention of *thomensis* in a separate genus of its own, *Dreptes*, until recently, had a good deal of justification.

A parallel to this is found in the Zosteropidae of the islands. The presumed first colonists (retained for a variety of reasons in a separate genus *Speirops*) have also greatly increased in size so that they are much bigger than any mainland zosterops, while the second arrivals on São Tomé and Principe, *Z. ficedulina*, are normal in size, and in colouring are distinguished mainly by their dullness. It is of the greatest interest, then, that on Annobon, where *Zosterops griseovirescens* is one of the only two passerines, it should be abnormally large, though no more divergent in colour than *Z. ficedulina*. (For details and discussion of the Gulf of Guinea Zosteropidae, see Moreau, 1957, pp. 384–391.) Much the same thing seems to have happened with the *Lamprotornis* starlings. On Principe, where a population of the mainland species *L. splendidus* is living alongside the first-wave colonist *L. ornatus*, they are not bigger than the *splendidus* of the mainland, but on Fernando Poo, where *L. splendidus* is the only starling of its genus, the individuals are distinctly larger.

The last and most remote of the four islands, Annobon, of less than 7 sq. miles, rises to 2000 ft. It carries a human population of 1400 but, surprisingly, "much of the original forest remains" according to Fry (1961). The same author found that the Moorhens *Gallinula chloropus* of the African subspecies *brachyptera*, which used to inhabit the crater lake, have gone; and he classes the little heron *Butorides striatus* and the kite *Milvus migrans*, which Amadon admitted as resident species, as merely vagrants. On the other hand, it seems from his evidence that the Emerald Cuckoo *Lampromorpha* (=*Chrysococcyx*) *cupreus* should be accepted as a breeding bird, which is all the more interesting because there are only two possible hosts, the local passerines being limited to the Paradise Flycatcher *Terpsiphone* and the endemic *Zosterops*. (A *Terpsiphone* has been recorded as a host of this cuckoo in the Congo.) It would be interesting to know whether the Annobon cuckoos are migrants as they seem to be elsewhere. The remaining land-birds are a Scops owl and two pigeons *Columba malherbii* and *Aplopelia larvata*. From their habits they would presumably not be in competition at all, but they seem to have fluctuated wildly in relative abundance. *Aplopelia* has developed an endemic subspecies, so it must have been on the island for some time.

The status of endemism in the birds of the Gulf of Guinea islands is summarized in Table XXX. As noted earlier, the three endemic (monotypic) genera are so odd that allocation to (sub-)families is uncertain. The relatively low endemism in the Fernando Poo bird fauna and in the groups A–C birds in all islands is emphasized. It will be seen that the great majority of the Principe and the São Tomé passerines differ from continental birds and most of them at higher than the subspecific level. In fact those three on São Tomé which do not are estril-

TABLE XXX

Endemism in the Gulf of Guinea islands

	Fernando Poo	Principe	São Tomé	Annobon
Groups A–C				
Total no. of species	13	3	12	2
Endemic species of uncertain superspecies	—	—	—	—
Other endemic species and semispecies	—	—	1	—
Endemic subspecies	1	1	3	—
Group D				
Total no. of species	39	10	10	3
Endemic species of uncertain superspecies	—	—	—	—
Other endemic species and semispecies	—	2*	3*	1*
Endemic subspecies	6	4	2	2
Group E				
Total no. of species	95	13	18	2
Endemic genera (monotypic)	—	1	2	—
Endemic species of uncertain superspecies	1	2†	4	1
Other endemic species and semispecies	—	4‡	7‡	—
Endemic subspecies	30	2	2	1

Figures marked (*) include one endemic pigeon occurring unchanged on all three islands. Figures marked (†) include an endemic finch shared between, and subspecifically separable on, São Tomé and Principe, and those marked (‡) two other birds to which the same applies.

dines, an *Estrilda*, a *Lonchura* and their parasite *Vidua*, which are likely to have established themselves only recently since clearings became important on the island. The same species account for three of the four non-endemic forms on Principe, the fourth being *Lamprocolius splendidus*, the interesting circumstances of which have been referred to already.

As will be seen from the footnotes to Table XXX, there are four birds, besides the *Cyanomitra olivacea* subsp., common to Fernando Poo and Principe, which seem to point to the colonization of one island from another, but the possibility cannot be ruled out that the present situation has arisen by colonization of all the islands concerned direct from the mainland, with subsequent extirpation of the stock there.

Trends of differentiation to the Gulf of Guinea islands have been fully discussed by Amadon. The colour differences from mainland forms are very variable, but perhaps that towards loss of lipochrome, as in the endemic oriole, the zosterops, the *Speirops* spp. and *Ploceus st. thomae*, is the most noteworthy. There are several cases of gigantism, in sunbirds, *Zosterops* and *Ploceus* especially. It is noteworthy that in the first of these cases (as in some others) the increased size may be correlated with the existence of a congeneric species on the same island, while in the last two species it is birds lacking such potential competition that are abnormally large. There is also a tendency for beaks of the insular birds to be bigger than those of mainland birds, even when general size (as measured by wing-length) is unchanged. Fernando Poo is affected no less than the outer islands, and no reason can be suggested.

17

The Bird Fauna of Madagascar

For the basic data on this island we can use the work of Rand (1936), who gives valuable descriptions of the country, a discussion of distribution and a complete list of bird species, with field-notes. Berlioz (1948) discussed the composition of the bird fauna. Paulian (1961) provided a wide-ranging discussion of the zoogeography of Madagascar and the neighbouring islands, but birds are mentioned only incidentally. I am greatly indebted to Dr Rand for reading and discussing my first draft of this chapter. The responsibility for its final form is of course mine.

Madagascar, lying between about 12° and 25° 30′ S., with an area of about 240 000 sq. miles, is separated from Africa by a channel nowhere less than 250 miles wide and deepening throughout its length to at least 2000 m (nearly 7000 ft). The persistent statements in zoological literature that the island has been separated since the Miocene seem to rest on nothing more than questionable inferences from the biogeography. It can now be accepted on geological grounds that its isolation dates from the Secondary (Millot, 1952). Hence it can be assumed that its bird fauna has arrived without the help of a landbridge, as Simpson (1940, 1952) has been prepared to accept also for its mammals.[*] The existence of the giant non-flying elephant birds, the Aepyornithidae, on Madagascar raises no great difficulty. Their prototype was presumably evolved *in situ* from a flying ancestor (cf. de Beer, 1964), as now seems to be accepted for the moas of New Zealand (Falla, 1964) and must certainly be accepted for the Dodo and the Solitaires of the Mascarene Islands.

Madagascar is a very compact island, with no important marine intrusions and with no evidence that it has at any stage in relevant geological time been dissected by the sea. It is simple in internal topography, with a single line of highland running down it west of centre (Fig. 61). It rises to 9000 ft but little is above 5000 ft and no real montane bird fauna is developed. As Rand says, "one finds no definite change in the bird life as one ascends the mountains" (1936, p. 220); by implication from p. 244 he is thinking of at least some 6000 ft. Rainfall is

---

[*] In his opinion (1940) the insectivore stock dates from the Palaeocene, the primitive primate stock from the Eocene, the carnivore stock from the Oligocene, the rodent stock from the Miocene. It is true that successful oversea colonization by a mammal, presumably by rafting, is what Simpson calls "an improbable event". But the present gap of about 250 miles may have been materially reduced at different stages; and on the time-scale concerned the "improbable" need have come off "about a dozen or fewer" times in considerably over 60 million years (Simpson, 1952).

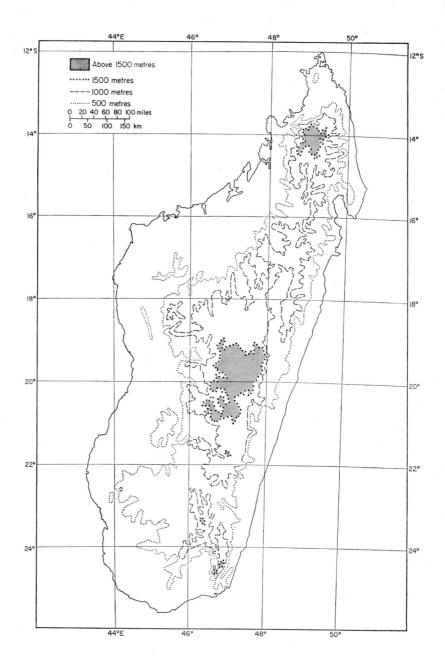

FIG. 61. The surface relief of Madagascar. (From Rand, 1936.)

very heavy on the east coast (facing the monsoons), up to 150 in. in places, but in most of the western three-quarters of the island it is under 50 in., with a long dry season, and in the south-west it is under 20 in. Following some centuries of occupation by agricultural people,* only about one-eighth of the surface is regarded as being under original vegetation today. The island is, however, floristically rich; it has, for example, over 1000 species of orchids (Millot, 1952). Primary evergreen forest still occupies some 12 000 sq. miles (Fig. 62), while the dry south-west, the "Subdesert", carries some highly character-istic xerophytic plants which show it to be no recent phenomenon. By far the greater part of the island is covered with a more or less wooded savanna, mainly deciduous, which degenerates into empty eroded grass-land on the central plateau, and it is probably here that the worst habitat destruction has taken place. However, on the whole one would think that big enough areas of natural vegetation remain for the bird fauna we see today to give no misleading impression of its structure. This is not to deny that a long succession of extinctions has taken place in the past. The giant Aepyornithidae of course have gone, perhaps helped to their doom by the first human colonists. Then again the ancestors of the extinct Raphidae of the Mascarene islands, the Dodo and Solitaire, must on geographical grounds have reached Mauritius and Rodriguez origin-ally from Madagascar; but there is no existing species in Madagascar that suggests relationship to the Raphidae on the conventional view that they are Columbiformes or Ralliformes, as argued by Lüttschwager (1961). In historically recent times there is little evidence of extinction in Madagascar, although increasing agricultural development must have favoured it; there appear from Rand's account to be not more than two or three species of birds that are known only from specimens collected in the nineteenth century and not found since. Hence, prima facie, when we discuss the bird fauna of Madagascar, it would seem that we are not merely dealing with ruins as we should be in the Mascarenes. Later, however, we shall see that there may perhaps be reason to revise this impression.

On geographical grounds the avian stock of Madagascar is much more likely to have come from Africa than from anywhere else. Gond-wanaland, the Indian Ocean land-mass, however real prior to the Tertiary, is irrelevant because too far back in time. Some elements of affinity with Asia rather than with Africa certainly exist. However, while a number of African families are absent from Madagascar (and some of them also from Asia), all the non-endemic families of Madagascar have representatives in Africa, and also all the eighty-seven non-endemic genera (see Table XXXI) except three, namely *Ninox* (owl), which is

---

* Amplifying a comment in his book, Paulian (*in litt.*) informs me that "the earliest traces of human occupation in Madagascar dated by radiocarbon are from around A.D. 800. Earlier traces have been found, such as animal bones showing marks of tools, but these are made by iron instruments, a fact which precludes any really remote occupation. Up to now no traces of a stone age have been recorded."

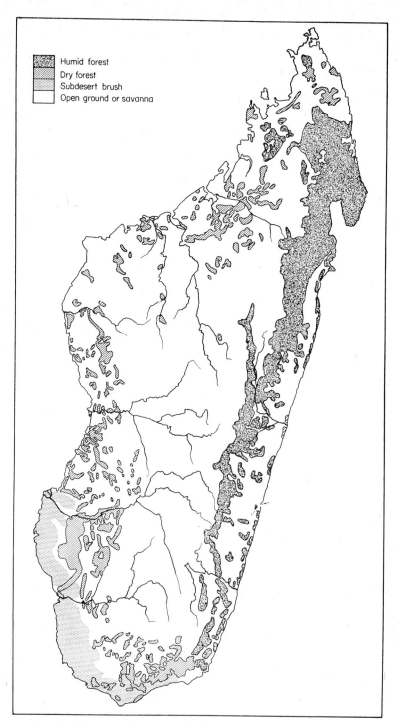

FIG. 62. The vegetation zones of Madagascar. (From Rand, 1936.)

mainly Australian and Malaysian, *Hypsipetes* (=*Microscelis*=*Ixocincla*) (bulbul), and *Copsychus* (turdine), both of which are typically Asiatic.* The bulbul *Hypsipetes madagascariensis* is peculiar in that various sub-species of it breed on the Comoros, on Aldabra Island and over a great part of southern Asia, yet in the Seychelles, which are geographically intermediate, another species of *Hypsipetes, crassirostris*, is in possession. This bird is, however, a member of the same superspecies (Rand and Deignan, 1960), the implication of which is that differentiation has proceeded more rapidly in the Seychelles than in other stations, whether insular or not, inhabited by bulbuls of this stock. The distribution of *H. madagascariensis* could be a relict of a more extensive one involving the African continent, like that of the coucal *Centropus toulou*, which inhabits much of Africa, in addition to Madagascar, the Comoros, Aldabra Island and India. Some support for such a view is given by the fact that a subspecies of the owl *Otus rutilus*, otherwise inhabiting Mada-gascar and the Comoro Islands, appears on Pemba Island, close to the East African coast and 500 miles from the Comoros, but has not been found on the intervening part of the continent nor on the intervening islands of Zanzibar and Mafia, which appear ecologically suitable. As a suggestion of the extinction that may have taken place in Africa dur-ing the long history of Madagascar we have the evidence of Darlington (1957, p. 521; and personal communication) that the closest relatives of some Madagascar reptiles and also beetles are found today in South America, with no link in Africa.

The other Madagascar species that breeds also in Asia (in northern India and China) but not in Africa, is the typically parasitic Little Cuckoo *Cuculus poliocephalus*. The two breeding populations are separated by some 3000 miles and there is nothing definite to show which was the original stock. However, since they differ only in wing-length, the separation of the stock has probably taken place only lately, and the habits of both of them suggest possible means by which the present position came about. Madagascar birds today appear in Africa in the off-season as far north as Kenya and Uganda (Fig. 52, p. 250): if they represent the original stock they might conceivably have travelled much farther north at some stage in the past when the country around the northern Indian Ocean and Arabian Sea was less forbiddingly arid, to

* As being of Asiatic affinity Rand (1936, p. 298) also mentions the Madagascar species of *Amaurornis*, but this genus is usually merged in *Porzana*, which is equally African, and he men-tions *Collocalia*, but the status of these swifts in Madagascar seems problematical. Since Rand wrote, the endemic Madagascar starling genus *Hartlaubius* has tentatively been combined by Amadon (1956) with the Asiatic *Saroglossa*; and it seems that, in fact, both are relatives of *Cinnyricinclus* (which is African as well as Oriental). Further, the suggestion has been made by Salomonsen and also by Deignan that *Neomixis* of Madagascar (three species) is a timaliine group allied to the Asiatic *Stachyris* rather than a sylviine of uncertain affinities, as hitherto classified.

Among the zoological puzzles presented by Madagascar is the fact that all the nineteen species of ticks are endemics belonging to Indo-Malaysian species-groups (H. Hoogstraal, personal communication).

TABLE XXXI

Endemism in the Madagascar bird fauna

Group	Genera In endemic families and subfamilies	In other families	Endemics in "other families" No.	Proportion	Species In endemic genera	In other genera	Endemics in "other genera" No.	Proportion	Subspecies Endemics among non-endemic species No.	Proportion	"Undifferentiated" species*
A	—	41	3	0·07	3	52	14	0·27	14	0·37	24
B	—	16	2	0·12	2	18	11	0·61	5	0·71	2
C	2	5	1	0·20	4	5	3	0·60	—	—	2
D	4	20	4†	0·20	18	19	10	0·53	8	0·88	1
E	11	34	19	0·56	47	16	13	0·81	2	0·66	1
	17	116	29		74	110	51				
	133				184						

* Non-endemic species of which the Madagascar representatives are not differentiated from mainland birds.

† *Cochlothraustes delalandei* is included in *Coua*, as was done by Edwards and Grandidier (1879–81) and also by Milon (1952).

leave representatives in Asia. On the other hand, Indian *C. poliocephalus* have been found migrating in large numbers through coastal Tanganyika. Their winter quarters are still not accurately known, but if, as may be supposed, they are in south-eastern Africa, an over-shooting to Madagascar could have colonized the island. The case is a special one because whichever of the two populations of this cuckoo was secondary it had to come to terms not only with the local environment in general, but also with the local birds that must serve as its hosts. It may be added that Madagascar has only one other parasitic bird, the cuckoo *Pachycoccyx audeberti*. Nothing is known of the hosts of either the *Cuculus* or the *Pachycoccyx*; it would be interesting to know whether there is overlap in this respect, and whether the present rarity of *Pachycoccyx* owes anything to competition from *Cuculus*.

The composition of the Madagascar bird fauna, by families, is given in Table XXVIII and its degree of endemism is analysed in Table XXXI. The basis of both these tables is of course Rand's (1936) work; in a few species, mainly of water-birds, breeding has been presumed in the absence of proof. For the present purpose I have omitted sea-fowl, the introduced myna *Acridotheses tristis* and the guinea fowl *Numida meleagris*, which (belonging to an African subspecies) is certain to have been introduced. Rand's classification has been followed in the main, since there has been no subsequent general revision. It may be borne in mind that in the non-endemic genera some of the birds that are classified as endemic species doubtless form superspecies with African birds, but pending examination they are accepted at their 1936 rating. Table XXXI is one of the few cases in which I give statistics for genera, not because I regard this most labile and subjective of taxonomic categories as having special validity in the Madagascar context, but because the distinctness of a high proportion of the species is thereby emphasized.

It will be seen from Table XXVIII that a number of African families, for which Madagascar seems ecologically suitable, are absent from the island, especially the cranes, turacos, colies, wood hoopoes, hornbills, barbets, woodpeckers, honey-guides, broadbills, pittas, shrikes (including helmet shrikes) and finches. On the other hand, Madagascar has three endemic families that are universally so classified and three others that have only recently been down-graded from that status. The first three consist of the Mesitornithidae, three species in two genera of rail-like birds inhabiting dry bush, the sub-oscine Philepittidae, two genera each of two species, and the Vangidae, twelve species in ten genera. Both the last two families are discussed separately below. The second group of endemics of supra-generic status consist of the former Leptosomatidae (one species) and Brachypteraciidae (five species), both now assigned as subfamilies to the Coraciidae, and the nuthatch-like Hypositttidae (one species), now provisionally made a subfamily of the Salpornithidae. Typical examples of both the Coraciidae and the Salpornithidae are absent from Madagascar.

It will be seen from Table XXXI that the breeding bird fauna of Madagascar consists of 133 genera, of which forty-six are endemic, including seventeen in endemic families or subfamilies. The species number 184, of which 125 are endemic, including seventy-four in endemic genera.* The overall proportion of two-thirds endemic species conceals, however, some very diverse elements: in water-birds only seventeen species are endemic out of fifty-five, only a little over one-third, while in passerines sixty are endemic out of sixty-three, nearly nine-tenths. Comformably, of the thirty species on the Madagascar list which are not differentiated from mainland birds even subspecifically, no less than twenty-four are water-birds and only one is a passerine. This situation will be discussed later. Meanwhile it may be noted that, as shown in Table XXXII, the mean number of species to the genus is

TABLE XXXII

The average numbers of species per genus in Madagascar (by groups)

Group	Endemic genera			Non-endemic genera		
	No.	No. of constituent species	Mean	No.	No. of constituent species	Mean
A	3	3	1·0	38	52	1·3
B	2	2	1·0	14	18	1·3
C	3	4	1·3	4	5	1·2
D	8	18	2·2	16	19	1·2
E	30	47	1·6	15	16	1·1

low, less than 1·4 in all groups and in both endemic and other genera, except the passerines and the group D non-passerines in the endemic genera. There the averages rise to 1·6 and 2·2 species respectively, owing in the latter case to the fact that the genus *Coua* contains ten species, far more than any other in Madagascar. The Madagascar averages compare with 3·5 for African passerines and 3·3 for group D birds as calculated on Sclater's (1928–30) classification, constructed in much the same taxonomic climate, as regards recognition of small genera, as was Rand's for Madagascar. Two factors can have contributed to this result: in the island a very limited degree of radiation within the genus, or, if radiation has not been so limited, considerable extinction.

* Carcasson (1964) gives statistics for the butterflies of the "Malagasy Sub-region", which includes the Mascarenes, Comoros and Seychelles as well as Madagascar. It appears that in these insects the endemism at the generic level is much lower than in birds, only in the proportion 0·15 compared with 0·34 (twelve endemics out of eighty-one, compared with forty-six out of 133), but at the specific level is actually higher, 0·77 against 0·67 (233 out of 301, compared with 125 out of 184). It would be interesting to know to what extent these differences are influenced by differences in taxonomic practice between these two groups of winged animals or in the case of the butterfly species by the prevalence of vicariant species on the various islands comprising the sub-region.

Aside from the monstrous phenomenon of the Aepyornithidae—since when there is no evidence of marked gigantism—it is clear that in some respects evolution in the birds of Madagascar has been vigorous and has proceeded to considerable lengths. The most remarkable manifestations are those of the endemic family of the shrike-like Vangidae, the twelve species of which are allocated among ten different genera, and the endemic genus *Coua* (belonging to the centropine, non-parasitic cuckoos), with ten species. Knowledge of the Vangidae in life seems inadequate for comparing their ecology, especially as those we know anything about are all arboreal and mainly or wholly insectivorous (see, however, Dorst, 1960). They are all stoutly built birds and most of them have strong rather shrike-like beaks. *Euryceros* has an enormous blue beak, deeper than the skull, but it is not known to have any correspondingly abnormal feeding habits. In *Falculea* the long curved beak resembles that of a wood hoopoe *Phoeniculus* and is said by Edwards and Grandidier (1879–81) to be used "à la recherche des insectes et des larves . . . dans les fissures du bois" (cf. van Someren, 1947), though Rand (1936, and *in litt.*) has no evidence of this.

The couas can be discussed more particularly by combining the information in Milon (1952) with that of Rand and bringing in for the apparently extinct *delalandei* the information given by Edwards and Grandidier (1879–81). In habits the couas fall into two main groups, for which Rand uses the terms "arboreal" and "terrestrial". Milon prefers "grimpeurs" and "coureurs", for he points out that the "arboreal" species may frequent bushes and the "terrestrial" often sun themselves on trees and run along horizontal branches. However, these terms of Rand's are adequate in so far as they relate to the foraging habits of the birds. It will be seen from Table XXXIII that only three of the ten couas are

TABLE XXXIII

Ecology of the Coua *spp.*

Species and feeding zone	Main habitat	Food
Arboreal		
C. caerulea	Forest	Insects, some fruit
C. cristata	Savanna	Insects, some fruit
C. verreauxi	Subdesert	?
Terrestrial		
C. coquereli	Savanna	Insects, some fruit
C. gigas	Savanna	Insects
C. reynaudi	Forest	Insects, some fruit
C. serriana	Forest	Fruit, some insects
C. delalandii	Forest	Molluscs
C. cursor	Subdesert	(?) insects
C. ruficeps	Subdesert and savanna	Insects

arboreal and that these three are all segregated ecologically. The seven terrestrial species comprise three belonging to forest, two to savanna and two at least partly to subdesert. The two pairs that are bracketed, *coquereli* with *gigas* and *reynaudi* with *serriana*, are described as definitely overlapping in range, but here also there is some evidence of segregation. In the first of these pairs the species differ considerably in size, which suggests a difference in the food taken; in the second pair *reynaudi* is said to take more insects than fruit and *serriana* the opposite. The third forest species is *hors concours*; it is (or was) a very large bird living entirely on molluscs, which moreover it smashes on a stone with sideways movements of the head, like a thrush. Of the last two species in Table XXXIII, it seems that *ruficeps* is in potential competition for insects with *cursor* in the subdesert and with *coquereli*, if not also with *gigas*, in the savanna. The swiftly-moving species of the subdesert, one of them appropriately called *cursor*, recall the Roadrunner *Geococcyx californicus* of a similar habitat in America. On the whole, then, it is evident even on the present imperfect information that the ecological segregation between the ten *Coua* spp. of Madagascar is considerable. It may be added that the birds show no particular morphological specialization and their beaks are all much alike.

No group of Madagascar birds shows more ecological radiation than the couas, and that is clearly not great. It is, however, possible that at some stage in the past at least one far more spectacular radiation took place. In Madagascar the sub-oscines, represented by the Philepittidae, form only an insignificant percentage of the bird fauna, as they do in Africa, and belong to a suborder, the Tyranni, otherwise represented in the Old World only by the Pittidae. The Philepittidae are comprised of two pairs which are very unlike in appearance and in habits. According to Rand, the two *Philepitta* spp. are "plump sluggish" birds of the forest living on fruit. The two *Neodrepanis* spp. are so like sunbirds, with long beaks and metallic gloss, that they were classified in the Nectariniidae until in 1951 Amadon discovered that their throat musculature is sub-oscine; and it is now assumed that it is their external resemblance to sunbirds rather than their musculature which is due to convergence.* They feed on insects more than on nectar, but certainly frequent flowers a good deal (Edwards and Grandidier, 1879–81). These two widely divergent branches of the Philepittidae may well be the sole relicts of an early radiation of this family in Madagascar, and the stock comment on the situation would be that the other sub-oscines have presumably succumbed in competition with oscine newcomers to the island. Caution is, however, necessary here in face of the history of the Lemuridae in Madagascar. They radiated remarkably but suffered a good deal of extinction in "comparatively recent times", in the absence of competi-

---

* Since this was written the *Neodrepanis* spp. have been reconsidered by Salomonsen (1965). While remaining satisfied that they are Philepittidae, he has extended the list of convergences with sunbirds.

tion from other mammals and apparently by no means all as a result of human activity (Hill, 1953).

Birds have presumably been arriving in Madagascar from Africa since before the beginning of the Tertiary epoch, but it may be inferred that at any rate latterly such birds have been able to establish themselves in the island only infrequently. It is true that most of the water-birds (thirty-eight out of fifty-five) are conspecific with African; they could on the whole cross the wide strait with less difficulty than many other birds. Individual species of them are more widely dispersed over the world in general, and as species they appear to be older (see also Chapter 6). Water-birds provide the only two cases of double invasion into Madagascar, by two Little Grebes, *Podiceps rufolavatus* (endemic) and *P. ruficollis*, and by two squacco herons, *Ardeola idae* (endemic) and *A. ralloides*. In neither of the non-endemic species is the insular population differentiated from the African; and *P. ruficollis* seems to have become established on the island only very recently indeed.*

While nearly three-quarters of Madagascar water-birds are conspecific with African ones, the proportion is far smaller in the other groups. It may be noted parenthetically that the two undifferentiated group C birds on the Madagascar list (Table XXXI) are *Coturnix* quails, both of which are known to move widely and erratically in Africa (see Chapter 13) and hence are the more likely to reach Madagascar from time to time.

The passerines are the group with the fewest non-endemic species; out of a total of sixty-three on the insular list only three are definitely African. This can be taken as an index of how rarely members of this group of birds have successfully established themselves in the island in the course of the late Pleistocene. The three exceptions are the crow *Corvus albus*, martin *Hirundo paludicola* and Stonechat *Saxicola torquata*. The last two have both been in Madagascar long enough to develop subspecies there but this does not necessarily mean more than a few thousands or even hundreds of years. However, the Stonechat has also colonized Réunion, 400 miles to the east, which it could hardly have reached except from Madagascar, and has again subspeciated there. The crow might prima facie have had a shorter tenure in Madagascar than either the martin or the stonechat, since its population there is undifferentiated: but anyway this crow does not vary throughout its great continental range, which extends from Senegal to Abyssinia and the Cape.

The great paucity of non-endemic passerines in Madagascar could

---

* Since this was written the result of a field study has been published (Voous and Payne, 1965). It has been ascertained that *P. rufolavatus* and *P. ruficollis* interbreed in Madagascar on such a scale that both may be superseded there by a stabilized hybrid population; that the latest arrival of the three local grebes, *ruficollies*, is at present the commonest, and that the endemic species *pelzelni*, presumed from its degree of distinctness to have much the longest insular history, is the rarest. This suggests a complete reversal in population status in the course of the last thirty years or so.

result from difficulty in making the crossing or from the fact that the bird fauna is too saturated to admit newcomers readily. It seems un-unlikely that the latter is true, notwithstanding the evolutionary potential that Madagascar has shown and the fact that it has had the entire Tertiary epoch to work in; for in some respects the bird fauna today seems unbalanced, at least by African standards. In the first place, for an area of a quarter of a million square miles, with a range of habitats from subdesert to the most humid of evergreen forests, a total of 184 species is not impressive (details in Table XXXI). This total is, for example, little more than one-third of that for Kenya, which has about the same area; it is only three-quarters of that for Gambia, a mere 4000 sq. miles with much narrower range of habitats. It is facile to attribute the relative poverty of Madagascar to "insular effect" (and after all Australia with 3 million sq. miles of area musters only 531 species, substantially less than Kenya, and no more than Zambia); but it is not clear how this can have operated in Madagascar. It is no ordinary island: although not connected with the continent during relevant geological time, it has repeatedly been replenished from there, and it has not, like so many others, been available as a stage for evolution merely since the Pliocene but through most of the lifetime of the class Aves. It is true, the geographical isolation precedent to speciation would seem difficult of attainment in the island, as discussed below, but the couas and vangids show that this could be overcome in a spectacular manner.

The next point is that the proportions contributed by the different groups of birds to the total bird fauna diverge greatly in Madagascar from those prevalent in Africa. For convenience of reference we may draw on the statistics for territorial bird faunas (excluding Somaliland and the Gambia) given in Table XIV (p. 130). The results are shown in Table XXXIV. It will be seen that the large proportion of water-

TABLE XXXIV

The proportional composition of the bird faunas of Madagascar and of African territories (mean and range)

	Group A	Group B	Group C	Group D	Group E
Madagascar	0·30	0·11	0·05	0·20	0·34
Africa	0·16	0·10	0·07	0·18	0·48
	(0·12–0·20)	(0·09–0·13)	(0·05–0·98)	(0·14–0·20)	(0·44–0·51)

birds and the small proportion of passerines in Madagascar are right outside the ranges of the African statistics and are respectively about double and three-quarters of the African means. Moreover, so far as the continental passerines are concerned, Table XXXIV considerably understates the African case for it is based only on non-forest birds. To make the African figures fully comparable with those for Madagascar we need to bring in the birds of the African forests, which vary enorm-ously in area and faunistic importance in the different territories. However, important or not, since the forest birds so largely belong to groups D and E, as shown in Chapter 5, in whatever African territory they occur they will increase the proportion of those groups, accentuate the Madagascar deficiency, and reverse the very slight superiority of Madagascar in group D birds.

There seems to be no ecological reason for Madagascar's wealth of water-birds. It has more breeding Anatidae and Rallidae than West Africa and the Sudan together and nearly as many Ardeidae. Ease of access by water-birds from the continent is presumably a factor, but diversity of habitat, facilitating multiplication of niches, must also be important. A comparative study of the Madagascar habitats would be worth while. Among the group B birds it may be noted that there are no vultures and no eagles except the so-called "serpent eagle" *Eutriorchis astur* of the forest. These deficiencies are no doubt connected with the peculiar state of the mammal fauna, which lacks large species to provide carrion, but the lemurs of Madagascar might have been expected to support some powerful eagle, such as *Stephanoaetus* of the African forests or *Pithecophaga* of the Philippines.

Among the smaller raptors it is remarkable that, while Madagascar has its resident peregrine and two species of kestrel, it is occupied during the summer, from about November to April, by both *Falco eleanorae* and *F. concolor*. These are the two falcons, discussed in Chapter 4, which in the Mediterranean, the Sahara and the Red Sea islands have their anomalous late breeding season adjusted to take advantage of the autumn migration of Palaearctic birds into Africa. Thereafter *F. eleanorae* is known to vacate its breeding grounds completely and it probably spends the off-season nowhere but in Madagascar. Its world population has reliably been estimated as only about 4000 (Vaughan, 1961). For *F. concolor* the information is more inadequate, but the fact that Rand recorded them so often, and from eight to fifteen at once, suggests that all the birds of this species also may congregate in Madagascar. Such scanty evidence as there is suggests that both the immigrant falcons are largely insectivorous, depending during their stay much upon Orthoptera, as do the two resident insular kestrels. It is virtually certain that the two immigrants reach Madagascar by way of East Africa and it may be postulated that they would not have preferred Madagascar to the continent if they could accommodate themselves in the latter. Here they would be in potential competition not only with the resident falcons but also with large numbers of wintering *F. naumanni*, *F. subbuteo*, *F. vespertinus* and *F. amurensis*.

For the relative paucity of passerine birds in Madagascar I can make no suggestions at all. As will be seen from Table XXVIII, Madagascar has only half as many species as Fernando Poo and only three-quarters as many as Zanzibar. Both these islands, being only a small fraction of the size of Madagascar and with a much narrower range of habitats, are prima facie far poorer ecologically than Madagascar. What is more, over one-third of the Madagascar passerines are made up of eleven Sylviidae (one *Cisticola* and ten species belonging to endemic genera) and the twelve Vangidae (some of which apparently replace the "missing" laniid shrikes). A noteworthy gap is the absence of pipits, which are otherwise so widespread in the Old World, and, compared with a dozen or more in most African territories, there are only two hirundines, the "sand-martin" *Hirundo paludicola* and *Phedina borbonica*, which seems to be typically a rock-nester. Since neither of these species is common, it is difficult to believe that the Hirundinidae are making full use of their feeding opportunities in Madagascar; and there is no obvious reason to think that in general their opportunities for nesting sites would differ much from those in the same latitudes of Africa. Conceivably there may be competition with swifts, in which, with four species, Madagascar does not do so badly.

By African standards Madagascar is also poorly off for fruit-eaters and for seed-eaters, though there is no reason to suppose that the wide range of Madagascar habitats is deficient in fruit-bearing plants or that there is any particular seasonal shortage; and as regards seeds, at least three of the main genera of important food-grasses of Africa, *Sorghum*, *Rotboellia* and *Echinochloa*, are regarded as native in Madagascar, apart from other species. Of fruit-eating birds the island wholly lacks barbets, hornbills, turacos and orioles and it has only a single starling compared with the dozen that are the complement of most African territories. In fact in Madagascar, besides the starling, only seven birds seem to be mainly fruit-eaters, namely, the fruit-pigeons, *Treron* and *Alectroenas*, the two *Philepitta* spp., one of the bulbuls, and one of the *Coracopsis* parrots. Another parrot and a thrush *Pseudocossyphus* seem to be the only other species partly dependent on fruit to an important extent, apart from the couas, which evidently eat only fruit that has fallen on the ground. As seed-eaters two columbids, three ploceine weavers, one estrildine and one little parrot *Agapornis* qualify as "mainly" and one galline bird *Margaritoperdix*, one sand-grouse, one button-quail, one quail and one lark as "partly". Using the same method of calculation as in Table XXVI, this means that among Madagascar birds of groups C–E only about 0·07 of the feeding attention is directed to fruit, which is only one-half to one-fifth of that in any of the African evergreen forests and actually lower than in most samples of the acacia steppe. The proportion of attention to seeds in Madagascar is 0·09, one-third of that in the African acacia steppe but rather more than the average in the African forests, which are so scantily supplied with grasses.

Madagascar possesses no representative of the woodpecker family (Picidae)* and no wood hoopoe (Phoeniculidae), the other African

---

* It is true that the woodpeckers in general appear to be typically sedentary, but there are some remarkable instances of what seem to be great opportunities awaiting them. For example, there are no woodpeckers in the Australasian Region and not even in Ireland. Yet one species has colonized the Canary Islands.

family that is typically dependent on insect food found in and on the branches and trunks of trees. However, Madagascar has evolved a passerine bird, *Falculea palliata*, attributed to the Vangidae, which shows remarkable convergence to the wood hoopoes "in shape with their long curved bill and bodily appearance, and in habit by running up tree-trunks and along branches, poking into crevices and hanging upside down by their claws, chattering loudly the whole time" (van Someren, 1947). Madagascar possesses also *Hypositta corallina*, a monotypic endemic genus (allied to the African *Salpornis*), which is "very creeper-like in its habits" (Rand). Still, together the *Falculea* and the *Hypositta* would seem very inadequately to fill the niches made available by a wide variety of woodland and forest of types that are occupied by several species of woodpecker in most parts of the African continent.

In Rand's list there are seventeen group D non-passerines and forty-four passerines which he specifically associates with evergreen forest, so that nearly two-thirds of the Madagascar species in these groups are found in evergreen forest. This contrasts with the situation in Africa, where only one-quarter of the birds occur in evergreen forest of all types put together. Actually, as Rand indicates (1936, p. 257), only a minority of the species are found throughout the Madagascar forests and the bird community in any one locality is probably a good deal smaller than the above figures suggest. On the other hand, on his data the Madagascar forest birds are by no means so limited in their habitat tolerances as they are in Africa, and in the island the distinction between forest and non-forest species goes nowhere so deep. For example, of *Newtonia brunneicauda* Rand writes: "equally at home in the heavy forest and secondary bush of the Oriental [Province of Madagascar], the deciduous woodland of the Occidental and the bush of the Sub-desert". Nothing like this could be written of any bird in Africa except the highly polytypic *Zosterops senegalensis*. There, as shown in Table V (p. 87), only 3% of the species are not allocable to one or other of these two main ecological categories. In Madagascar, of the sixty-one species Rand associates with the "Humid Forest" his text shows that at least twenty-six are also regular in the non-forest, with or without subspecific differentiation, while seventeen of the species actually extend into the subdesert. This involves a breadth of ecological tolerance by individual species that is without parallel on the African continent.

Rand has an interesting discussion of the two parrots *Coracopsis vasa* and *C. nigra*, "very much alike in colour, differing chiefly in size, and of which the ranges more or less coincide". One seems to be more frugivorous than the other, but each extends from the forests of the Humid East right through to the Subdesert and each has subspeciated. In general, Rand's view is, however, that the birds now in the humid forests extended over the whole central plateau before it was deforested and the "western fauna", i.e. non-forest, "probably developed in a rather restricted area and under strong influence from the east, as is also indicated by the paucity of the endemic fauna in the west" (p. 234). However, it seems doubtful whether under the existing climatic regime evergreen forest can have been well developed over the west of the island since most of it would be in the rain-shadow of the highlands (see Fig. 62) which catch the easterly monsoons. It is true that the wide ecological spread of Madagascar species seems for the most part to have been achieved recently because, while a few of the ecologically tolerant species are represented by different subspecies in habitats of different humidity, most of them are not. These latter could be regarded as evidence for recent penetration of non-forest habitats by typically forest species whose area has recently been reduced by human agency; but in Africa it has been extremely rare for any forest species to show such adaptability. Moreover, if the human factor is thus invoked it must be borne in mind that, as indicated on p. 329, it is unlikely to have been effective on a large scale for more than about a thousand years; so that those species which have differentiated subspecifically between forest and other habitats must have accomplished it within this period. However, as mentioned in Chapter 1, such a speed of evolution is known to have been exceeded in birds elsewhere.

If anything, the discussion in the foregoing paragraphs makes more difficult the whole problem of speciation in Madagascar by birds. It is true that, as Mayr (1963) remarks, phenotypically distinct populations within a few kilometres of each other can occur in mice and in small winged insects, but birds are individually far more mobile. Today in Madagascar there is plenty of subspecific differentiation in birds but few species appear to have populations, whether distinctive or not, that are effectively isolated from each other. Yet somehow some groups have speciated repeatedly, apparently entirely within the island, to culminate in the twelve species of the Vangidae and the ten of the couas. These centropine cuckoos are admittedly not the most mobile of birds, but even in them the process of their evolution through geographical isolation within Madagascar is extremely difficult to envisage. In an area of no more than 1000 × 300 miles at its widest part no factor of geographical distance can be invoked to help; geographical barriers are not impressive, except for the line of highlands under the conditions of a glaciation; and there has been no dissection by the sea. We have presumably then to rely on ecological barriers. Although it is easy to imagine the existence of some ecological islands and concurrent local ecological barriers at any given time, it is essential to postulate that such barriers must have persisted without intermission long enough for genetic incompatibility to become established in the birds involved. Given the rapidity with which climatic vicissitudes evidently succeeded each other on the continent in the same latitudes as Madagascar, at any rate in the latest stages of the Pleistocene (Chapter 3), I do not see how the requirements of long-term ecological isolation within the island can have been met, unless full genetical isolation was attainable by sundered populations much more rapidly than would be conceded by most evolutionists (but see p. 11). Furthermore, from the wide ecological tolerance shown by some Madagascar species (and even by some subspecies), it is clear that in general ecological barriers would need to be more rather than less marked than we should need to postulate in Africa.

What may be inferred of the Pleistocene history of the island by analogy with the African sequence does not help. It can be accepted that, no less than the African continent, it shared in the fluctuations in temperature and humidity associated with the glaciations. At the coolest stages—and down to 18 000 years ago—most of the area above 500 m in Fig. 61 would have been "montane" (yet there is no sign today that any real montane bird fauna developed). Increase in humidity would have enlarged the forest and tended to produce a more continuous block. But neither the topography nor the run of the isohyets give any reason to believe that even the western lowland "savanna" area was cut into separate blocks by the intrusion of either forest or subdesert. In the subdesert itself, though confined to quite a small part in the south of the island, the presence of endemic species of plants and animals shows that it has had a continuous existence for a long time.

Clearly, the evolutionary problems presented by the Madagascar bird fauna have not diminished now that Rand has given us so many more facts. In part the two things that must worry us most are contradictory. It is difficult to see how speciation among the birds could have proceeded in the island to the extent it did; but, since evolution has had unlimited time and has showed such remarkable achievements in certain groups of birds, one might have expected the bird fauna to be much richer than it is. The most likely explanation is that there have been widespread extinctions, not only of the sub-oscines, now represented only by the *Neodrepanis* and the *Philepitta* pairs, but also of the "modern" birds, especially of the passerines.* Several features of the bird fauna today point to this conclusion. The first is its poverty and unbalance compared with African areas. The second is the low average number of species per genus, which, compared with the African figures, suggests that in Madagascar more than twice as many species as now exist there may have recently been lost. Thirdly, the bird species in Madagascar

---

* In this connexion an interesting point arises with regard to the sole species of *Zosterops* in Madagascar, *Z. maderaspatana*. It is a thoroughly "ordinary" member of its genus, looking very like some continental populations of both India and Africa, though not particularly like its nearest neighbour across the Mozambique Channel. Now, on geographical grounds, it is virtually certain that Madagascar was the source of the *Zosterops* stock that originally colonized the Mascarene Islands, to the east. Yet the two species there, each represented in different subspecific forms on both Mauritius and Réunion, both show eccentric developments, right outside the range of variation in either Africa or India. Therefore, either the ancestral stock or stocks of these birds on Madagascar has been superseded there by *Z. maderaspatana* or else this species has remained relatively stable, while very distinct specific characters, together with genetic incompatibility, have been developed twice in the Mascarene Islands, with subspeciation between Mauritius and Réunion thereafter (see discussion in Moreau, 1957, p. 403).

are so imperfectly segregated into each of the main habitats, namely forest, wooded savanna, and "subdesert" that a considerable proportion of them occupy both forest and also one or both of the drier zones. This is a degree of adaptability extremely rare in Africa (Chapter 5), and it is difficult to imagine how it can have come about in Madagascar except by such casualties in the bird fauna of one or other of the habitats that numerous niches became vacant.

Unless human agency can be held responsible, which I doubt very much, the cause of the extinctions must have been climatic, a dry period so prolonged and all-embracing that it massacred the forest species extensively, or a similar humid period that swamped the dry-country species, or perhaps one immoderate period after the other. But whatever happened, the water-birds do not seem to have been so badly hit as other groups and the passerines seem to have suffered most. Also, whatever the climatic vicissitudes, the flora is, as noted above, still considered to be a rich one. Moreover, it is difficult to imagine that any such vicissitudes that affected the island would not have been felt strongly also on the opposite mainland. We know something about the climatic changes in these parts (Chapter 3), but not nearly enough. In any case, it is difficult to imagine that such an access of aridity as that which extended Kalahari conditions east to the Victoria Falls and north through Angola around 12000 years ago would not have suffered much diminution in severity as it approached the maritime conditions of the Mozambique Channel, let alone on the eastern side of it. Clearly the postulate of climatic changes as responsible for the poverty of the Madagascar bird fauna is not without difficulties.

18

The Bird Faunas of the Other East-coast Islands

COMORO ISLANDS

Although various collections have been made in the Comoro islands (Fig. 63), the first systematic investigation was that of the expedition under the leadership of Benson (1960a, b) in 1958. The bird fauna of the archipelago can now be regarded as almost definitely known, and he has provided an exhaustive description and discussion, without which the following summary would have been impossible.

The archipelago consists of four islands, all of volcanic material and each surrounded by deep water, within an area of about 100×140 miles located in the 400-mile strait between Africa and the northern end of Madagascar. From the point of view of geographical distance the prospects of colonization from either source are about equal, but since the prevalent winds are north-easterly for part of the year and south-easterly for the rest, the chances are against arrivals from Africa; and in fact of the fifty-two presumably native Comoro birds twenty-eight are of Madagascar origin against twelve from Africa, while the remainder cannot be ascribed definitely to either source. Three species, a crow *Corvus albus*, an estrildine *Lonchura cucullatus* and a Ploceine weaver *Foudia madagascariensis*, all typically hangers-on of man and his agriculture, are probably recent arrivals. The first two come from Africa and occupy the same position in the Gulf of Guinea islands. Four species that can be accepted as introduced are a guinea fowl *Numida*, lovebird *Agapornis*, mynah *Acridotheres* and sparrow *Passer*.

From west to east the islands and their most relevant particulars are as shown in Table XXXV. Group A birds are excluded because the uneven distribution of watery habitats in the islands vitiates the water-bird statistics.

TABLE XXXV

Statistics of the Comoro Islands

	Area (sq. km.)	Greatest height (m)	Humans per sq. km.	Bird species excluding water-birds
Grand Comoro	950	2560	87	34
Moheli	216	810	23	31
Anjouan	378	1578	164	29
Mayotte	370	660	54	25

On the whole the islands are very humid, though there is great local and altitudinal variation and the months May–October are drier than the other half of the year. Evergreen forest probably covered almost the whole of the islands originally. Man is understood to have been practis-

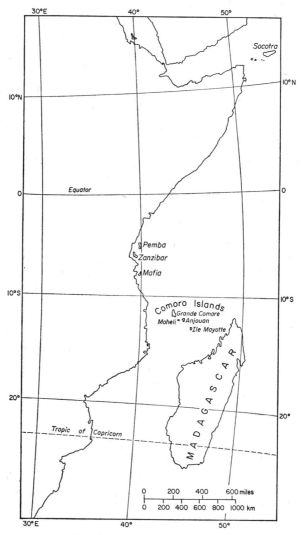

FIG. 63. The east-coast islands.

ing agriculture of a sort on the islands for over 1000 years, but except on Anjouan habitat destruction does not seem calamitous and Benson concluded that "on the whole the Comoro avifauna has so far been remarkably fortunate, in comparison with other small islands of the Indian Ocean, especially in largely escaping the effects of European exploitation of the land". While, however, of the birds reported in the nineteenth century Benson failed to find only one on a single island, there are gaps in distribution from which extinctions can be inferred. Thus, the coucal *Centropus toulou* lives in both Africa and Madagascar, but not in the Comoros, though its poor powers of flight suggest that it used these islands as intermediate stages in its expansion of range and there is nothing in the ecology of the Comoros to suggest that they would be unsuitable for the bird. Within the archipelago there are several anomalies of distribution; for example, five species, such as *Cyanolanius madagascariensis*, which there is every reason to accept as originating in Madagascar, are found only on one or other of the two more westerly islands and not on either of those nearer the presumed source of origin. Such absences may of course be the result of competition rather than of human intervention, though there are no grounds for thinking so; but this factor can hardly be invoked to account for the difference in number of species on the four islands. Originally the ecology of all four must have been very similar, except in prevalence of watery habitats and except that the biggest, Grand Comoro, is the only one high enough to carry a distinct montane environment. Omitting water-birds, and limiting consideration to groups B–E, it will be seen from Table XXXV that Moheli, the smallest island but the one with lowest human density, has thirty-one species, compared with only twenty-five and twenty-nine respectively on Anjouan and Mayotte, which are both bigger, and with thirty-four on Grand Comoro, which is four times as big and also more lofty. Unless differential extinction had occurred elsewhere as a result of "development", Moheli would surely not have so rich a bird fauna in comparison with the other islands.

As will be seen from Table XXXVI, endemism at the specific and subspecific levels is high in the insular bird fauna as a whole. The

TABLE XXXVI

Endemism in the Comoros bird fauna

Endemics	A	B	Group C	D	E
Genera (monotypic)	–	–	–	–	1
Species of uncertain affinities	–	–	–	–	3
Species allocable to superspecies	–	–	–	–	6
Subspecies*	1	2	–	10	11
Undifferentiated	5	4	1	4	4
Total	6	6	1	14	25

* Based on the number of species in which at least one of the island populations is distinct from populations outside the archipelago.

differences between the bird groups are instructive. No more than sub-specific differentiation has taken place in groups A–C and that to a limited extent, namely in three out of thirteen species, but much more, in ten out of fourteen species, in group D, the "other non-passerines". By contrast, out of twenty-five passerines we have one endemic genus, three species of uncertain affinities, five endemic species which appear to be members of superspecies and eleven endemic subspecies. The only four species that show no insular evolution are the three hangers-on of man already mentioned, *Corvus albus*, *Lonchura cucullatus*, *Foudia madagascariensis* and the bulbul *Hypsipetes madagascariensis*. The first of these species shows no variation in its vast African range and the second very little, so that they are less likely than some other species to do so in the islands, even if they are not recent arrivals. The *Hypsipetes* shows slight variation on Grand Comoro but is not accepted as a "good subspecies" there.

The differences between the islands to some extent repeat the same pattern and are of much interest. In the first place, twenty-three of the twenty-seven non-passerines occur on more than one island, but only three have differentiated within the archipelago; and it is at first sight surprising that one of these should be so potentially mobile a bird as the goshawk *Accipiter francesii* (also of Madagascar), a forest-edge bird that has developed three subspecies within the archipelago. Compare the subspeciation of the kestrel *Falco tinnunculus* in the Cape Verdes.) By contrast, only twelve out of the twenty-five passerine species occur on more than one island and of these seven have differentiated between two or more of the islands—five times the proportion of the non-passerines. The fact that more than half the passerines species occur on only one or other of the islands may well be because intergeneric competition reduces the instances in which more than one species could co-exist. Table XXXVII shows the distribution in the islands of those genera which have more than one species there, distinct species being shown by different letters.

TABLE XXXVII

Distribution of congeneric species within the Comoros

	Grand Comoro	Moheli	Anjouan	Mayotte
Dicrurus	a	–	b	c
Foudia	a + b	a + b	a + b	a + b
Hypsipetes	a	a + b	a	a
Nectarinia	a + b	a + b	c	d
Nesillas	a	a + b	a	–
Zosterops	a + b	c	c	d

The genera *Foudia* (Ploceine), *Hypsipetes* (bulbul) and *Nesillas* (warbler) have no representatives in Africa. The ancestors of both the Comoro *Foudia* spp. and both the *Nesillas* spp. presumably reached the archipelago from Madagascar. There is no reason to suppose that the two *Nesillas* spp. were not derived from double invasions of the same Madagascar stock but the same does not apply to the Foudias. The ancestral stock of the endemic *Foudia eminentissima* may have been allied to *F. omissa* of Madagascar rather than to that of the second invader *F. madagascariensis* (Moreau, 1960b). The *Hypsipetes* situation is more complicated and has in part already been referred to in discussing the origin of the Madagascar bird fauna. The *H. madagascariensis*, that occupy all four islands, are regarded as belonging to the same subspecies as the Madagascar ones and are presumably derived from there, but other subspecies occur on Aldabra and Gloriosa Islands just to the north, and also widely in southern Asia. Yet as already noted in Chapter 17, *H. madagascariensis* does not occur in the Seychelles archipelago which intervenes geographically. For this reason the possibility has been suggested that at one time this species ranged down the coast of Africa, in which case the Comoros birds would presumably have been derived from there and have themselves been the source of the Madagascar population. However, on the Seychelles, which lie some 800 miles north-east of the Comoros, the only bulbul is *H. crassirostris*, which is the second species of *Hypsipetes* inhabiting Moheli (as a different subspecies) but is not known from the intervening islands or from anywhere else. It is impossible to say whether the Comoros *crassirostris* is derived from the Seychelles or the reverse; and if the reverse what was the source of the Comoros stock. Neither is it clear which species colonized Moheli (and presumably the rest of the Comoros) first, but it rather looks as if *madagascariensis* had been the second and more successful colonist.

The two *Foudia* species are well segregated ecologically; *eminentissima*, which is subspecifically different on each of the four islands and no doubt the senior species there, is a bird of the forests, eating mainly insects and fruit; *madagascariensis*, evidently a recent arrival, depends on man-made habitats, "scrubland and cultivation", and lives mainly on seeds. Between the two *Hypsipetes* spp. and between the two *Nesillas* spp. no ecological segregation is obvious on the present scanty information. It is noteworthy that in both cases Moheli is the island showing the successful double invasion. This is, as previously noted, the island where the original environment is least disturbed, but it is in the west of the group and it is difficult to believe that the second species reached there without establishing itself for a time on Anjouan and Mayotte, which lie on the way from Madagascar.

The other three genera included in Table XXXVII are represented in the archipelago by stock from both Africa and Madagascar. The simplest case is that of the drongos, with a different species on each of the three islands they occupy. In *Nectarinia*, while the two eastern islands each have a distinct species, two other species co-exist on each of the two western islands; these two birds differ in size but nevertheless often feed together (Benson, 1960a, p. 93). In *Zosterops*, with four species in the archipelago, segregation is well marked. Grand Comoro is the only island to have two species, apparently as a result of double invasion from Africa; these are segregated by altitude, a situation exactly like that on some islands in Indonesia (Stresemann, 1939). Incidentally, the higher-level bird, *Z. mouroniensis*, has the typical characters of montane *Zosterops* throughout Africa, larger size and more melanin in the plumage (cf. Moreau, 1957). On the whole, therefore, the possibilities of inter-generic competition in any island of the Comoros are at the present time limited to a pair of sunbirds, a pair of bulbuls and a pair of sylviines (*Nesillas*).

There are some striking absences from the archipelago as a whole; in particular it has no member of the cuckoo, woodpecker, swallow, shrike or starling families. The absence of the cuckoo is particularly difficult to understand since *Centropus* occupies both Africa and Madagascar and in the latter *Coua* has flourished so greatly. There seems nothing whatever to take the woodpecker niche in the Comoros. Hirundines are prima facie unlikely to have been squeezed out by swifts, since so many species of both co-exist elsewhere; but Benson (1960a, p. 22) has noted that all three swift species in the Comoros "apparently" feed "close to the ground", which suggests that they may locally be occupying the niches of swallows and martins. The absence of shrikes is not compensated by the prevalence of Vangidae, as in Madagascar, except in one island, Moheli, which has *Cyanolanius* (Vangidae). This surely must at one time have existed also on the eastern islands; but it may perhaps be significant that Moheli is the only island that possesses no drongo, for these are aggressive birds which might perhaps be effective competitors. The absence of starlings (and turacos) leaves fruit-eating in the trees to pigeons and parrots, with, apparently, some help from the aboriginal *Foudia*. This situation is approached in Madagascar, but in the Comoros it is surprising that no African starling has established itself, especially as there would be no competition for nest-holes except from the small owl *Otus rutilus* and perhaps the smaller of the Vasa parrots.

MAFIA, ZANZIBAR AND PEMBA

The birds of these three islands have been listed, and discussed along with the rest of their vertebrate fauna, by Moreau and Pakenham (1941). A few records for Mafia have been added by Moreau (1944c), but this island has not had the same prolonged attention from resident ornithologists as have Zanzibar and Pemba.

The three islands have a great deal in common. They range in size only from 240 (Mafia) to 320 (Pemba) and 640 sq. miles (Zanzibar). None rises more than 100 m above the sea. All have a good rainfall, 1500–2000 mm, though with a more prolonged dry season in Mafia than on the other islands. Evergreen forest has probably been the dominant vegetation on all the islands, but it has suffered from prolonged human occupation. Apart from primitive agriculture, the islands have had trading relations with the north and east for some 2000 years. Now the dominant vegetation of the islands consists of coconut plantations, clove plantations (not in Mafia) and shifting cultivation for food crops. Apart from much secondary growth, little remains of evergreen forest except 3 sq. miles in Mafia, 12 sq. miles (of swamp forest) in Zanzibar and 2 sq. miles in Pemba (which may at one stage have been practically all cleared). Hence it is certain that the bird faunas have been to some extent artificially impoverished; an indication is provided by the fact that the forest on Zanzibar houses three species not known elsewhere on that island or on the others.

The most striking difference between the three islands is not obvious to the eye. Although none is more than 30 miles from the mainland, Zanzibar and Mafia are on the continental shelf and Pemba is not. According to the Admiralty charts it is possible to pass from Zanzibar to the mainland without encountering a depth of more than 19 fm (35 m). As in the case of Fernando Poo, it can be calculated from information we have on the post-glacial rise in the sea-level (Chapter 3) that Zanzibar was probably cut off from the mainland only after 10000 years ago. Similarly for Mafia, where the critical depth of the strait is only 10 fm (about 18 m), isolation took place after 8000 years ago. But for most of its separate existence Mafia has been less effectively isolated than Zanzibar, because the strait is narrower and full of banks which could have been used as "stepping-stones". Any coastal movement there was in this area seems to have been a sinking, so that the periods of isolation would, if anything, be shorter than those calculated above. By contrast, Pemba is isolated by a trough over 800 m deep. It is agreed that this is due to faulting but the postulated date of separation has varied from the Miocene to the mid-Pleistocene. The evidence from the vertebrates as a whole rather favours the more recent dating. Pemba has seven terrestrial mammals, six of which are subspecifically different from mainland forms but shared with another of the islands. They include a little antelope, *Cephalophus caeruleus*, most unlikely to have reached the island by rafting but conceivably introduced a few centuries ago. This antelope is shared with Mafia but not with Zanzibar, which intervenes geographically and is only 30 miles south of Pemba.

As shown in Table XXVIII Zanzibar has 105 species of birds, only about half the total for a similar narrow range of altitude and habitat on the opposite mainland coast (Moreau and Pakenham, 1941); but hardly any of the important families are missing and the Zanzibar bird fauna is as a whole a richer and better balanced one than that of either of the other islands. Pemba has only seventy-three breeding species, Mafia sixty-nine, the latter figure probably incomplete for water-birds and raptors. In these groups Pemba actually excels Zanzibar as well as

Mafia, but we get a very different impression if we concentrate on "other non-passerines" and passerines, with seventy-two in Zanzibar against forty-eight in Pemba and fifty-two in Mafia, and still more different if we consider passerines alone, forty-five against twenty and thirty-two. It is true that a part of this advantage is associated with the larger vestige of forest on Zanzibar than on the other islands, for eight of the Zanzibar passerines are typically forest species, compared with only two on Pemba and on Mafia. But each of these two latter islands wholly lacks a number of families that are prominent on the mainland and are represented also on Zanzibar. Among the most surprising absences on Pemba are babblers, barbets, bulbuls, drongos, orioles, thrushes, shrikes and woodpeckers, which between them total about forty-five species on the mainland coast only 30 miles away, and as many as sixteen on Zanzibar. By contrast, Mafia has three bulbuls, two shrikes, two thrushes, yet lacks babblers, drongos, woodpeckers, and also hornbills and parrots, which last two appear even on Pemba. Although it is probably that the bird fauna of Mafia is still not completely known, I do not believe that members of all these families, which are mostly conspicuous, would have been missed if they were present. All the same, if they really do not occur there it is impossible to suggest why.

The passerine bird fauna of Pemba, a mere 30 miles off shore, looks almost like that of an oceanic island, and it is extremely difficult to imagine why this should be so. Moreover, although it was well studied by Pakenham, he has, as he confirms to me, no evidence of what in the ecosystem can be taking the place of such normally important elements as the drongos or the shrikes or the thrushes. Two oddities there certainly are. Pemba alone among the islands possesses a *Zosterops*, a member of a genus that has reached many remote islands, some of them repeatedly, and on Pemba *Zosterops vaughani* is one of the commonest birds. In the absence of detailed information, there is no evidence that it is filling some particular niche left vacant on Pemba by a normal occupant, but there is a strong presumption that it may be, because on the actual coastal strip of neighbouring East Africa no *Zosterops* flourishes. This suggests that on the mainland coast there may be some successful competitor that is absent from Pemba. The other ecological peculiarity among the birds of Pemba is *Acrocephalus baeticatus*, which on the African continent behaves as a typical reed warbler but on the island is not particularly attached to freshwater vegetation and in a garden will "come and hop round after the manner of an English robin" (Vaughan, 1930). As already noted, the parallel with the acrocephaline on the Cape Verde Islands is close. It is also worth noting that, while Pemba and Zanzibar each have only a single small diurnal raptorial bird, on Pemba it is *Falco dickinsoni*, there opposite the northern limit of its mainland range, and on Zanzibar it is *Elanus caeruleus*, which has a vastly more extensive, intercontinental range.

Endemism is very low among the island birds, but slightly higher in

Pemba than in the other two islands. Mafia has not even a subspecies of its own, but shares one with Zanzibar, *Erythropygia barbata greenwayi*, differing in colour characters from *E. b. barbata* which occupies the mainland from Kenya to the Zambezi. I doubt whether the colouring of these birds in the two islands is due to convergence, and, especially since a bulge of the Tanganyika coast practically intervenes between them, it seems unlikely that one island population of *greenwayi* has been derived from the other, but rather that both are relicts of a former continuous mainland population of *greenwayi*. If this is correct, then birds with the characters of *greenwayi* have been superseded on the mainland coast in the course of about the last 7000 years. Apart from *E. b. greenwayi*, Zanzibar possesses three subspecies that are found nowhere else, and one other shared only with Pemba. The three are presumably the product of evolution over the 9000 years during which Zanzibar has been isolated. No doubt their evolution has been facilitated by the fact that they are of forest species, which are typically very sedentary, so that individuals are unlikely to have crossed the strait once it had been formed. Gene-flow would thus effectively have ceased. In fact one of the species, *Tauraco fischeri*, appears so poor a flyer that it is probably incapable of crossing 25 miles of sea. The subspecies of sunbird shared between Zanzibar and Pemba remains a peculiar problem. Whether the island populations are relicts of a mainland one or were evolved in one of the islands, individuals must have crossed the 30 miles of sea in or out of Pemba.

Pemba has one endemic subspecies that has never been claimed as anything higher and three other endemics that have; that is, omitting water-birds, less than one-tenth of the Pemba species are distinguishable from those of the opposite mainland, a proportion that, in view of the history of this island, seems surprisingly low. It is possible that intermittent gene-flow from the mainland may be responsible for the failure of most species to differentiate in Pemba, especially since hardly any of them are typically forest birds, a category that might be expected to be particularly sedentary.

The three more interesting endemics consist of a green pigeon *Treron pembaensis*, which can be regarded as forming a superspecies with *T. australis* of Africa and Madagascar (Husain, 1958), the owl *Otus pembaensis*, which forms a superspecies with the Madagascar *O. rutilus* (Benson, 1960a, p. 61) and the white-eye *Zosterops vaughani*. This bird is very distinct from African white-eyes, especially from any populations of *Z. senegalensis*, which occupy the mainland of East Africa. Probably, however, this Pemba *Zosterops* should, like the pigeon, be regarded as forming a superspecies with the mainland birds. In all three cases it seems that evolution has gone further on Pemba than it has on the younger islands of Zanzibar and Mafia and, as mentioned under Madagascar, the presence of a *rutilus*-type of owl on Pemba indicates that at one time it ranged up the East African coast.

Sokotra

Sokotra is an island approximately 70 miles long by 20 broad, lying about 150 miles east of Cape Guardafui and 150 from the south coast of Arabia. It seems to have been cut off from the Horn of Africa since about the end of the Pliocene and there are small arid intermediate islands, Abd-el-Kuri and the Brothers, with hardly any land-birds. Useful descriptions of Sokotra and its vegetation have recently been given by Botting (1958, 1960) and by Popov (1957). It is nearly all an extremely rugged limestone plateau at around 1500 ft above the sea, "particularly barren and waterless", but in the north-east granite peaks rising to nearly 5000 ft are "usually cloud-covered and damp, with small grassy plateaux, aromatic shrubs and small trees". Two-thirds of the plant species are endemic. During part of the year strong south-westers, which would help immigration from Somalia, are common. It is doubtful whether the scanty human population and domestic stock have had much effect on the island as a bird habitat.

The ornithology of Sokotra has been extraordinarily neglected. Very little has been published about it, the last contribution I can trace being that of Forbes (1903), reporting the results of an expedition in 1898–99, and incorporating preceding visitors' notes. In the subsequent sixty-five years no ornithologist seems to have studied there, but thirty-four specimens, collected in March 1934 at my instigation by an Arab working for Col. H. M. T. Boscawen, are now in the British Museum. All the seventeen species it includes were already known to Forbes. Finally in 1964 A. Forbes-Watson spent some months collecting on the island on behalf of the United States National Museum. As he has kindly informed me, he was able to add nothing to the list of resident birds compiled by the early workers except to collect the swift, at first believed to be a new species. It can therefore be taken that the bird fauna is known adequately for discussion. Since its features are so curious and its documentation is not very accessible, it is worth listing in detail below.

The birds of Sokotra known or presumed to be resident

E = Endemic

Group A (water-birds)

PHALACRACORIDAE. *Phalacrocorax nigrogularis.*

Group B (raptors, etc.)

AEGYPIIDAE. *Neophron percnopterus.*

AQUILIDAE. *Buteo buteo* subsp. (E). Proved to be a common breeding bird. The Sokotra population forms an undescribed subspecies (Benson and Irwin, 1963).

FALCONIDAE. *Falco tinnunculus archeri.* The kestrel of Somaliland. *Falco peregrinus* subsp. was observed by Forbes-Watson and is presumably resident, but *F. biarmicus* was not.

STRIGIDAE. *Otus scops socotranus* (E). Less streaked than either the African or the European Scops. Another, large, owl was seen by Ogilvie-Grant and Forbes.

GROUP C (ground-birds)

PTEROCLIDAE. *Pterocles l. lichtensteini.*

GROUP D (other non-passerines)

APODIDAE. A swift, *Apus** (E).

CAMPRIMULGIDAE. *Camprimulgus nubicus jonesi* (E). Greyer than the Somaliland birds.

COLUMBIDAE. *Oena capensis, Streptopelia* (=*Stigmatopelia*) *senegalensis, Treron* (= *Vinago*) *waalia.*

CUCULIDAE. *Centropus superciliosus sokotrae* (E). Greyer than Somaliland birds.

GROUP E (passerines)

ALAUDIDAE. *Eremopterix nigriceps melanauchen.* The Somaliland bird.

CORVIDAE. *Corvus ruficollis.* A bird of Arabia and the Sahara, not of Somaliland.

EMBERIZIDAE. *Fringillaria tahapisi insularis* (E). Paler than the Somaliland bird.

Fringillaria socotrana (E). There is a suggestion that this replaces the other bunting at high levels above 4000 ft, and if so its total population must be very small. Like *tahapisi, socotrana* is devoid of yellow, but it looks as if it may have been derived from different stock.

FRINGILLIDAE. *Rhynchostruthus s. socotranus* (E). Differs from both the Arabian and the Somaliland subspecies in colour pattern of the foreparts. The affinities of this monotypic genus with the big beak are uncertain.

HIRUNDINIDAE. *Hirundo* (=*Ptyonoprogne*) *obsoleta arabica.* Specimens collected in winter in Sokotra may well belong to a resident population, since this bird breeds in both Arabia and Somaliland.

LANIIDAE. *Lanius excubitor uncinatus* (E). With longer beak and less white on scapulars than Somaliland birds.

MOTACILLIDAE. *Anthus similis sokotrae* (E). More distinctly marked than Somaliland birds.

NECTARINIIDAE. *Nectarinia* (=*Cyanomitra*) *balfouri* (E). A sunbird devoid of any pigment except dusky melanin and yellow in pectoral tufts, but very unusual amongst male sunbirds in being streaky and mottled above and below. Affinities uncertain.

PLOCEIDAE. *Passer iagoensis insularis* (E). With grey instead of rufous rump as in the mearest African subspecies (the nearest point of which is extreme western Somaliland).

STURNIDAE. *Onychognathus blythii.* Known on the continent only from Eritrea to Somalia.

Onychognathus frater (E). Overlaps *blythii* in size. It may well be a derivative of *O. morio,* the rock starling widespread in eastern Africa but not found in Somalia.

SYLVIIDAE. *Cisticola juncidis haesitata* (E). Either to be regarded as a strongly marked insular subspecies (as C. M. N. White does), or as a distinct allied species (as Lynes did). Confined to a limited area of "thick bush-like grass' at a low altitude. The island birds are greyer and paler than those of the neighbouring mainland.

Incana incana (E). A monotypic endemic genus near *Cisticola.* Far more widespread in the island than *haestata,* namely in almost all bush-clad parts of the island at all elevations.

[TURDIDAE. The Isabelline and Desert Chats *Oenanthe* spp. recorded by Ogilvie-Grant and Forbes, and also obtained by Boscawen's collector, were probably only winter visitors.]

ZOSTEROPIDAE. *Zosterops abyssinica socotrana* (E). Brighter green than Somaliland birds.

* As we go to press this bird has been described not as a full species, but as *Apus pallidus berliozi* by Ripley (1965).

It will be seen from the foregoing that, with only twenty-seven species of land-bird, the bird fauna is thin. However, especially in passerines it compares not badly in numbers with the far more lush and varied habitats offered by the oceanic islands in the Gulf of Guinea and by those of the Comoros. Its history is of course different; it could conceivably retain birds whose ancestral stock occupied that part of the Horn of Africa before Sokotra was detached as an island. The species most likely to be of this status is the sylviine *Incana*, which would represent a proto-Cisticola. One of the most interesting resident local birds is the buzzard, not greatly different from the nearest population of *Buteo buteo*, namely *vulpinus* nearly 2000 miles to the north, in Asia Minor. This is the only clearly extra-Ethiopian bird resident in the island. It is probably a relict of a former wide southern extension of range by this species during a glacial period which may have led equally to its establishment, in the form *B. b. oreophilus*, on African mountains. Or, since Steppe Buzzards *B. b. vulpinus* migrate to Africa from Eurasia for the winter, the Sokotran and tropical African populations could be derived from passage migrants "stopping over". However, to these birds in Sokotra a remarkable parallel is provided by another endemic subspecies of *B. buteo* in the Cape Verde Islands at almost the same latitude, off the opposite side of Africa, where no European buzzards pass on migration. Hence an extension of the species' breeding range, southwards into the tropics during a glaciation, is the most likely means to account for the Sokotran population today.

Endemism in the fifteen Sokotra passerines is remarkably high, one monotypic genus, three species, two of which may belong to mainland superspecies, and six subspecies, a level which suggests that isolation from the continent is pretty complete, notwithstanding the favourable seasonal winds. In the twelve non-passerines endemism is far lower, only one species (the swift) and two subspecies, but most of the birds concerned show little or no evidence of plasticity elsewhere. As so often, the composition of the bird fauna is full of surprises. Among the non-passerines, given the nature of the environment, with so little arborescent growth, perhaps no other species could have been expected with any confidence except a bee-eater and a hoopoe. On the other hand, it is at first sight surprising that *Centropus superciliosus*, which always appears so incompetent a flyer, should have occupied the island. However, another of these non-parasitic cuckoos, *C. toulou*, has reached Madagascar and islands in its vicinity.

Among the Sokotra passerines it is curious that so large a proportion, six out of fifteen, should be made up of three pairs of species, the *Fringillaria* buntings (belonging to a family, the Emberizidae, not established on any other African island at all), the *Onychognathus* starlings and the *Cisticola/Incana* warblers; yet there is no flycatcher, no turdine, no tit, no babbler—to mention groups to members of which the environment seems suited—and a remarkably poor representation of larks and weavers.

Evolution in the island has taken no spectacular course. There is not a single clear abnormality in size. Other local variations have been diverse, but with a main tendency to pallor or dullness of plumage—a feature common in the birds of small islands. The local sunbird, lacking all bright colour except in its pectoral tufts, may perhaps be an extreme example of this. It is paralleled in this respect by the dingy sunbird of the Seychelles, but so it is by the mouse-coloured Sunbird *Nectarinia* (=*Cinnyris*) *veroxii* of the East African coast.

General Discussion of the Insular Bird Faunas

The insular bird faunas described in the preceding chapters may now be considered together, especially in relation to the size and the nature of the individual islands. There are of course several factors that in various degrees make the bird faunas only imperfectly comparable one with another: (1) the history of each island; (2) its degree of isolation; (3) the complexity of the habitats it carries; (4) the extent to which its habitats and its bird fauna have suffered from human interference, and (5) the completeness with which the bird fauna is known. The first factor can adequately be allowed for by dividing the islands into two categories: "continental" and "other". The latter, vaguer, term is preferred to "oceanic" because of the anomalous history of Pemba, already mentioned. The second factor, the degree of isolation, should I think be estimated in the case of an archipelago on the basis of the distance separating the closest inshore member of the group from a major landmass. This is as low as 60 miles in the case of the easternmost island of the Canaries, compared with 150 miles for Sokotra and 350 for the Cape Verdes. The significance of differences of this magnitude is uncertain, especially for such mobile animals as birds and in view of the dispersal that has occurred round New Zealand, referred to on p. 303. The third factor, complexity of the insular habitats, is of major importance. Ideally, degree of complexity should be assessed on a scale and an appropriate factor derived from it applied to the size of the island, but I cannot devise one. However, one thing that can be done to make the bird faunas more comparable for the present purpose is to exclude water-birds, since their habitat is lacking on some of the islands. Also it is useful to group together the four islands, Fuerteventura, Lanzarote, S. Thiego and Sokotra, which from aridity or extreme habitat destruction offer only a poor environment with few trees. At the other end of the scale Madagascar, with a range from semi-desert to luxuriant evergreen forest, is in a class by itself. But the remaining non-continental islands cannot be classified accurately. Except for those which are low and arid, the predominant natural vegetation was evergreen forest of one sort or another on both the Canaries and the tropical islands, but it is not easy to estimate how the different types of forest should rank in complexity. Other things being equal, an island that rises above a certain elevation will by virtue of that alone have a more complex habitat, and the native species that are absent from the lower altitudes in Tenerife and Grand Comoro show the practical effect of this; at the same time, in an archipelago the highest island tends to be the largest and the proportionate effect of height alone on number of species is not great. Moreover, human interference with natural vegetation has a double effect, as noted before. If it has not gone too far on a forested island it will not lead to extinctions, while cleared areas can accommodate non-forest species, as has happened for *Foudia madagascariensis* on the Comoros. In this respect the most favourable stage for abundance of bird species may have been reached now on Moheli and has almost certainly long passed on Pemba. Finally, the possibility that the bird faunas of some of the islands under discussion may be less adequately known than others can at this stage of ornithological exploration be dismissed as of minor importance.

Bearing in mind all the foregoing considerations, I think it is useful to group the insular bird faunas into four categories, each homogeneous in at least one or two of the most important characters, as listed in Table XXXVIII, and to plot them accordingly in Fig. 64. Purely for convenience of plotting, the size of the island is expressed as the square root of the area as measured in square miles, but, even so, Madagascar cannot be accommodated since it is eighty times as big as any other island. Excluding minor islands of the Cape Verdes, twenty-one bird faunas are available for discussion. The number of species for the individual Canary islands is taken from Volsøe (1955), each of the few which he lists with a query being counted as a half.

The first thing brought out by Fig. 64 is that, while the points plotted for the tropical and extra-tropical oceanic bird faunas in categories 2b and 2c intermingle, those for the continental islands and those for the "poor" islands are both segregated. Next, in each of the four categories of islands except the ecologically poor the tendency for number of species to rise with area is consistent. In categories 2b and 2c it can be calculated that the bird faunas increase by about one-half for an increase of eight times in area of island. (Darlington (1957), working on the herpetological faunas of Caribbean islands, tentatively reached the conclusion that number of species doubled with tenfold increase in area.) Extrapolation gives 2·25 times as many species in an island sixty-four times as big, 3·37 as many on an island 512 times as big. On this formula it is found that the groups B–E bird fauna of Madagascar, totalling 130 species and already noted as very deficient by African continental standards, lies almost exactly on the curve of expectation for the African oceanic islands. This is all the more remarkable because it is far older than any of the others here considered, has a far greater range of habitats, and has shown a great capacity for the production of species on its own surface, as witnessed by the Vangidae and the lemurs, for example. Furthermore, though eighty times bigger than the continental island of Fernando Poo and with a far bigger range of habitats, Madagascar has actually fewer species (apart from water-birds). All this gives additional grounds for regarding the Madagascar bird fauna as by no

TABLE XXXVIII

The insular bird faunas in relation to the areas of the islands

Island	Area (sq. miles)	√ area	Number of species in group*				Total species*	Ratio of group E to	
			B	C	D	E		group D	group B-E
1. Continental islands									
Fernando Poo	800	28	6	—	39	95	140	2·4	0·68
Mafia	240	15	4	2	20	32	58	1·6	0·55
Zanzibar	640	25	8	6	27	45	86	1·7	0·52
2. Other islands									
(a) Ecologically poor islands									
Fuerteventura	675	26	5½	4½	5	13	26½	2·4	0·45
Lanzarote	340	18	6	4½	4½	12	27	2·7	0·44
S. Thiego	360	19	4	2	3	9	18	3·0	0·50
Sokotra	1300	36	4	1	6	15	26	2·5	0·58
(b) Other tropical islands									
Anjouan	147	12	6	1	11	11	29	1·0	0·38
Annobon	7	3	1	—	3	2	6	0·7	0·33
Grand Comoro	370	19	6	1	11	16	34	1·5	0·47
Madagascar	240000	490	20	10	37	63	130	1·7	0·48
Mayotte	144	12	5	1	10	9	25	0·9	0·36
Moheli	84	9	4	1	11	15	31	1·4	0·45
Pemba	320	18	9	4	18	20	51	1·1	0·39
Principe	50	7	—	—	10	13	23	1·3	0·32
São Tomé	400	20	3	1	10	18	32	1·8	0·56
(c) Other extra-tropical islands									
Gomera	150	12	4½	2½	7	18½	32½	2·6	0·57
Gran Canaria	590	24	7½	4	6	21	38½	3·5	0·57
Hierro	110	10	3	2	4	17½	26½	4·4	0·66
Palma	280	17	4	2	7	20	33	2·9	0·61
Tenerife	790	28	7½	4	6	22	39½	3·7	0·55

* A species, the breeding of which is explicitly doubtful, is scored ½.

means coming up to expectation. As already mentioned, I doubt whether human destruction of habitat can be held accountable for most of this, and widespread extinction through climatic catastrophe must be postulated.

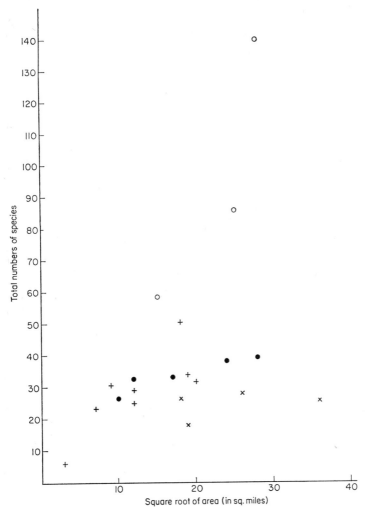

FIG. 64. Number of passerine species plotted against area of island. o, Category 1 bird faunas; ×, category 2a; +, category 2b; ●, category 2c.

As shown in Table XXXVIII, the three continental islands are by far the richest in species; in them the rise in number of species in relation to area of island is much steeper, for the number about doubles when the area is trebled. The relative richness of Fernando Poo may be due largely to the habitat factor, for this island has retained much of its forest bird fauna and gained more species in the clearings, whereas in Mafia and Zanzibar the natural vegetation, especially forest, has been greatly damaged. It should be noted also that Pemba, anomalous in its insular status, appears on Fig. 64 between the continental islands and the oceanic. Finally, in category 2a, the impoverished islands, the points plotted are unconformable, for, as shown in Table XXXVIII, the bird faunas of the dry islands of the Canaries are richer than those of S. Thiego and Sokotra. It is difficult to believe that the habitats on these last two islands are poorer than the others, so it may be that their greater isolation—by some six times and three times the sea-mileage respectively—is a significant factor.

Although, as noted above, certain regularities can be perceived when the totals for groups B–E in each bird fauna are plotted, it will be seen from Table XXVIII (pp. 304–307) that they conceal marked unconformities in the individual groups. One of the more extreme examples is in the ground-birds of the continental islands, among which Zanzibar has six, Mafia two and Fernando Poo none. Since the passerines are the most numerous group they are the most suitable for discussion. When the passerine totals are plotted against size of island, as in Fig. 65, the result is much more irregular than that for the total B–E. The points for the "other tropical islands" are now thoroughly mixed up with those for the Canary Islands and that for Pemba, instead of standing aside, comes among them. The extraordinary predominance of Fernando Poo is accentuated and, as shown in Table XXVIII, it actually possesses nearly one and a half times as many passerine species as Madagascar.

If now we consider the proportion of passerines in the B–E fauna of each island we find that it is lowest, under two-fifths, in Anjouan, Annobon, Mayotte and Pemba, and is highest, around two-thirds, in Hierro and Fernando Poo. Among the tropical non-continental islands (category 2b) the proportions actually range from 0·33 to 0·49, except for Principe and São Tomé, both 0·56. Yet of the bird fauna of Africa south of the Sahara (Table V, p. 84) passerines form nearly two-thirds (0·61) and of individual African communities (Tables XXIV and XXV, pp. 276 and 286) 0·48–0·80. It might be suggested that the anomalous composition of the insular bird faunas do not reflect their original natural state, but are the result of human interference. If this were true it would, however, be necessary to explain why passerines had suffered more than other birds from this cause. Moreover, Moheli, the Comoro island that has been quoted as having suffered comparatively little from "development", has only a low proportion of passerines, 0·47, actually lower than in any of the continental communities examined.

In attempting to analyse these anomalies a little more closely, it may be recalled that passerines encounter no competition at all (and in Africa not a great deal of predation) from group B birds, raptors and owls. Competition between passerines and group C ground-birds is surely slight, and practically limited to some overlap in the taking of seed off the ground. There is more potential competition between some passerines and some group D birds ("other non-passerines")—in seed-eating with some doves, in fruit-eating (starlings with barbets, turacos, fruit pigeons and parrots), in aerial feeding (swallows with swifts). Some of these possibilities have been mentioned incidentally under individual islands, and inspection of the insular lists in Table XXVIII suggests that competition between passerines and group D birds is not important. In fact the balance of numbers between these two groups is highly variable, as shown in Table XXXVIII. In the Canary Islands and the arid islands, together with Fernando Poo, there are two and a half to

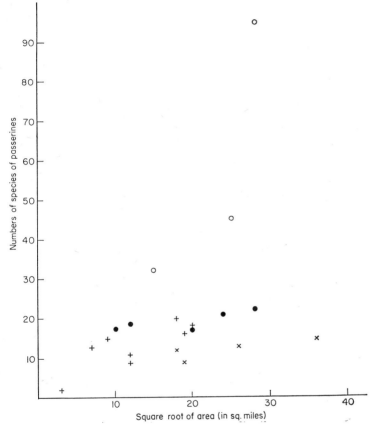

FIG. 65. Number of passerine species plotted against area of island. ○, category 1 bird faunas; ×, category 2a; +, category 2b; ●, category 2c.

over three times as many passerines as group D species; in none of the other islands are they as much as twice as numerous and in most of them not one and a half times. This compares with a figure of over three times in the continental bird fauna (2·8 for forest birds, 3·4 for the others, as can be calculated from Table V). I am at a loss to explain these anomalous results. They suggest that in most well-vegetated tropical islands (but not in Fernando Poo) opportunities are proportionately less for passerines than for group D birds, or else that, compared with group D birds, individual passerine species occupy on the average proportion-ately broader niches than they do on the continent. The Canary Islands and the Cape Verdes are the ones in which field-work has given most attention to this question. In the former archipelago two species in particular, the Blue Tit and the Chiffchaff, have been shown to occupy wider niches than they do on the continent, apparently as a result of the absence of other species of *Parus* and *Phylloscopus*. In the Cape Verdes there is evidence of considerable change and broadening of niches, especially in the three species of warbler. Of these the most remarkable is the Acrocephalus. A bird of the same genus in Pemba can be cited as the only outstanding case of the same thing in the other insular bird faunas. In most of the other islands special attention has not been given by field workers to the ecological relations of the species, but even where this has been done, by Benson (1960a) in the Comoros, no case has been found of obvious broadening of niche, as compared with that occupied by the same species in Madagascar or in Africa.

A contrary effect, namely a narrowing of niche, is likely to occur where a "double invasion" (Mayr, 1942) of related stock from the same source, whether the mainland or another island, has established itself. It is of course not always clear how far the limits of "related stock" can be stretched in this connexion. From one point of view, any two species in the same genus are of interest as potential competitors; but some of these pairs of species, such as the two *Vasa* parrots on the Comoros and the two *Streptopelia* doves on Zanzibar, may be assumed to be merely in the same mutual relations in the islands as they are in their presumed country of origin, in these cases Madagascar and the mainland of Africa respectively. The "double invasions" of real interest are those in which it seems possible to discern in the area of origin a species to which both invaders are more closely related than to any other and not merely one with which both are congeneric. On this definition the undoubted cases of double invasion in the numerous oceanic islands under review are only seven, as shown below, the more differentiated and so pre-sumably earlier invader on each island being listed first.

> Canary Islands: *Fringilla teydea* and *F. coelebs*, which occupy exclusive habitats.
> Principe: *Lamprotornis ornatus* and *L. splendidus*. The mutual relations of these starlings are not known. Snow (1950) found *L. ornatus* in both forest and plantations and failed to see the other species.
> São Tomé: *Cyanomitra thomensis* and *C. newtoni*, which differ greatly in size and feeding habits.
> Madagascar: *Podiceps pelzelni* and *P. ruficollis*; *Ardeola idae* and *A. ralloides*. The mutual relations within each pair are not known.
> Moheli: *Nesillas typicus* and *N. mariae*. Same comment.
> Grand Comoro: *Zosterops mouroniensis* and *Z. senegalensis*, which are segregated altitudinally.

Although the foregoing are the only cases that can strictly be regarded as double invasions a few others are worth mentioning, for they have the ecological characters of these phenomena, if not the taxonomic.

Fernando Poo, Principe and São Tomé each have two zosteropids, one larger than any mainland birds and highly modified (into what is usually regarded as a different genus), the other little changed. Different feeding habits are indicated (see also Snow, 1950), and in Fernando Poo there is also habitat segregation.

São Tomé: If *Neopiza concolor* is a member of the Fringillidae, not of the Ploceidae, it may be derived from the same stock as *Poliospiza rufobrunnea*, which also occurs on this island. The greater size of *Neospiza* and its far heavier beak presuppose different feeding habits in the two birds.

Moheli: *Hypsipetes madagascariensis* and *H. crassirostris* are probably derived immediately from different stocks, in Madagascar and the Seychelles area respectively. Both inhabit the evergreen forest on Moheli, but since the latter is larger, especially in beak, different food may be surmised.

Sokotra: *Fringillaria socotrana* and *F. tahapisi* seem to be separated altitudinally, but are doubtfully of the same stock. *Onychognathus blythii* and *O. frater* may be of the same stock, but their ecological relations are unknown.

Thus, considering that we have been dealing with twenty-one islands, the proportion of double invasions is low, for they total only seven that are strictly admissible, together with another eight interesting cases on the borderline. In two of the first seven pairs of species the birds occupy different habitats, and in a third pair (as also in the Gulf of Guinea Zosteropidae) the species have developed different sizes, and hence presumably different feeding habits (cf. Lack, 1944; Moreau, 1948).

It will be clear from the discussions on the various islands that the degree of endemism differs to an important extent. To arrive at some sort of numerical index for this, Table XXXIX has been compiled, using a scale that is admittedly arbitrary. Each endemic subspecies has been assessed as one point; an endemic species (or semi-species) that is merely an insular member of a superspecies, three points; an endemic species not allocable in this way, nine points; and an endemic genus, twenty-seven points. The points are then summed and divided by the total of B–E species in the bird fauna. (Water-birds are omitted from the calculations because they are not comparable on different islands.) The inclusion of the genus in this scheme is possible because all the genera in question are monotypic. Madagascar cannot be included in this comparison partly because of its polytypic genera and partly because of the necessity for making some allowance for its endemic families and sub-families, which would be difficult. Certainly, of course, its endemism is high (Table XXXI).

Table XXXIX brings out clearly the low status of the continental islands, with no index higher than 0·3; and Pemba also scores only 0·2.

TABLE XXXIX

Endemism in the bird faunas (groups B–E) of the African islands

	Total no. of species	Number of endemic				Total points*	Proportionate endemism†
		subspecies	species that are members of superspecies	other species	genera (monotypic)		
Continental islands							
Fernando Poo	140	37	—	1	—	46	0·3
Mafia	67	—	—	—	—	—	0·00
Zanzibar	105	3	—	—	—	3	0·03
Other islands							
Canaries	49	36	4	2	—	66	1·3
Cape Verdes	26	11	—	2	—	29	1·1
Comoros	52	24	6	2	1	96	1·8
Pemba	72	1	2	1	—	16	0·2
Principe	26	7	6	2	1	70	2·7
São Tomé	40	7	11	4	2	130	3·2
Sokotra	26	10	(?) 1	2	1	58	2·2

* Assessed at 1 for a subspecies, 3 for a species that is a member of a superspecies, 9 for any other species, unless it forms a monotypic genus, in which case it counts as 27.

† Calculated by dividing the score in column 7 by the number of B–E species in the bird fauna.

It may also be borne in mind that in these in-shore islands not all the endemics may have been evolved in isolation, but some may be relicts. Specific instances have been cited. Of the other islands the next lowest scores are those of Cape Verdes (1·1) and Canaries (1·3), while the highest are Comoros (1·8), Sokotra (2·2), Principe (2·7) and Sao Tomé (3·2).* Thus degree of isolation does not seem to be of overwhelming importance. If it were, the Cape Verdes, which are more remote from the continent than any other of the oceanic islands, would not rank lowest among them. The same applies, with less force, to Sokotra, which is some 150 miles off shore compared with only 140 miles for Principe and 170 for São Tomé. These two islands certainly have the most peculiar bird faunas of all, for their scores in Table XXXIX, high as they are, make no allowance for the fact that none of their three endemic genera can certainly be allocated to a family; and moreover the gigantism and/ or oddity of such other species as *Nectarinia thomensis, Speirops* spp. and *Ploceus st. thomae* can hardly be paralleled in any of the other islands, Madagascar of course excepted. In other words, evolution has gone further on the Gulf of Guinea islands than on any of the others under discussion, including, in particular, the Comoros, and it is most difficult to put forward any suggestion why this should be so.

Mayr (1965) has discussed "the nature of colonizations in birds". He has pointed out incidentally that at any rate in the western Pacific it is possible to designate "colonizing families" and that they will be found to constitute a higher percentage of the bird fauna in islands further from the prime source of colonists. Owing to the arrangement of the islands round Africa, all peripheral to the continent, and to their heterogeneous natures these ideas cannot be applied with much effect in this part of the world. The recent continental islands of Fernando Poo, Zanzibar and Mafia have to be omitted from consideration since their birds have presumably not been derived in the main by "colonization" over water and the figures for the Madagascar species in individual families cannot for the present purpose be taken at their face value because in some families the representation has been so much increased by speciation within the island. There is the further caveat that, owing to their situation, the Canary Islands would have had little or no chance to acquire members of any purely tropical family of birds. One final point must be borne in mind, and that is that the colonists an island can expect to get will be affected by the nature of the neighbouring coast of the mainland, as well as by the habitats the island itself provides. For example, the Gulf of Guinea islands are ecologically unsuitable for larks and so are the shores of Nigeria, Cameroon and Gabon. Hence the absence of larks in the Gulf of Guinea islands is not necessarily a mark against the colonizing ability of the Alaudidae. On the other hand, the

* If the scale 1, 2, 4, 8 is adopted (instead of 1, 3, 9, 27), thus in effect giving more weight to subspecies, this order is unchanged, except that Sokotra then scores relatively less than Principe and São Tomé and equal with the Canaries and the Comoros.

relatively remote Cape Verde Islands, which have open country and look across the sea to lark country in Senegal, have been colonized by four species.

From the figures for the island bird faunas in Table XXVIII, it can be said that the raptors, the owls, the swifts, the parrots, the sunbirds, the estrildines and other weavers, the flycatchers, the warblers, and the white-eyes are the families that have distributed themselves into most islands. I hesitate to include the thrushes here because, although they appear on most islands, they hardly live up to the expectations aroused by the fact that their continental representatives include species adapted to every kind of habitat. This list contains some illustrations of the remark by Mayr (1965) that "one cannot predict on the basis of morphological, ecological or physiological characteristics whether or not a given species will be a successful colonizer". In particular, neither the sunbirds nor the white-eyes could have been expected to get around in the way they have (though the latter are notably sociable); nor, turning for a moment to a smaller group than the family, a cisticola (grass warbler) to reach Sokotra and Madagascar, if no other islands. In this respect no single species (-group) is more remarkable than the Paradise Flycatchers, which have colonized even Annobon, as well as Madagascar and neighbouring islands. On the opposite side of this problem, while no one with experience of a trogon or a turaco in the wild state would be surprised that they fail to appear on oceanic islands, the reverse applies to the hirundine and shrike families. Both include notable long-distance migrants—the former even within the African continent—possessed moreover of the habit of moving in parties; yet both make a very poor show in Table XXVIII.

In conclusion, it may be noted that among the African islands there is no evolutionary situation resembling those of the Galápagos Islands (in the Geospizidae; see especially Lack, 1947) and of the Hawaiian Islands (in the Drepaniidae; Amadon, 1950), except in Madagascar. There adaptive radiation has been considerable in the Vangidae and has probably been so also in the sub-oscine Philepittidae, though these are now poorly represented by only a few survivors. Also at lower level, within the genus, the couas of Madagascar have radiated more in habits than in morphology. These examples are the more remarkable because they have been achieved on a single land-mass of no great size—less than one-tenth that of Europe—without the aid of exchange between islands.

It is beyond the scope of this book to attempt any wide range of comparisons with insular bird faunas elsewhere, but it is suggested that the data here presented may be useful in a world-wide discussion of island bird faunas, which should take into account the ideas of MacArthur and Wilson (1963) and which it would be supremely interesting to have at this stage of zoology.

20

Conclusion

The discussion in the foregoing chapters has shown, all through, the extent to which geographical and ecological problems are intermingled. The discoveries that have illuminated the conditions of the Late Pleistocene have made it easier to understand and to explain the main difficulties of discontinuous distribution in Africa, but they have if anything made it harder to explain the extent to which the biota north and south of the Sahara differ from each other in most of their elements. In this connexion a remarkable unconformity discloses itself. In the main outlines of their distribution, the birds of Africa have close parallels in the plants and also in the butterflies. In all these three classes the species of the montane forests are nearly all different from those of the lowland, so that one is fully justified in speaking of a distinct montane fauna and flora; and again the species inhabiting the Maghreb are to a great extent different from those inhabiting the southern side of the Sahara. But in these respects the mammals differ strongly: no distinct montane fauna is recognized and, except for the bats, the mammals of the Maghreb are predominalty Ethiopian. Between the birds and the butterflies the parallels extend indeed into such details as the peculiar and unexpected status of the fauna of each on Marsabit Mountain, the faunal relations of the East African coastal forests to the West African, and the affinities of the relict montane forests, Kungwe and Ufipa, on the east side of Lake Tanganyika. It would be of great interest if other classes of organism could be brought into this comparison.

It may here be useful to recapitulate some of the points brought out in the course of this book, and they may conveniently be summarized chapter by chapter.

As indicated in Chapter 1, it is essential for fruitful zoogeographical discussion that the birds of evergreen forest, which are highly specific to that habitat, should be dealt with separately from the others. This is a proviso probably applicable also to other tropical areas besides Africa. Of special importance in Africa, because of its high relief, is the distinction between typically lowland and typically montane species, the division between them being, in the main, at around 1500 m above the sea in the tropics. Also the significance of discussion is greatly enhanced if the birds are divided into five groups, which are, broadly speaking, (A) water-bird families, (B) raptors and owls, (C) game and other ground-birds, (D) other non-passerines, (E) passerines. One more preliminary consideration is the virtual certainty that evolutionary change may be far more rapid than has hitherto been supposed, at any rate among passerine birds. An exceptionally high proportion of the birds in this group belong to superspecies and there are other reasons for supposing that passerine species are on average "younger" than others.

Chapter 2 describes essential features of the climate and vegetation and also the seasonality of different parts of Africa, with especial reference to the breeding seasons of the birds. Most of the surface of tropical Africa is covered with savanna or wooded steppe where a rainy season is succeeded by an extremely well-marked dry season, which to all appearances is ecologically very unfavourable. Yet appearances must be to a considerable extent deceptive. On the one hand, in some of the evergreen forests, which certainly look as if they had no markedly unfavourable season, nearly all the birds concentrate their breeding into a few months of the year. On the other hand, in areas where the dry season appears to be very severe and great scarcity of food would be postulated at that time, the great majority of the bird species manage to persist throughout the year and moreover, north of the Equator, Palaearctic migrants are actually accommodated throughout the dry season.

Chapter 3 supplies the information about changing state of Africa in the Late Pleistocene. In the most (and last) severe phase of the Last Glaciation the temperature was so much reduced that a continuous block of montane conditions extended from Abyssinia to the Cape and round the northern edge of the Congo basin to the Cameroon highlands. South of the Sahara, species intolerant of montane conditions must have been confined to West Africa, to the centre of the Congo basin, to part of the southern Sudan and to a strip of country round the coast. The glaciation did not begin to relax until about 18 000 years ago, and thereafter temperatures rose to a little above the present around 6000 years ago. Thus it can be accepted that, as regards the balance between montane and lowland, conditions did not approximate to those of the present until after about 12 000 years ago.

It is not known to what extent evergreen forest expanded beyond its present limits during the cool period of the glaciation but it is known to have suffered great vicissitudes as a result of climatic changes, the causes of which are unknown. Prior to about 70 000 years ago most of the Congo forest was ruined during an access of aridity spreading from the south, while at an uncertain, perhaps later, date aridity that spread from the north across West Africa must greatly have reduced the forests there. In the southern tropics around 12 000 years ago Kalahari conditions spread again north to the lower Congo and east down the Zambezi Valley, doubtless reducing and cutting up what is now the great block of miombo (brachystegia) woodland.

By contrast, what is now the Sahara was greatly mitigated during much of the Late Pleistocene. A woodland of typically Palaearctic species extended at least as far south as Ahaggar and more xerophytic Mediterranean vegetation practically to the site of the present Lake

Chad, which was greatly expanded. Whether at the same time the birds so far south were typically European is not known. There can have been no great barrier to the movement of animals across what is now the western Sahara; but about 5000 years ago, before the end of the Neolithic, conditions changed quickly and catastrophically and the present extreme desert soon supervened.

Chapter 4 discusses the bird faunas of Mediterranean Africa and of the Sahara. The first is nearly all Palaearctic (European) while the species inhabiting the Sahara are neither typically European nor typically Ethiopian. Presumably following the post-Neolithic desiccation the desert bird fauna spread from a refuge further east. The oases hold a scanty bird fauna, in the north comprised of mainly Palaearctic species. In Egypt the Ethiopian component of the bird fauna differs much in composition from that in the Maghreb, partly under the influence of the Nile.

In Chapter 5 the (Ethiopian) bird fauna south of the Sahara is analysed between forest and other species and between typically montane and typically lowland. Using to the full the polytypic species concept, the total arrived at for species breeding on the continent south of the Sahara is 1481, and so sharp is the distinction between forest and other species that less than fifty cannot be allocated to one category or the other. The total is made up as in Table XL.

TABLE XL

The species-composition of the Ethiopian bird fauna

Forest			Non-forest			Indeter-minate
Typically lowland	Typically montane	Both	Typically lowland	Typically montane	Both	
250	120	39	887	74	69	42

The montane birds are now confined to a small fraction of the area open to them as recently as 12 000 years ago; by contrast, for a long period prior to that date the lowland species were confined to a small fraction of their present area. The present fragmentation of the montane area is conducive to speciation today, while the state of Africa during the glaciation was conducive to speciation in lowland birds. The montane zone has probably always supported a much poorer variety of habitats than the lowland and at present the montane non-forest bird fauna has a smaller proportion of species peculiar to it than any other, perhaps in part because its habitats have been so much interfered with by man. A special evolutionary problem is posed by a number of forest species confined to secondary growth and forest edges; the hypothesis is that such birds, evolved elsewhere, have acquired their habitat specificity as a result of competition.

On average, African group A birds (water) and group B (raptors and owls) have more extensive geographical ranges both within the continent and beyond it, than others, and especially the passerines.

Chapter 6 discusses the Ethiopian bird fauna in terms of its endemism and its affinities with the bird faunas of other continents. With South America they are almost confined to group A birds, with Europe and with India they are strongest in group A and group B birds and weakest in passerines. So far as species are concerned, less than one-twentieth of those which breed in the Ethiopian Region breed also in Europe and a proportion slightly lower than that of Ethiopian species breeding in India. This is the more noteworthy because from Europe about 150 species travel to south of the Sahara every year on migration, while there is hardly any such annual interchange with India. The rarity with which European species have established themselves probably means that the Ethiopian communities are too "saturated" easily to admit other birds to breed. Some of the species shared with India do not occur in Arabia, from which they have probably disappeared recently as a consequence of post-Neolithic desiccation.

In Chapter 7 the non-forest bird faunas of a number of African territories are compared with each other and in relation to the range of habitats available. Area *per se* is of comparatively little importance in this respect. Forest birds are omitted because their habitat is deficient or absent in some territories. While Kenya is the richest territory, the most important outcome of the comparison is to show that West Africa, with 1 million sq. miles and the full range of habitats from steppe-desert to humid savanna, is inhabited by slightly fewer (non-forest) species than Malawi, 37000 sq. miles and with a limited range of habitats. This comparative poverty of West Africa in non-forest birds seems to be only one aspect of a general biological poverty of the northern tropics of Africa compared with the southern. Contributory causes may be the fact that the northern tropical biomes form comparatively narrow latitudinal strips unstable in time, and that the topography is very simple, so that chances of isolation, conducive to speciation, are poor. Tropical West Africa has moreover been to a great extent isolated from the rest of the Ethiopian Region by ecological barriers. At the other end of the Region the Cape Province too has a notably poor bird fauna considering its very wide range of habitats. It is specially deficient in group D birds, the "other non-passerines", comprised largely of families dependent on fruit and on large insects.

In Chapter 8 some comparisons are made between Ethiopian bird faunas and others elsewhere. The Ethiopian as a whole is much poorer than that of South America, mainly in forest birds. This may be a consequence of the ecological history; the severe disturbances suffered by the Ethiopian forests in the Late Pleistocene is not known to have had a counterpart in South America.

When the bird fauna of the Gambia Valley is compared with that

of the Thames, and the bird fauna of an arid African territory, Somaliland, with Morocco and with Arizona, the African bird faunas are
shown to be the richer. One main factor is that the group D birds, the
"other non-passerines" are much more richly represented in the tropical
bird faunas than at higher latitudes. In this group of birds, species
dependent all the year round on fruit or on large insects are especially
well represented; and these are foods that outside the tropics are seasonally unobtainable.

Chapter 9 discusses the comparatively poor bird faunas of the lowland rain-forest of Africa. The present break in the Guinea forests by
the intrusion of savanna through the "Dahomey Gap" is not associated
with any important change in the forest birds on either side of it and
hence is presumed to be comparatively modern. A number of species
appear to be limited to the north-east corner of the Congo forest, a cir-
cumstance for which no reason can be found. By contrast, several
western species or superspecies of the lowland forest which extend east
to the Lake Victoria basin reappear on or near the east coast of Africa,
separated there from their relatives by the mass of the Kenya highlands
and the dry country of the coastal hinterland. To account for their
extension it seems necessary to postulate a period that was both warm
enough and wet enough for a connexion of lowland forest—not an easy
concept.

Chapter 10 discusses the lowland non-forest species. There are three
main foci of endemism, all connected with dry country, namely the dry
south-western corner of the continent, Somaliland with a prolongation south through the acacia country to northern Tanganyika, and the
strip south of the Sahara from Senegal to Abyssinia. Although Lake
Chad has been so potentially important a barrier to east-and-west communication in the recent past, especially in conjunction with the forests
of the Cameroons to the south, endemism to the west of this line is low.
There seem to be no species peculiar to the belt of *Isoberlinia* woodland
of the northern tropics whereas the corresponding formation in the
south has twenty-three endemic species—another aspect of the biological
poverty of West Africa. About forty species of birds (and also some
mammals), the habitat of which is dry acacia country, have a discontinuous range, inhabiting the Somali area and then after an interval,
which includes more or less of Tanganyika and Zambia, reappearing
in the dry south-west of the continent. This postulates the existence of
a corridor of dry acacia country right through East Africa sometime in
the Late Pleistocene, which may have been either before or after the
wet period postulated in Chapter 9.

Chapter 11 discusses the distribution and differentiation of the
specialized montane forest bird fauna. Sedentary as these birds appear
to be individually, it must be inferred that a few species cross some 80
miles of extremely dry country to Marsabit Mountain frequently
enough to inhibit subspeciation there, while three species are common

to the mountains of both Africa and the Yemen, implying a hop of some 300 miles. However, in the mountain forests of the Cameroons, isolated by 1200 miles from mountains to the east, half the birds are East African and presumably reached there by way of a forest connexion round the northern rim of the Congo basin when the climate was at last 5° cooler than it is now. The varied degree to which the Cameroon birds are differentiated points to more than one period of intercommunication. Again, the birds of the relict forests of the Angolan highlands, nearly as isolated today as are those of the Cameroons, all belong to East African species.

The affinities of the Kenya montane bird fauna with the East Congo are what might be expected from their connexion until probably only some 12 000 years ago, and the Kenya endemism is very low. That in the bird fauna of the Tanganyika–Nyasa mountains is far higher and an unaccountable feature is the high degree of subspeciation in mountains near the Kenya–Tanganyika border, often in sight of one another. Another anomaly is that the forests of the Abyssinian plateau are extremely deficient in bird species. At the other end of the continent the number of montane forest species is much reduced south of the Zambezi. An example of "multiple" invasion, that would be remarkable in a marine island, is afforded by the four species of *Cryptospiza* (estrildine weavers) inhabiting the forests on Ruwenzori Mountain.

In the non-forest montane birds, discussed in Chapter 12, a main point of interest is that the Abyssinian highlands are rich in endemics, the converse of their status with regard to forest birds. Special interest attaches to the afroalpine areas, limited to above the timber-line on a few East African mountains. Most of the scanty resident bird faunas, which vary much from one mountain to another, are derived from the lower montane levels, without subspeciation, but there are three endemic species, the highly discontinuous ranges of which raise special problems.

Chapter 13 describes the regular migration of African birds within the continent. As with Palaearctic migrants, insectivorous species predominate. Also, most of the movement is between breeding grounds at higher latitudes and wintering grounds nearer the Equator, but a few species cross the Equator. A feature unknown in the Palaearctic is that during the (summer) rains the strip on the south side of the Sahara is sought not only by birds coming there to breed but also by birds arriving for the "off" season.

The migrants from the Palaearctic (Chapter 14), which visit Africa in enormous numbers, distribute themselves very differently from the migrants of the Americas, in that they appear almost completely to ignore evergreen forest and they occupy the continent to its southern extremity. (In South America few of the land-birds cross the Equator; in Africa nearly half the migrant species reach nearly to the southern edge of the tropics (Rhodesia) and many enter Cape Province.) A remarkable feature is that those migrants which winter in the northern tropics of Africa are, unlike those in the southern, subsisting through the dry season, which has every appearance of being the bad season of the year for food. Ecological relations in this connexion are by no means understood, especially in relation to the resident African species. Moreover prior to departure in spring the migrants, faced with the 1000-mile crossing of the Sahara, must put on much fat, an operation which in the birds wintering in the drought of the northern tropics seems to be achieved in a brief time under conditions that have every appearance of being unfavourable.

Because of the ecological changes that have taken place on a great scale in both Europe and Africa in the Late Pleistocene the present pattern of migration to and from the Palaearctic has never occurred before and the striking adaptations of the birds concerned must be very recent.

Chapter 15 compares the bird faunas of sample areas of different African vegetation types. Evergreen forest at low altitudes is the richest in species (and has the highest proportion of passerine birds), but rapidly becomes impoverished with increasing altitude. By comparison, deciduous woodland of various types is remarkably rich, except mopane. While no information on the avian biomass exists, it is possible to give numerical expression, purely on the basis of the species present, to the amount of "feeding attention" given to each of the main types of bird food (invertebrates, fruit and seeds), and thus to show how the proportionate "attention" differs from one vegetation type to another.

Chapters 16 to 19 describe the land-bird faunas of the islands round Africa, in relation to their evolution and their ecology, though discussion is hampered by the uncertain, but often gross, damage done to the insular habitats by human occupation. The Canary Islands and to some extent the Cape Verdes are peculiar in that they evidently owe their present stock in part to immigration during a glaciation. For the assessment of the degree of endemism a simple numerical formula has been devised which can be applied to all the islands except Madagascar, and this shows how much further evolution has proceeded in the islands of the Gulf of Guinea (except the "continental" Fernando Poo) than in the others. When the islands are divided into ecological categories a clear correlation between area and number of species emerges. There is no evidence of strong adaptive radiation in any of the archipelagos, but it has evidently taken place within Madagascar. This great island poses a number of unique problems: its bird fauna is far poorer than might have been expected on ecological and evolutionary grounds; and for several reasons it must be postulated that there has been considerable recent extinction, not the result of human interference. In Madagascar the proportions of the total bird fauna contributed by each of the five "groups" of birds differ markedly from those on the African continent; and a peculiarity shared by Madagascar with nearly all the other islands is the low proportion of passerine species.

In general, reviewing all the discussions in this book, it will be seen that two main themes have reappeared incidentally again and again. One is the mutability of the whole biological scene and the other is the problem of the number and the nature of species comprising a bird fauna, whether of a continental area, of an island or of a vegetation community. The mutability is a matter that exercised the minds of the older biologists little, except in terms of geological epochs. For us, the discoveries of the past few years on one front and another have provided an entirely new perspective of the evolutionary potential and of the extent to which the face of Africa has changed, not merely in the course of the Pleistocene but most notably in the last few thousand years. The period spanning the last major change, that from the height of the last phase of the Last Glaciation, around 20000 years ago, to the hypsi-thermal ("climatic optimum") around 6000 years ago is so short that until recently it would have seemed to most people almost negligible on the evolutionary time-scale. We now know that it is possible for what any conventional bird taxonomist would recognize as a "good sub-species", meriting a trinomial, to be developed in much less than 100 generations, and hence it seems probable that, given geographical isolation and of course that selection pressure which it is difficult to conceive as ever wholly lacking, full species can be evolved in a few thousand years, at least in passerine birds.

Thus we have to face the probability that headlong changes in the avifaunal situation have taken place in Africa; and because there have been not one but at least four glaciations, comprising various degrees of intensity, we must envisage repeated changes of comparable magnitude in the course of the Pleistocene. The most sweeping of all the changes is that within the space of some 10000 years, from the stage when most of tropical Africa was dominated by those montane bird faunas now persisting only in relatively small ecological islands, to the present pattern in which the lowland bird faunas, merely peripheral during the glaciation, have flooded nearly the whole of tropical Africa. Changes that are smaller, but still on a vast scale, are the repeated reduction and fragmentation of the lowland evergreen forests, with contemporary spread of the arid areas of the continent. Mediterranean vegetation had extended well into the northern tropics very late in the Pleistocene, to be overtaken by the desiccation of the Sahara. Between about 5000 and 2000 years ago (that is, starting some 200 human generations before the present) over 2 million sq. miles of Africa became uninhabitable except by a few highly specialized plants and animals. In the face of such mighty and such rapid changes, what becomes of that conventionally accepted stability of the tropical environments which is invoked as the ultimate reason why so many more species are found in low latitudes than in high? Moreover, it must always be borne in mind that by far the greater part of Africa is occupied by vegetation types that change greatly in the course of each year, for in the wet season they are lush and green, in the dry season they are desiccated and the biomass of the vegetation is greatly diminished by leaf-fall and by fire.

The difference in "stability" between the ecology of the high latitudes and that of the tropics, though real, has surely been exaggerated. I suspect that for all too many people "the tropics" has connoted tropical rain-forest and that conventional ideas on the subject have ultimately been coloured by Bates' account of the Amazons (1863). The truth is that, in Africa, evergreen forest does not bulk so large as it does in South America, and so far as the lowlands are concerned it probably never has done. In Africa, then, it is the vegetation types showing strong seasonal changes that are typical of "the tropics" and that should be kept in mind when comparisons are being made with a temperate zone. Moreover, within the Temperate Zone itself (and I have in mind the north temperate because biologically it is so much better known and also more important than the southern), the difference in seasonal "stability" between a typical Mediterranean area and an area that regularly experiences winter snowfall with continued frost is greater than that between the latter and a tropical monsoon area, where up to eight months of unbroken drought succeed the rainy season. In this connexion it should be borne in mind that the Cape Province of South Africa, which is extra-tropical, is poor in bird species compared with territories at lower latitudes in the continent, though it has a full range of habitats and a remarkably rich flora.

Much the same comment applies to the supposed ecological stability of low latitudes compared with high in the past. In tropical Africa the vegetation zones have expanded and contracted, been split, moved their locations, again and again in the course of the Pleistocene. The higher latitudes of the Temperate Zone, repeatedly locked in ice or shivering in a peri-glacial climate, have indeed undergone worse vicissitudes during the Pleistocene, but this does not apply to much of the country bordering the western Mediterranean, especially on the south, where the fluctuations in vegetation have been no more than between deciduous broad-leaved and sclerophyllous woodland. Yet, when tropical and extra-tropical areas that seem broadly comparable are examined together, the tropical is the richer in species every time. This happens even when the extremities of Africa itself are brought into the comparison, and even when the tropical samples come from north of the Equator, which are so much poorer in species than are those from the south.

Consider now some of the problems that the bird faunas of Africa and its islands present us with. The continent as a whole is poorer in bird species than South America (as it is in plants); and we suppose with some justification that, although the temperature changes associated with the glaciations would have been the same in both continents, the biological effects would not have been so severe in South America because of its different topography. Next, in Africa the number of species of birds and indeed also of plants and mammals, if not also of other groups, north of the Equator is less than in equivalent areas to the south. A combination of factors has been invoked to account for limited local speciation in the north, and interchange with the southern tropics of Africa has been restricted for savanna animals by the presence in the equatorial belt of the Congo forest in the west, the humid Lake Victoria basin, the Kenya highlands and the arid coastal hinterland in the east.

Turning now to the communities in equatorial and southern tropical Africa, the lowland forest is indeed the richest in species, with fully one-third more species than the deciduous brachystegia woodland, which looks as if it must hold much poorer resources and which, moreover, there is reason to suppose was reduced and dissected very late in the Pleistocene, as was the evergreen forest. Moreover, the number of forest species falls off so greatly with altitude that, notwithstanding the vastly greater biomass of the vegetation per unit of area in the montane forests, the number of bird species they hold is only about the same as in the poorest deciduous woodland, which is mopane. Indeed both brachystegia woodland and acacia steppe are richer in bird species than is montane forest, not only in total, but also if comparison is restricted to those which depend on the ecological category best represented in both types of community, namely the birds dependent mainly on invertebrate food. It might be supposed that in the deciduous habitats there would be a great seasonal fluctuation in the bird population; in numbers there may be—that remains to be studied—but only a small minority of the species seem to evacuate brachystegia or acacia or mopane when conditions look worst, in the dry season. There may perhaps be less of total than of partial migration, which tends to escape record, but against this—and against what seems probable to the casual observer—migrant birds from the Palaearctic come into what appear to be among the poorest habitats, the acacia woodland and steppe of the northern tropics, and are able to support themselves there throughout the dry season, presumably the worst part of the year. Evergreen forest, whether lowland or montane, experiences no such invasion, perhaps in part because its communities are too saturated and its population, like its conditions, is too constant for there to be any room for incomers. Yet the awkward fact remains that, with all its great biomass of vegetation on a unit of area, the number of bird species in the montane forest is as low as it is in any climax arboreal community and much lower than in most.

When we turn to the islands off Africa, making allowance for the effects of human interference, certain anomalies emerge from comparative examination. One is that in nearly all of them, including Madagascar, the proportion of passerine species, is lower than in the continental bird faunas. It seems impossible to rationalize this result at present, all the more because the two islands that have the highest proportion of passerines, Hierro and Fernando Poo, are otherwise so dissimilar that it is difficult to believe that the same factor can have been responsible for this result in both. The other major anomaly is that, apart from water-birds, the two largest islands, Fernando Poo and Madagascar, have the same number of species, although the latter is 300 times the bigger and is far richer in its range of habitats. True, Fernando Poo is a continental island and Madagascar is not; but Madagascar has received repeated infusions from the continent and evolution on the most extensive scale is known to have taken place on its surface. The present numerical equality in species must be due to coincidence, resulting from a severe degree of extinction in Madagascar in circumstances that are obscure. Incidentally, the fact that in certain groups vigorous adaptive radiation has taken place on this unbroken land-surface of only 250 000 sq. miles, where it is difficult to postulate ecological barriers of long duration, suggests that the conditions of that isolation which is accepted as necessarily precedent to speciation are more easily fulfilled than is usually supposed.

If what has been said in this book has done something to elucidate some problems of bird distribution, the same cannot be said of problems of species-number. Here all that has been achieved is to multiply challenging examples. The truth is that we know altogether too little that is significant about the ecology of our material. We know virtually nothing about the succession through the year of any category of bird food in any habitat, either quantitatively or even qualitatively. We know hardly anything of the numbers and biomass and, in detail, of the diet of the birds comprising any African community. There are indirect indications that in even the habitats with the strongest seasonal fluctuations food may not be as scarce at the "worst" time of year as our impressions suggest. Consequently nowhere can we have any but the crudest idea of the niche each species fills. We do not even know in which species food supply is the critical factor in determining presence or abundance and in which it is not. It is perhaps not too much to hope that the preliminary survey in this book, with its presentation of problems and of opportunities, will help to stimulate that research without which understanding is impossible.

References

Acocks, J. P. H. (1953). Veld types of South Africa. *Bot. Surv. S. Afr.*, Mem. 28.

Agron, S. L. (1962). Evolution of bird navigation and the earth's axial precession. *Evolution, Lancaster, Pa.* **16**, 524–527.

Allen, S. S. (1864). Remarks on Dr A. Leith Adams's "Notes and observations on the birds of Egypt and Nubia". *Ibis* **6**, 233–243.

Amadon, D. (1950). The Hawaiian honeycreepers (Aves, Drepaniidae). *Bull. Am. Mus. nat. Hist.* **95** (4), 153–262.

Amadon, D. (1951). Le pseudo-souimanga de Madagascar. *Oiseau Revue fr. Orn.* **21**, 59–63.

Amadon, D. (1953). Avian systematics and evolution in the Gulf of Guinea. *Bull. Am. Mus. nat. Hist.* **100** (3).

Amadon, D. (1956). Remarks on the starlings, family Sturnidae. *Am. Mus. Novit.* **1803**, 1–41.

Amadon, D. (1957). Remarks on the classification of the perching birds (order Passeriformes). *Proc. zool. Soc., Calcutta*, Mookerjee Memorial Volume, pp. 259–268.

Antevs, E. (1954). Climate of New Mexico during the last glaciation. *J. Geol.* **62**, 182–197.

Arambourg, C. (1962). Les faunes mammalogiques du Pleistocène circum-méditerranéen. *Quaternaria* **6**, 97–109.

Archer, G. and Godman, E. M. (1937–61). "The Birds of British Somaliland". Vols. 1 and 2, Gurney and Jackson, London; Vols. 3 and 4, Oliver and Boyd, Edinburgh.

Arkell, A. J. (1964). "Wanyanga." Oxford University Press, London.

Ashmole (née Goodacre), M. J. (1962). The migration of European thrushes: a comparative study based on ringing recoveries. *Ibis* **104**, 314–346, 522–559.

Ashmole, N. P. (1963a). The extinct avifauna of St Helena Island. *Ibis* **103b**, 390–408.

Ashmole, N. P. (1963b). Sub-fossil bird remains on Ascension Island. *Ibis* **103b**, 382–389.

Bagnold, R. A. (1931). Journeys in the Libyan Desert, 1929 and 1930. *Geogrl J.* **78**, 13–39, 524–535.

Baker, R. H. (1951). The avifauna of Micronesia, its origin, evolution and distribution. *Univ. Kans. Publs Mus. nat. Hist.* **3**.

Bakker, E. M. van Zinderen (1962a). "Palynology in Africa." Seventh Report. University of the Orange Free State, Bloemfontein.

Bakker, E. M. van Zinderen (1962b). Botanical evidence for quaternary climates in Africa. *Ann. Cape Prov. Mus.* **2**, 16–31.

Bakker, E. M. van Zinderen (1963). Palaeobotanical studies. *S. Afr. J. Sci.* **59**, 332–340.

Bakker, E. M. van Zinderen (1964a). "Palynology in Africa." Eighth Report. University of the Orange Free State, Bloemfontein.

Bakker, E. M. van Zinderen (1964b). A pollen diagram from equatorial Africa, Cherengani, Kenya. *Geologie Mijnb.* **43**, 123–128.

Bakker, E. M. van Zinderen (1964c). Pollen analysis and its contribution to the palaeoecology of the Pleistocene in southern Africa. *In* "Ecological Studies in Southern Africa" (D. H. S. Davis, ed.), pp. 24–34. W. Junk, The Hague.

Bally, P. R. O. (1964). Quoted in *Oryx* **7**, 210.

Bannerman, D. A. (1914–15). Report on the birds collected by the late Mr Boyd Alexander (Rifle Brigade) during his last expedition to Africa. Part 1. *Ibis* (10) **2**, 596–631. Part 2. *Ibis* (10) **3**, 89–121.

Bannerman, D. A. (1930–51). "The Birds of West Tropical Africa." 8 vols. Crown Agents, London.

Bannerman, D. A. (1953). "The Birds of West and Equatorial Africa." 2 vols. Oliver and Boyd, Edinburgh and London.

Bannerman, D. A. (1963). "Birds of the Atlantic Islands." Vol. 1. Oliver and Boyd, Edinburgh and London.

Basilio, A. (1963). "Aves de la Fernando Po." Coculsa, Madrid.

Bates, H. W. (1863). "The Naturalist on the River Amazons." Murray, London.

Beadle, L. C. (1962). The evolution of species in the lakes of East Africa. *Uganda J.* **26**, 44–54.

Beebe, W. (1917). "Tropical wild life in British Guiana." New York Zoological Society, New York.

Beebe, W. (1947). Avian migration at Rancho Grande in north-central Venezuela. *Zoologica, N.Y.* **32**, 153–168.

Beer, G. de (1964). Phylogeny of the ratites. *In* "New Dictionary of Birds" (A. Landsborough Thomson, ed.), pp. 681–685. Nelson, London.

Benson, C. W. (1948). Evergreen forests near Blantyre. Comparative variety of bird species. *Nyasald J.* **1**, 45–52.

Benson, C. W. (1951). A roosting site of the Eastern Red-footed Falcon *Falco amurensis*. *Ibis* **93**, 467–468.

Benson, C. W. (1952). Notes from Nyasaland. *Ostrich* **23**, 144–159.

Benson, C. W. (1953). "A Check List of the Birds of Nyasaland." Nyasaland Society, Blantyre and Lusaka.

Benson, C. W. (1957). Migrants at the south end of Lake Tanganyika. *Bull. Br. Orn. Club* **77**, 88.

Benson, C. W. (1960a). The birds of the Comoro Islands. *Ibis* **103b**, 5–106.

Benson, C. W. (1960b). Les origines de l'avifaune de l'archipel des Comores. *Mém. Inst. scient. Madagascar* (A) **14**, 173–204.

Benson, C. W. (1962). Some additions and corrections to "A Check List of the Birds of Northern Rhodesia". *Occ. Pap. natn. Sth. Rhod.* **26B**, 631–652.

Benson, C. W. (1964). Some intra-African migratory birds. *Puku Occ. Pap. Dep. Game N. Rhod.* **2**, 53–66.

Benson, C. W. and Benson, F. M. (1949). Notes on birds from northern Nyasaland and adjacent Tanganyika Territory. *Ann. Transv. Mus.* **21**, 155–177.

Benson, C. W. and Irwin, M. P. S. (1963). Some comments on the "Atlas of European birds" from the Ethiopian aspect. *Ardea* **51**, 212–229.

Benson, C. W. and Irwin, M. P. S. (1964). The migrations of the pitta of Eastern Africa. *Nth Rhod. J.* **5**, 465–475.

Benson, C. W. and Irwin, M. P. S. (1966). The brachystegia avifauna. *Ostrich*. (In press.)

Benson, C. W. and White, C. M. N. (1957). "Check List of the Birds of Northern Rhodesia." Govt. Printer, Lusaka.

Benson, C. W. and White, C. M. N. (1962). Discontinuous distributions (Aves). Proc. First Fed. Sci. Congr. Salisbury [Rhodesia], May 18–22, 1960, pp. 195–216.

Benson, C. W., Irwin, M. P. S. and White, C. M. N. (1962). The significance of valleys as avian zoogeographical barriers. Ann. Cape Prov. Mus. 2, 155–189.

Benson, C. W., Brooke, R. K. and Vernon, C. J. (1964). Bird breeding data for the Rhodesias and Nyasaland. Occ. Pap. natn. Mus. Sth. Rhod. 27B, 30–105.

Berlioz, J. (1948). Le peuplement de Madagascar en oiseaux. Mém. Inst. scient. Madagascar (A) 1, 181–192.

Bernard, E. A. (1962). Interprétation astronomique des pluviaux et interpluviaux du Quaternaire africaine. Annls Mus. r. Congo belge Ser. 8vo 40 (1), 67–95 (see also pp. 127–130).

Beucher, F. (1963). Flores quaternaires au Sahara nord occidental, d'après l'analyse pollinique de sédiments prélevés à Hassi-Zguilma (Saoura). C. r. hebd. Séanc. Acad. Sci. 256, 2205–2208.

Bond, G. (1963). Pleistocene environments in Southern Africa. In "African Ecology and Human Evolution" (F. C. Howell and F. Bourlière, eds.), pp. 308–334. Viking Fund Publications in Anthropology, No. 36, New York.

Booth, A. H. (1958). The zoogeography of West African primates: a review. Bull. Inst. fr. Afr. noire 20, 587–622.

Booth, B. D. McD. (1961). Breeding of the Sooty Falcon in the Libyan desert. Ibis 103A, 129–130.

Botting, D. S. (1958). The Oxford University Expedition to Socotra. Geogrl J. 124, 200–207.

Botting, D. S. (1960). "Island of the Dragon's Blood." Hodder and Stoughton, London.

Bourne, W. R. P. (1955). The birds of the Cape Verde Islands. Ibis 97, 508–556.

Brain, C. K. (1962). In discussion. Ann. Cape Prov. Mus. 2, 312.

Brodkorb, P. (1960). How many species of birds have evolved? Bull. Fla St. Mus. 5 (3), 41–53.

Broekhuysen, G. J. (1959). The biology of the Cape Sugar bird Promerops cafer. Ostrich, Suppl. 3, 180–221.

Broekhuysen, G. J. (1964). The status and movements of the European Swallow, Hirundo rustica, in the most southern part of Africa. Ardea 52, 140–165.

Brooke, R. K. (1963). Birds round Salisbury [Rhodesia], then and now. S. Afr. Fauna, Ser. 9.

Brooks, C. E. P. and Mirlees, S. T. A. (1932). A study of the atmospheric circulation over tropical Africa. Met. Off. Geophys. Mem. No. 55.

Brosset, A. (1961). Ecologie des oiseaux du Maroc oriental. Trav. Inst. scient. chérif. Ser. Biol. 22.

Brown, L. (1959). "The Mystery of the Flamingos." Country Life, London.

Brown, L. H. (1965). "Africa. A Natural History." Hamish Hamilton, London.

Bruneau de Miré, Ph. (1957). Observations sur la faune avienne du Massif de l'Aïr. Bull. Mus. natn. Hist. nat. Paris 29, 130–135.

Büdel, J. (1951). Die Klimazonen des Eiszeitalters. Eiszeitalter Gegenw. 1, 16–26.

Büdel, J. (1963). Die pliozänen und quantären Pluvialzeiten der Sahara. *Eiszeitalter Gegenw.* **14,** 161–187.

Bury, G. W. (1915). "Arabia Infelix." Macmillan, London.

Butzer, K. W. (1958). Studien zum vor- und frühgeschichtlichen Landschaftswandels der Sahara und Levante seit klassischen Altertum. II. Das ökologische Problem der Neolitischen Felsbilder der östlichen Sahara. *Abh. math.-naturw. Kl. Akad. Wiss. Mainz,* pp. 1–49.

Butzer, K. W. (1959a). Die Naturlandschaft Ägyptens während der Vorgeschichte und der dynastischen Zeit. *Abh. math.-naturw. Kl. Akad. Wiss. Mainz* **2,** 43–122.

Butzer, K. W. (1959b). Environment and human ecology in Egypt during predynastic and early dynastic times. *Bull. Soc. Géogr. Egypte* **32,** 43–87.

Butzer, K. W. (1961). Climatic change in arid regions since the Pliocene. *In* "A History of Land Use in Arid Regions" (L. Dudley Stamp, ed.). UNESCO, Paris.

Capot-Rey, R. (1953). "Le Sahara français." Paris.

Capot-Rey, R. (1961). Borkou et Ouninga. *Mém Inst. Rech. sahar.,* No. 5.

Carcasson, R. H. (1964). A preliminary survey of the zoogeography of African butterflies. *E. Afr. Wildl. J.* **2,** 122–157.

Carr, A. F. (1950). Outline for a classification of animal habits in Honduras. *Bull. Am. Mus. nat. Hist.* **94** (10), 566–594.

Cave, F. O. and Macdonald, J. D. (1955). "Birds of the Sudan." Oliver and Boyd, Edinburgh and London.

Cawkell, E. M. (1964). The utilization of mangroves by African birds. *Ibis* **106,** 251–253.

Cawkell, E. M. and Moreau, R. E. (1963). Notes on birds in the Gambia. *Ibis* **105,** 156–178.

Chapin, J. P. (1923). Ecological aspects of bird distribution in tropical Africa. *Am. Nat.* **57,** 106–125.

Chapin, J. P. (1932–54). The birds of the Belgian Congo. (4 vols.) *Bull. Am. Mus. nat. Hist.* **65, 75, 75A, 75B.**

Chapman, F. M. (1926). The distribution of bird life in Ecuador. Part 1. *Bull. Am. Mus. nat. Hist.* **55,** 133–144.

Church, R. J. H., Clarke, J. I., Clarke, P. J. H. and Henderson, H. J. R. (1964). "Africa and the Islands." Longmans, London.

Clancey, P. A. (1964). "The Birds of Natal and Zululand." Oliver and Boyd, Edinburgh and London.

Clapham, C. S. (1964). The birds of the Dahlac archipelago. *Ibis* **106,** 376–388.

Clark, J. D. (1954). "Prehistoric Cultures of the Horn of Africa." Cambridge University Press.

Clark, J. D. (1957). A review of prehistoric research in Northern Rhodesia and Nyasaland. Third Pan-Afr. Congr. Prehist. Livingstone, 1955, pp. 412–432.

Clark, J. D. (1962a). Carbon 14 chronology in Africa south of the Sahara. *Annls Mus. r. Congo belge Ser. 8vo.* **40** (2), 303–314.

Clark, J. D. (1962b). The spread of food production in sub-Saharan Africa. *J. Afr. Hist.* **3,** 211–228.

Clark, J. D. (1963). "Prehistoric Cultures of Northeast Angola and their Significance in Tropical Africa." Companhia de Diamantes de Angola, Lisboa.

Clark, J. D. (1964). The prehistoric origins of African culture. *J. Afr. Hist.* **5,** 161–183.

Clarke, J. D. and Fagan, B. (1965). Charcoals, sands and Channel decorated pottery from Northern Rhodesia. *Am. Anthrop.* **67**, 354–371.

Coetzee, J. A. (1964). Evidence for a considerable depression of the vegetation belts during the Upper Pleistocene on the East African mountains. *Nature, Lond.* **204**, 564–566.

Cole, M. M. (1963). Vegetation and geomorphology in Northern Rhodesia. *Geogrl J.* **129**, 290–310.

Cooke, H. B. S. (1958). Observations relating to Quaternary environments in east and southern Africa. *Trans. geol. Soc. S. Afr.* **60** (annexure).

Cooke, H. B. S. (1962). The Pleistocene environment in Southern Africa. Hypothetical vegetation in Southern Africa during the Pleistocene. *Ann. Cape Prov. Mus.* **2**, 11–15.

Cooke, H. B. S. (1963). Pleistocene mammal faunas of Africa, with particular reference to southern Africa. *In* "African Ecology and Human Evolution" (F. C. Howell and F. Bourlière, eds.), pp. 64–116. Viking Fund Publications in Anthropology, No. 36, New York.

Cooke, H. B. S. (1964). The Pleistocene environment in southern Africa. *In* "Ecological Studies in Southern Africa" (D. H. S. Davis, ed.), pp. 1–23. W. Junk, The Hague.

Cullen, J. M., Guiton, P. E., Horridge, G. A. and Peirson, J. (1952). Birds on Palma and Gomera (Canary Islands). *Ibis* **94**, 68–84.

Curry-Lindahl, K. (1958). Internal timer and spring migration in an equatorial migrant, the Yellow Wagtail (*Motacilla flava*). *Ark. Zool.* (2) **11**, 541–557.

Darlington, P. J. (1957). "Zoogeography." Wiley, New York.

Darlington, P. J. (1959). Area, climate and evolution. *Evolution, Lancaster, Pa.* **13**, 488–510.

Davies, O. (1963). Deep-ocean cores for Pleistocene dating. *Ghana J. Sci.* **3**, 98–110.

Davis, P. (1965). Recoveries of swallows ringed in Britain and Ireland. *Bird Study* **12**, 151–169.

Delacour, J. and Vaurie, C. (1957). A classification of the Oscines (Aves). *Contr. Sci.*, No. 16.

Delibrias, G. and Hugot, H. J. (1962). Datation par la methode dite "de C 14" du Néolithique de l'Adrar Bous (Ténéréen). *In* "Documents scientifiques Mission Berliet-Ténéré-Tchad" (H. J. Hugot, ed.), pp. 71–72. Arts et Metiers Graphiques, Paris.

Delibrias, G., Hugot, H. and Quezel, P. (1959). Trois datations de sédiments sahariens récents par le radio-carbone. *Lybica* **5**, 267–270.

Diesselhorst, G. (1959). Die geographische Variabilität von *Melocichla mentalis* (Fraser). *Opusc. zool., Münch.* No. 36.

Dobbs, K. A. (1959). Some birds of Sokoto, Northern Nigeria. *Niger. Fld* **24**, 102–119.

Dobzhansky, T. (1959). Evolution in the tropics. *Am. Sci.* **38**, 209–221.

Dorst, J. (1957). The Puya stands of the Peruvian high plateaux as a bird habitat. *Ibis* **99**, 594–599.

Dorst, J. (1960). Considerations sur les passereaux de la famille des Vangidés. *Proc. XII int. Orn. Congr., Helsinki* **1**, 173–177.

Dorst, J. (1962a). "The Migrations of Birds." Heinemann, London.

Dorst, J. (1962b). Considérations sur l'hivernage des canards et limicoles paléarctiques en Afrique tropicale. *Terre et Vie* 1962, 183–192.

Douaud, J. (1957). Les migrations au Togo (Afrique occidentale). *Alauda* **24**, 146–147.

Edwards, A. Milne and Grandidier, A. (1879–81). "Histoire physique, naturelle et politique de Madagascar." Vols 12–15. Histoire naturelle des oiseaux. Imprimerie Nationale, Paris.

Eggeling, W. J. (1947). Observations on the ecology of the Budongo rain forest. *J. Ecol.* **34**, 20–87.

Eisentraut, M. (1963). "Die Wirbeltiere des Kamerungebirges." Paul Parey, Hamburg and Berlin.

Elgood, J. H. (1959). Bird migration at Ibadan, Nigeria. *Ostrich*, Suppl. 3, 306–316.

Elgood, J. H. and Sibley, F. C. (1964). The tropical forest edge avifauna of Ibadan, Nigeria. *Ibis* **106**, 221–248.

Elgood, J. H., Sharland, R. E. and Ward, P. (1966). Palaearctic migrants in Nigeria. *Ibis* **108**, 84–116.

Elliott, H. F. I. (1940). An island in Lake Victoria. *Tanganyika Notes Rec.* **10**, 28–40.

Elliott, H. F. I. and Fuggles-Couchman, N. R. (1948). An ecological survey of the birds of the Crater Highlands and Rift Lakes, northern Tanganyika Territory. *Ibis* **90**, 394–425.

Emlen, J. T. (1956). A method for describing and comparing avian habitats. *Ibis* **98**, 565–576.

Etchécopar, R. D. and Hüe, F. (1964). "Les oiseaux du Nord de l'Afrique de la Mer Rouge aux Canaries." Boubée, Paris.

Fairbridge, R. W. (1961). Convergence of evidence on climatic change and ice ages. *Ann. N.Y. Acad. Sci.* **95**, 542–579.

Fairbridge, R. W. (1962). World sea-level and climatic changes. *Quaternaria* **6**, 111–134.

Fairbridge, R. W. (1963). Nile sedimentation above Wadi Halfa during the last 20,000 years. *Kush* **11**, 96–107.

Falla, R. A. (1964). Moa. *In* "New Dictionary of Birds" (A. Landsborough Thomson, ed.), pp. 477–479. Nelson, London.

Fanshawe, D. B. (1959). Floristic composition of miombo woodland. *Publs Cons. scient. Afr. S. Sahara* **52**, 55–57.

Faure, H., Manguin, E. and Nydal, R. (1963). Formations lacustres du quaternaire supérieur du Niger oriental. Diatomites et ages absolus. *Bull. Serv. Rech. géol. min.* **3**, 42–63.

Firbas, F. (1951). Die quartäre Vegetationsentwickelung zwischen den Alpen und der Nord- und Ostsee. *Erdkunde* **5**, 6–15.

Fischer, A. G. (1960). Latitudinal variations in organic diversity. *Evolution, Lancaster, Pa.* **14**, 64–81.

Fisher, J. (1964). Extinct birds. *In* "New Dictionary of Birds" (A. Landsborough Thomson, ed.), pp. 259–264. Nelson, London.

Fisher, J. and Peterson, R. T. (1964). "The World of Birds." Macdonald, London.

Flint, R. F. (1959a). Pleistocene climates in eastern and southern Africa. *Bull. geol. Soc. Am.* **70**, 343–374.

Flint, R. F. (1959b). On the basis of Pleistocene correlation in East Africa. *Geol. Mag.* **96**, 265–284.

Flint, R. F. (1963a). Status of the Pleistocene Wisconsin stage in central North America. *Science, N.Y.* **139**, 402–404.

Flint, R. F. (1963b). Pleistocene climates in low latitudes. *Geogrl Rev.* **53**, 123–129.

Flint, R. F. and Brandtner, F. (1961). Outline of climatic fluctuation since the last interglacial age. *Ann. N.Y. Acad. Sci.* **95**, 457–460.

Flower, S. S. (1932). Notes on the recent mammals of Egypt, with a list of species recorded from that kingdom. *Proc. zool. Soc. Lond.* 1932, 369–450.

Forbes, H. O. (1903). "The Natural History of Sokotra and Abd-el-Kuri." Free Public Museums, Liverpool.

Fournier, P. and D'Hoore, J. (1962). Carte du danger d'érosion en Afrique au sud du Sahara. 1:10000000. Commission de Co-operation Technique en Afrique, Paris and London.

Friedmann, H. (1955). The honey-guides. *Bull. U.S. natn. Mus.*, No. 208.

Friedmann, H. (1964). Evolutionary trends in the genus *Clamator*. *Smithson. misc. Collns* **146** (4).

Fry, C. H. (1961). Notes on the birds of Annobon and the other islands in the Gulf of Guinea. *Ibis* **103a**, 267–276.

Fuggles-Couchman, N. R. (1953). The ornithology of Mt Hanang, in northern-central Tanganyika Territory. *Ibis* **95**, 458–482.

Geyr von Schweppenburg, H. (1917). Vogelzug in der westlichen Sahara. *J. Orn., Lpz.* **65**, 43–65.

Gibb, J. (1954). Feeding ecology of tits, with notes on Treecreeper and Goldcrest. *Ibis* **96**, 513–543.

Gillett, J. B. (1955). The relation between the highland floras of Ethiopia and British East Africa. [Also discussion.] *Webbia* **11**, 459–466, 489–496.

Gillet, H. (1960). Observations sur l'avifaune du massif de l'Ennedi (Tchad). *Oiseau Revue fr. Orn.* **30**, 44–82, 99–134.

Gillman, C. (1949). A vegetation-types map of Tanganyika Territory. *Geogrl Rev.* **39**, 7–37.

Godwin, H. (1956). "The History of the British Flora." Cambridge University Press.

Godwin, H. (1960). Radiocarbon dating and quaternary history in Britain. *Proc. R. Soc.* B, **153**, 287–320.

Godwin, H., Suggate, R. P. and Willis, E. H. (1958). Radiocarbon dating of the eustatic rise in ocean-level. *Nature, Lond.* **181**, 1518–1519.

Good, R. (1953). "The Geography of the Flowering Plants", 2nd ed. Longmans, Green, London.

Goodwin, D. (1959). Taxonomy of the genus *Columba*. *Bull. Br. Mus. nat. Hist.* Zool. **6** (1), 1–23.

Grove, A. T. (1958). The ancient erg of Hausaland, and similar formations on the south side of the Sahara. *Geogrl J.* **124**, 528–533.

Grove, A. T. and Pullan, R. A. (1963). Some aspects of the Pleistocene palaeogeography of the Chad basin. *In* "African Ecology and Human Evolution" (F. C. Howell and F. Bourlière, eds.), pp. 230–245. Viking Fund Publications in Anthropology, No. 36, New York.

Guichard, K. M. (1955a). Habitats of the Desert Locust in Western Libya and Tibesti. *Anti-Locust Bull.* **51**.

Guichard, K. M. (1955b). The birds of Fezzan and Tibesti. *Ibis* **97**, 393–424.

Haldane, L. A. (1956). Birds of the Njombe District. *Tanganyika Notes Rec.* **44**, 1–27.

Hall, B. P. (1960a). The ecology and taxonomy of some Angola birds. *Bull. Br. Mus. nat. Hist.* Zool. **6**, 369–453.

Hall, B. P. (1960b). The faunistic importance of the scarp of Angola. *Ibis* **102**, 420–442.

Hall, B. P. (1963a). The francolins, a study in speciation. *Bull. Br. Mus. nat. Hist. Zool.* **10**, 107–204.

Hall, B. P. (1963b). The status of *Spizocorys razae* Alexander. *Bull. Br. Orn. Club* **83**, 133–134.

Hall, B. P. and Moreau, R. E. (1962). A study of the rare birds of Africa. *Bull. Br. Mus. nat. Hist. Zool.* **8** (7), 316–378.

Hallett, A. F. and Brown, A. R. (1964). A method of trapping European Swallows. *Ostrich* **35**, 293–296.

Hamilton, T. H. and Armstrong, N. E. (1965). Environmental determination of insular variation in bird species abundance in the Gulf of Guinea. *Nature, Lond.* **207**, 148–151.

Hammen, R. van der, and Gonzalez, E. (1960a). Upper Pleistocene and Holocene climate and vegetation of the "Sabana de Bogotá" (Columbia, South America). *Leid. geol. Meded.* **25**, 261–315.

Hammen, T. van der, and Gonzalez, E. (1960b). Holocene and Late Glacial climate and vegetation of Páramo de Palacio (Eastern Cordillera, Columbia, South America). *Geologie Mijnb.* **39**, 737–745.

Harrison, J. L. (1962). The distribution of feeding habits among animals in a tropical rain forest. *J. Anim. Ecol.* **31**, 53–63.

Harroy, J.-P. (1949). "Afrique: terre qui meurt." Marcel Hayez, Bruxelles.

Haumann, L. (1955). La "region afroalpine" en phytogeographie centro-africaine. *Webbia* **11**, 467–469.

Hedberg, O. (1951). Vegetation belts of the East African mountains. *Svensk bot. Tidskr.* **45**, 140–202.

Hedberg, O. (1961). The phytogeographical position of the afroalpine flora. "Recent Advances in Botany," pp. 914–919. University of Toronto Press.

Hedberg, O. (1962). Mountain plants from southern Ethiopia, collected by Dr John Eriksson. *Ark. Bot.* (2) **4**, 421–435.

Hedberg, O. (1964). Features of afroalpine plant ecology. *Acta phytogeogr. suec.* **49**.

Heim de Balsac, H. (1936). Biogéographie des mammifères et des oiseaux de l'Afrique du nord. *Bull. biol.* Suppl. 21.

Heim de Balsac, H. and Mayaud, N. (1962). "Les oiseaux du nord-ouest de l'Afrique." Lechevalier, Paris.

Heinzelin, J. de. (1963). Observations on the absolute chronology of the Upper Pleistocene. *In* "African Ecology and Human Evolution" (F. C. Howell and F. Bourlière, eds.), pp. 285–303. Viking Fund Publications in Anthropology, No. 36, New York.

Hellmayr, C. E. and Conover, B. (1932–49). "Catalogue of Birds of the Americas." 11 vols. Field Museum of Natural History, Chicago.

Hensley, M. M. (1954). Ecological relations of the breeding bird population of the desert biome in Arizona. *Ecol. Monogr.* **24**, 185–207.

Hesse, R., Allee, W. C. and Schmidt, K. P. (1937). "Ecological Animal Geography." Wiley, New York.

Heu, R. (1961). Observations ornithologiques au Ténéré. *Oiseau Revue fr. Orn.* **31**, 214–239.

Hill, W. C. O. (1953). "Primates: Comparative Anatomy and Taxonomy. I. Strepsithini." Edinburgh University Press.

Hopwood, A. T. (1954). Notes on the recent and fossil mammalian faunas of Africa. *Proc. linn. Soc. Lond.* **165**, 1952–53: 46–49.

Hoyle. (1955). [In discussion.] *Webbia* **11**, 493.

Hugot, H. J., Quézel, P. and Martinez, O. (Ed.) (1962). Documents scientifiques. Mission Berliet Ténéré–Tchad, Paris.

Husain, K. Z. (1958). Subdivisions and zoogeography of the genus *Treron* (green fruit-pigeons). *Ibis* **100**, 334–348.

Jackson, F. J. (1938). "The Birds of Kenya Colony and the Uganda Protectorate." 3 vols. Gurney and Jackson, London.

Jackson, F. J. and Sclater, W. L. (1938). "The Birds of Kenya Colony and the Uganda Protectorate." 3 vols. Gurney and Jackson, London.

Jackson, S. P. (1961). "Climatological Atlas of Africa." CCTA/CSA, Lagos–Nairobi.

Jacot-Guillemard, C. (1963). Catalogue of the birds of Basutoland. *S. Afr. Orn.* Ser. 8.

Jany, E. (1960). An Brutplätzen des Lannerfalken (*Falco biarmicus biarmicus* Kleinschmidt) in einer Kieswüste der inneren Sahara (Nordrand des Serir Tibesti) zur zeit des Frühjahrzugs. *Proc. XII int. orn. Congr.*, pp. 343–352.

Jany, E. (1963). Salma Kebir-Kufra-Djabel al-Uweinat. Ein Reisebericht aus der östlichen Sahara. *Erde, Berl.* **94**, 334–362.

Johnston, R. F. and Selander, R. K. (1964). House sparrows: rapid evolution of races in North America. *Science, N.Y.* **144**, 548–550.

Keast, A. (1961). Bird speciation on the Australian continent. *Bull. Mus. comp. Zool. Harv.* **123**, 303–495.

Keay, R. W. J. (1955). Montane vegetation and flora in the British Cameroons. *Proc. Linn. Soc. Lond.* **165** (2), 140–143.

Keay, R. W. J. (Ed.) (1959). "Vegetation Map of Africa." Oxford University Press, London.

Kendeigh, S. C. (1961). "Animal Ecology." Prentice-Hall, Englewood Cliffs, New Jersey.

Kendrew, W. G. (1961). "The Climates of the Continents," 5th ed. Clarendon Press, Oxford.

Klopfer, P. H. and MacArthur, R. A. (1961). On the causes of tropical species diversity, niche overlap. *Am. Nat.* **95**, 223–226.

Knecht, S. (1960). Ein Beitrag zur Erforschung der kanarischen Vogelwelt, inbesondere der Brutvögel, unter hauptsächlicher Berücksichtigung der drei westlichen Inseln Teneriffa, Gomera und La Palma. *Anz. orn. Ges. Bayern* **5**, 525–556.

Lack, D. (1944). Ecological aspects of species formation in passerine birds. *Ibis* **86**, 260–286.

Lack, D. (1947). "Darwin's Finches." Cambridge University Press.

Lack, D. (1956). The species of *Apus. Ibis* **98**, 34–62.

Lack, D. and Southern, H. N. (1949). Birds on Tenerife. *Ibis* **91**, 607–626.

Lamb, H. H. (1963). On the nature of certain climatic epochs which differed from the modern (1900–39) normal. *Arid Zone Res.* **22**, 125–150.

Lamprey, H. F. (1963). The Tarangire game reserve. *Tanganyika Notes Rec.* **60**, 10–22.

Lawson, W. J. (1963). Geographical variation in the southern African populations of the Dusky Flycatcher *Muscicapa adusta* (Boie). *Bull. Br. Orn. Club* **83**, 4–7.

Lawton, R. M. (1963). Palaeoecological and ecological studies in the Northern Province of Northern Rhodesia. *Kirkia* **3**, 46-77.

Leakey, L. S. B. (1964). Prehistoric man in the tropical environment. *In* "The Ecology of Man in the Tropical Environment", pp. 24–29. International Union for the Conservation of Nature Publications, Series 4.

Liversidge, R. (1959). Tropical mountain birds south of the Zambesi. *Ostrich*, Suppl. 3, 68–78.

Livingstone, D. A. (1962). Age of deglaciation in the Ruwenzori Range, Uganda. *Nature, Lond.* **194,** 859–860.

Logan, W. E. M. (1946). An introduction to the forests of central and southern Ethiopia. *Inst. Pap. Commonw. For. Inst.* No. 24.

Lüttschwager, J. (1961). "Die Drontevögel." Ziemsen Verlag, Wittenberg-Lutherstadt.

Lynes, H. (1925). On the birds of North and Central Darfur. VI. *Ibis* (12) **1,** 757–797.

Lynes, H. (1930). Review of the genus *Cisticola*. *Ibis* (12) **6,** Suppl.

MacArthur, R. H. and MacArthur, J. W. (1961). On bird species diversity. *Ecol. Monogr.* **42,** 594–598.

MacArthur, R. H. and Wilson, E. O. (1963). An equilibrium theory of insular zoogeography. *Evolution, Lancaster, Pa.* **17,** 373–387.

MacArthur, R. H., MacArthur, J. W. and Preer, J. (1962). On bird species diversity. II. Prediction of bird census from habitat measurements. *Am. Nat.* **96,** 167–174.

Mackworth-Praed, C. W. and Grant, C. H. B. (1952–60), "African Handbook of Birds." Ser. 1. Birds of eastern and north-eastern Africa. 1st ed. Vol. 1, 1952, Vol. 2, 1955; 2nd ed. Vol. 1, 1957, Vol. 2, 1960. Longmans, Green, London.

McLachlan, G. R. and Liversidge, R. (1957). "Roberts' Birds of South Africa." Central News Agency [Johannesburg].

MacLeod, J. G. R., Murray, C. d'C. and Murray, E. M. (1953). Death of many migrants at Somerset West. *Ostrich* **24,** 118–120.

Malbrant, R. (1954). Contribution à l'étude des oiseaux du Borkou–Ennedi–Tibesti. *Oiseau Revue fr. Orn.* **24,** 1–47.

Marchant, S. (1953). Notes on the birds of south-eastern Nigeria. *Ibis* **95,** 38–69.

Marchant, S. (1954). The relationship of the southern Nigerian avifauna to those of Upper and Lower Guinea. *Ibis* **96,** 371–379.

Marchant, S. (1963). Migration in Iraq. *Ibis* **105,** 369–398.

Martin, E., Martin, R. and Robinson, J. (1962). European Stork breeding in the Bredasdorp District. *Ostrich* **33,** 26.

Martin, P. S. and Harrell, B. E. (1957). The Pleistocene history of temperate biotas in Mexico and eastern United States. *Ecology* **38,** 468–480.

Mauny, R. (1956). La grande 'faune ethiopienne' du Nord-Ouest Africain du Paléolithique à nos jours. *Bull. Inst. fr. Afr. noire* (A) **18,** 246–279.

Mayr, E. (1942). "Systematics and the Origin of Species." Columbia University Press, New York.

Mayr, E. (1952). Conclusion [to "Symposium on the South Atlantic basin"]. *Bull. Am. Mus. nat. Hist.* **99** (3), 255–258.

Mayr, E. (1963). "Animal Species and Evolution." Belknap Press, Harvard, Cambridge, Mass.

Mayr, E. (1965). The nature of colonization in birds. *In* "The Genetics of the Colonizing Species" (H. G. Baker and G. L. Stebbins, eds.), pp. 29–47. Academic Press, New York.

Mayr, E. and Amadon, D. (1951). A classification of recent birds. *Am. Mus. Novit.*, 1496.

Mayr, E., Linsley, E. G. and Usinger, R. L. (1953). "Methods and Principles of Systematic Zoology." McGraw-Hill, New York.

Meinertzhagen, R. (1930). "Nicoll's Birds of Egypt." 2 vols. Hugh Rees Ltd., London.

Meinertzhagen, R. (1934). The biogeographical status of the Ahaggar Plateau in the central Sahara, with special reference to birds. *Ibis* (13) **4**, 528–571.

Meinertzhagen, R. (1954). "Birds of Arabia." Oliver and Boyd, Edinburgh.

Merikallio, E. (1958). Finnish birds: their distribution and numbers. *Fauna fenn.* **5**.

Miller, A. H. (1963). Seasonal activity and ecology of the avifauna of an American equatorial cloud forest. *Univ. Calif. Publs Zool.* **66**, 1–78.

Millot, J. (1952). La faune malgache et le mythe gondwanien. *Mem. Inst. scient. Madagascar* (A), No. 7.

Milne, G. (1947). A soil reconnaissance journey through parts of Tanganyika Territory. *J. Ecol.* **35**, 192–265.

Milon, Ph. (1952). Notes sur le genre *Coua. Oiseau Revue fr. Orn.* **22**, 75–90.

Milon, Ph. (1959). Sur la migration à Madagascar du *Cuculus poliocephalus rochii. Ostrich*, Suppl. 3, 242–249.

Monod, T. (1957). "Les grandes divisions chorologiques de l'Afrique." Scientific Council for Africa south of the Sahara. Publ. 24. London.

Monod, T. (1958). Majâbat al-Koubra. Contribution à l'étude de l' "Empty Quarter" ouest-saharien. *Mem. Inst. fr. Afr. noire*, No. 52.

Monod, T. (1963). The late Tertiary and Pleistocene in the Sahara. *In* "African Ecology and Human Evolution" (F. C. Howell and F. Bourlière, eds.), pp. 117–229. Viking Fund Publications in Anthropology, No. 36, New York.

Moreau, R. E. (1927). Migration as seen in Egypt. *Ibis* (12) **3**, 443–468.

Moreau, R. E. (1930). On the age of some races of birds. *Ibis* (12) **6**, 229–239.

Moreau, R. E. (1933). A note on the distribution of *Gypohierax angolensis. J. Anim. Ecol.* **2**, 179–183.

Moreau, R. E. (1934a). A contribution to the ornithology of the Libyan desert. *Ibis* (13) **4**, 595–632.

Moreau, R. E. (1934b). A contribution to tropical African bird-ecology. *J. Anim. Ecol.* **3**, 41–69.

Moreau, R. E. (1935a). A synecological study of Usambara, Tanganyika Territory, with particular reference to birds. *J. Ecol.* **23**, 1–43.

Moreau, R. E. (1935b). Some eco-climatic data for closed evergreen forest in tropical Africa. *J. Linn. Soc. Zool.* **39**, 285–293.

Moreau, R. E. (1936a). A contribution to the ornithology of Kilimanjaro and Mount Meru. *Proc. zool. Soc. Lond.* 1935, 843–891.

Moreau, R. E. (1936b). Breeding seasons of birds in East African evergreen forest. *Proc. zool. Soc. Lond.* 1936, 631–653.

Moreau, R. E. (1938). Bird-migration over the north-western part of the Indian Ocean, the Red Sea, and the Mediterranean. *Proc. zool. Soc. Lond.* (A) **108**, 1–26.

Moreau, R. E. (1939). A supplementary contribution to the ornithology of Kilimanjaro. *Revue Zool. Bot. Afr.* **33**, 1–15.

Moreau, R. E. (1941). The ornithology of Siwa Oasis, with particular reference to the results of the Armstrong College Expedition, 1935. *Bull. Inst. Égypte* **23**, 247–261.

Moreau, R. E. (1943). A contribution to the ornithology of the east side of Lake Tanganyika. *Ibis* **85**, 377–412.

Moreau, R. E. (1944a). Clutch-size: a comparative study, with special reference to African birds. *Ibis* **86**, 286–347.

Moreau, R. E. (1944b). Kilimanjaro and Mount Kenya: some comparisons with special reference to the mammals and birds. *Tanganyika Notes Rec.* **18**, 28–68.

Moreau, R. E. (1944c). Additions to the ornithology of the Mafia group of islands. *Ibis* **86**, 33–37.

Moreau, R. E. (1945). Mount Kenya: a contribution to the biology and bibliography. *Jl E. Africa nat. Hist. Soc.* **18**, 61–92.

Moreau, R. E. (1948). Ecological isolation in a rich tropical avifauna. *J. Anim. Ecol.* **17**, 113–126.

Moreau, R. E. (1950). The breeding seasons of African birds—1. Land birds. *Ibis* **92**, 223–267.

Moreau, R. E. (1951). The British status of the quail and some problems of its biology. *Br. Birds* **44**, 257–276.

Moreau, R. E. (1952a). Africa since the Mesozoic: with particular reference to certain biological problems. *Proc. zool. Soc. Lond.* **121**, 869–913.

Moreau, R. E. (1952b). The place of Africa in the Palaearctic migration system. *J. Anim. Ecol.* **21**, 250–271.

Moreau, R. E. (1953). Migration in the Mediterranean area. *Ibis* **95**, 329–364.

Moreau, R. E. (1954). The main vicissitude of the European avifauna since the Pliocene. *Ibis* **96**, 411–431.

Moreau, R. E. (1957). Variation in the western Zosteropidae (Aves). *Bull. Br. Mus. nat. Hist. Zool.* **4** (7), 309–433.

Moreau, R. E. (1958a). The *Malimbus* spp. as an evolutionary problem. *Rev. Zool. Bot. Afr.* **57**, 241–255.

Moreau, R. E. (1958b). Some aspects of the Musophagidae. *Ibis* **100**, 67–112, 238–270.

Moreau, R. E. (1960a). Conspectus and classification of the Ploceine weaver-birds. *Ibis* **102**, 298–321.

Moreau, R. E. (1960b). The Ploceine weavers of the Indian Ocean. *J. Orn., Lpz.* **101**, 29–49.

Moreau, R. E. (1961). Problems of Mediterranean–Saharan migration. *Ibis* **103a**, 373–427, 580–623.

Moreau, R. E. (1963). Vicissitudes of the African biomes in the Late Pleistocene. *Proc. zool. Soc. Lond.* **141**, 395–421.

Moreau, R. E. (1964a). The re-discovery of an African owl *Bubo vosseleri*. *Bull. Br. Orn. Club* **84**, 47–52.

Moreau, R. E. (1964b). Breeding season. *In* "A New Dictionary of Birds" (A. Landsborough Thomson, ed.), pp. 106–108. Nelson, London.

Moreau, R. E. (1966). On estimates of the past numbers and of the average longevity of avian species. *Auk* **83**, 403–415.

Moreau, R. E. (1967). Water birds over Sahara. *Ibis* **109** (In press.)

Moreau, R. E. and Pakenham, R. H. W. (1941). The land vertebrates of Pemba, Zanzibar and Mafia. *Proc. zool. Soc. Lond.* (A) **110,** 97–128.

Moreau, R. E. and Sclater, W. L. (1937). The avifauna of the mountains along the Rift Valley in north central Tanganyika Territory (Mbulu District). *Ibis* **(14)** 1, 760–786.

Morel, G. and Bourlière, F. (1955). Recherches écologiques sur *Quelea qu. quelea* L. de la vallée du Sénégal. Part 1. *Bull. Inst. fr. Afr. noire* **17** A, 617–663.

Morel, G. and Bourlière, F. (1956). Recherches écologiques sur *Quelea qu. quelea* L. de la vallée du Sénégal. Part 2. *Alauda* **24,** 97–122.

Morel, G. and Bourlière, F. (1962). Relations écologiques des avifaunes sédentaires et migratrices dans une savane sahelienne du bas Sénégal. *Terre et Vie* **4,** 371–393.

Morel, G. and Morel, M. Y. (1962). La reproduction des oiseaux dans une région semi-aride: la valée du Sénégal. *Alauda* **30,** 161–203, 241–269.

Morel, G. and Roux, F. (1962). Données nouvelles sur l'avifaune du Sénégal. *Oiseau Revue fr. Orn.* **32,** 28–56.

Morison, C. G. T., Hoyle, A. C. and Hope-Simpson, J. F. (1948). Tropical soil-vegetation catenas and mosaics. *J. Ecol.* **36,** 1–84.

Mortelmans, G. and Monteyne, R. (1962). Le quaternaire du Congo occidental et sa chronologie. *Annls Mus. r. Congo belge Ser. 8vo* **40** (1), 97–126.

Morton, J. K. (1961). The upland floras of West Africa—their composition, distribution and significance in relation to climatic changes. C. r. IV. Réunion plénière de l'AETFAT. Junta de Investigações do Ultramar, Lisboa.

Newbold, D. and Shaw, W. B. K. (1928). An exploration in the south Libyan Desert. *Sudan Notes Rec.* **9,** 103–194.

Niethammer, G. (1959). Die Rolle der Auslese durch Feinde bei Wüstenvögeln. *Bonn. zool. Beitr.* **10,** 179–197.

Nisbet, A. C. T. (1963). Weight-loss during migration. Part 2. *Bird-Banding* **34,** 139–159.

Oakley, K. P. (1964). "Frameworks for Dating Fossil Man." Weidenfeld and Nicolson, London.

Oatley, T. B. (1966). Competition and local migration in some African Turdidae. *Ostrich.* (In press.)

Odum, E. P. (1960). Lipid deposition in nocturnal migrant birds. *Proc. XII int. Orn. Congr.,* 1958, pp. 563–575.

Paulian, R. (1961). "Faune de Madagascar." 13. La zoogéographie de Madagascar et des îles voisines. Institut Recherche Scientifique, Tananarive.

Peters, J. L. (1931). "Check-list of Birds of the World." Vol. 1. Harvard University Press, Cambridge, Mass.

Phillips, A., Marshall, J. and Monson, G. (1964). "The Birds of Arizona." University of Arizona Press, Tucson.

Phillips, J. (1959). "Agriculture and Ecology in Africa." Faber, London.

Phillips, J. F. V. (1931). Forest succession and ecology in the Knysna region. *Bot. Juno. S. Afr. Mem.* 14.

Pinto, A. A. da R. (1959). Um esboço da avifauna sedentaria da Região da Gorongoza (Moçambique). *Ostrich* Suppl. 3, 98–125.

Pitt-Schenkel, C. J. W. (1938). Some important communities of warm temperate rain forest at Magamba, West Usambaras, Tanganyika Territory. *J. Ecol.* **25,** 50–81.

Popov, G. B. (1957). The vegetation of Socotra. *J. Linn. Soc. Lond.* Zool. **55** (362), 706–720.

Prigogine, N. (1960). La faune ornithologique du massif du Mont Kabobo. *Annls Mus. r. Congo belge Ser. 8vo.* **85,** 5–46.

Quézel, P. (1957). "Peuplement végétal des hautes montagnes de l'Afrique du Nord." Lechevalier, Paris.

Quézel, P. (1960). Flore et palynologie sahariennes, leur signification bioclimatique et paléoclimatique. *Bull. Inst. fr. Afr. noire* (A) **22,** 353–360.

Quézel, P. and Martinez, O. (1958). Étude palynologique de deux diatomites de Borkou (Territoire du Tchad). *Bull. Soc. Hist. nat. Afr.* **49,** 230–244.

Quézel, P. and Martinez, C. (1962). Premiers resultats de l'analyse palynologique. *In* "Documents scientifiques Mission Berliet Ténéré–Tchad" (H. J. Hugot, ed.), pp. 313–327. Arts et Metiers Graphiques, Paris.

Rand, A. L. (1936). The distribution and habits of Madagascar birds. *Bull. Am. Mus. nat. Hist.* **72,** 143–499.

Rand, A. L. and Deignan, H. G. (1960). Family Pycnonotidae. *In* "Check-list of Birds of the World," Vol. 9, pp. 221–230. Museum of Comparative Zoology, Cambridge, Mass.

Rattray, S. M. (1960). "The Grass Cover of Africa." FAO, No. 49. Rome.

Rayner, M. (1962). Palaearctic birds in southern Nigeria. *Ibis* **104,** 415–416.

Reichenow, A. (1900–03). "Die Vogel Afrikas." 3 vols. J. Neumann, Neudamm.

Reynolds, V. (1965). "Budongo. A Forest and its Chimpanzees." Methuen, London.

Richards, P. W. (1952). "The Tropical Rain Forest. An Ecological Study." Cambridge University Press.

Richards, P. W. (1963). What the tropics can contribute to ecology. *J. Ecol.* **51,** 231–241.

Ripley, S. D. (1953). Considerations on the origin of the Indian bird fauna. *Bull. natn. Inst. Sci. India* **7,** 269–275.

Ripley, S. D. (1954). Comments on the biogeography of Arabia with particular reference to birds. *J. Bombay nat. Hist. Soc.* **52,** 241–248.

Ripley, S. D. (1961). "A Synopsis of the Birds of India and Pakistan." Bombay Natural History Society, Bombay.

Ripley, S. D. (1965). Le martinet pale de Socotra (*Apus pallidus berliozi*). *Oiseau Revue fr. Orn.* **35,** no. spec. 101–102.

Rosen, B. von (1962). Träskostork i Fornegypten. *Fauna Flora Upps.* 1962, 174–178.

Roux, F. (1959). Quelques données sur les Anatidés et Charadriidés paléarctiques hivernants dans la basse vallée du Sénégal et sur leur écologie. *Terre et Vie* **106,** 315–321.

Rudebeck, G. (1955). Some observations at a roost of European Swallows and other birds in the south-eastern Transvaal. *Ibis* **97,** 572–580.

Ruwet, J. C. (1964). La périodicité de la reproduction chez les oiseaux du Katanga. *Gerfaut* **54,** 84–110.

Salomonsen, F. (1965). Notes on the sunbird-asitys (*Neodrepanis*). *Oiseau Revue fr. Orn.* **35,** no. spec. 103–111.

Salt, G. (1954). A contribution to the ecology of upper Kilimanjaro. *J. Ecol.* **42,** 375–423.

Schmidt-Nielsen, K. (1964). "Desert Animals. Physiological Problems of Heat and Water." Clarendon Press, Oxford.

Schnell, R. (1950). "La Forêt Mense." Paul Chevalier, Paris.

Schouteden, H. (1948–56). De Vogel van Belgisch Congo en van Ruanda-Urundi. *Annls Mus. r. Congo belge Sér. 8vo* C (4) 2, 3, 4, 5.

Schüz, E. (1952). "Vom Vogelzug." Paul Schöps, Frankfurt am Main.

Sclater, W. L. (1911). On the birds collected by Mr Claude H. B. Grant at various localities in South Africa. Parts 2 and 3. *Ibis* (9) **5**, 405–437, 695–741.

Sclater, W. L. (1928–30). "Systema Avium Aethiopicarum." 2 vols. Taylor and Francis, London.

Scott, H. H. (1942). "In the High Yemen." John Murray, London.

Scott, H. H. (1952). Journey to the Gughé Highlands (southern Ethiopia), 1948–49. *Proc. Linn. Soc. Lond.* **163**, 85–189.

Scott, H. H. (1955). Journey to the High Simien district, northern Ethiopia, 1952–53. *Webbia* **11**, 425–450.

Serle, W. (1950). A contribution to the ornithology of the British Cameroons. *Ibis* **92**, 342–376, 602–638.

Serle, W. (1954). A second contribution to the ornithology of the British Cameroons. *Ibis* **96**, 47–80.

Serle, W. (1957). A contribution to the ornithology of the Eastern Region of Nigeria. *Ibis* **99**, 371–418, 628–685.

Serle, W. (1964). The lower altitudinal limit of the montane forest birds of the Cameroon Mountain, West Africa. *Bull. Br. Orn. Club* **84**, 87–91.

Serle, W. (1965). A third contribution to the ornithology of the British Cameroons. *Ibis* **107**, 60–94, 230–246.

Shantz, H. L. and Marbut, C. F. (1923). "The Vegetation and Soils of Africa." American Geographical Society, Res. Ser. 13, New York.

Shaw, H. K. A. (1947). The vegetation of Angola. *J. Ecol.* **35**, 23–48.

Shepard, F. P. (1964). Sea level changes in the past 6000 years: possible archeological significance. *Science, N.Y.* **143**, 574–576.

Simon, P. (1965). Synthèse de l'avifaune du massif montagneux du Tibesti. *Gerfaut* **55**, 26–71.

Simpson, G. G. (1940). Mammals and land bridges. *J. Wash. Acad. Sci.*, 137–163.

Simpson, G. G. (1952). Probabilities of dispersal in geologic time. *Bull. Am. Mus. nat. Hist.* **99** (3), 163–176.

Simpson, G. G. (1953). "The Major Features of Evolution." Columbia University Press, New York.

Slobodkin, L. B. (1962). "Growth and Regulation of Animal Populations." Holt, Rinehart and Winston, New York.

Slud, P. (1960). The birds of Finca "La Selva", Costa Rica: a tropical wet forest locality. *Bull. Am. Mus. nat. Hist.* **121** (2), 51–148.

Smith, K. D. (1955). The winter breeding season of land-birds in eastern Eritrea. *Ibis* **97**, 480–507.

Smith, K. D. (1957). An annotated check list of the birds of Eritrea. *Ibis* **99**, 1–26, 307–337.

Smithers, R. H. N., Irwin, M. P. S. and Paterson, M. L. (1957). "A Check List of the Birds of Southern Rhodesia." Rhodesia Ornithological Society.

Smythies, B. E. (1960). "The Birds of Borneo." Oliver and Boyd, Edinburgh.

Snow, D. W. (1950). The birds of São Tomé and Principe in the Gulf of Guinea. *Ibis* **92**, 579–595.

Snow, D. W. (1952). A contribution to the ornithology of north-west Africa. *Ibis* **94**, 473–498.

Snow, D. W. (1954). Trends in geographical variation in Palaearctic members of the genus *Parus*. *Evolution, Lancaster, Pa.* **8**, 19–28.

Someren, R. A. L. van and van Someren, V. G. L. (1911). "Studies of Bird-life in Uganda." John Bale and Danielson, London.

Someren, V. D. van (1947). Field notes on some Madagascar birds. *Ibis* **89**, 235–267.

Someren, V. G. L. van (1922). Notes on the birds of East Africa. *Novit. zool.* **29**, 1–246.

Someren, V. G. L. van (1932). Birds of Kenya and Uganda, being addenda and corrigenda to my previous paper in *Novitates Zoologicae*, **29** (1922). *Novit. zool.* **37**, 252–380.

Someren, V. G. L. van (1939). Museum's expedition to Chyulu Hills. *Jl E. Africa nat. Hist. Soc.* **14**, 1–151.

Someren V. G. L. van and van Someren, G. R. C. (1949). The birds of Bwamba (Bwamba County, Toro District, Uganda). *Uganda J.* Spec. Suppl. 13.

Storer, R. W. (1960). Adaptive radiation in birds. *In* "Biology and Comparative Physiology of Birds" (A. J. Marshall, ed.), Vol. 1, pp. 15–55. Academic Press, New York and London.

Stresemann, E. (1927–34). "Handbuch der Zoologie", Bd. 7. Aves. Walter de Gruyter, Berlin and Leipzig.

Stresemann, E. (1939). *Zosterops siamensis* Blyth—ein gelbbauchige Rasse von *Zosterops palpebrosa*. *J. Orn., Lpz.* **87**, 156–164.

Stresemann, E. (1954). Zur Frage der Wanderungen des Eleonorenfalken. *Vogelwarte* **17**, 182–183.

Summers, R. (1960). Environment and culture in Southern Rhodesia. *Proc. Am. phil. Soc.* **104**, 266–292.

Täckholm, V., Draa, M. and Abdel Fadeel, A. A. (1956). "Students' Flora of Egypt." Anglo-Egyptian Bookshop, Cairo.

Taylor, J. S. (1949). Notes on the martins, swallows and swifts: Fort Beaufort, C.P. *Ostrich* **20**, 26–28.

Tennent, J. R. M. (1964). The birds of Endau Mountain in the Kitui District of Kenya. *Ibis* **106**, 1–6.

Thomas, A. S. (1943). The vegetation of the Karamojo District, Uganda. *J. Ecol.* **31**, 149–177.

Thomson, A. L. (1936). "Bird Migration." Witherby, London.

Traylor, M. A. (1963). "Check-list of Angolan Birds." Companhia de Diamantes de Angola, Lisboa.

Troughton, K. G. (1959). The marsupial fauna: its origin and radiation. *In* "Biogeography and Ecology in Australia." (Keast *et al.*, eds.). W. Junk, The Hague.

Tuck, R. F. (1959). Observations dans le massif du Tibesti en 1957. *Alauda* **27**, 199–204.

Udvardy, M. F. D. (1959). Notes on the ecological concepts of habitat, biotope and niche. *Ecology* **40**, 725–728.

Ulfstrand, S. and Lamprey, H. (1960). On the birds of the Kungwe-Mahari area in western Tanganyika. *Jl E. Africa nat. Hist. Soc.* **23**, 223–232.

Undeland, R. E. (1964). Verreaux's Eagle *Aquila verreauxii* in Sinai. *Ibis* **106**, 258.

Valverde, J. A. (1957). "Aves de Sahara español." Consejo Superior de Investigaciones Cientificas, Madrid.

Vaughan, J. H. (1930). The birds of Zanzibar and Pemba. Part II. *Ibis* **(12)** 6, 1–48.

Vaughan, R. (1961). *Falco eleanorae. Ibis* **103a**, 114–128.

Vaurie, C. (1957). Systematic notes on Palaearctic birds, no. 29. *Am. Mus. Novit.* 1854.

Vaurie, C. (1959). The birds of the Palaearctic fauna. H. F. & G. Witherby, London.

Verdcourt, B. (1963). New Mollusca from Mt. Kulal, Kenya. *Arch. Molluskenk.* **92**, 237–245.

Volsøe, H. (1951). The breeding birds of the Canary Islands. *Vidensk. Meddr. dansk. naturh. Foren.* **113**, 1–153.

Volsøe, H. (1955). The breeding birds of the Canary Islands. *Vidensk. Meddr. dansk. naturh. Foren.* **117**, 117–177.

Voous, K. H. (1959). The relationship of the European and Aethiopian avifaunas. *Ostrich* Suppl. 3, 40–46.

Voous, K. H. (1960). "Atlas of European Birds." Nelson, London.

Voous, K. H. and Payne, H. A. W. (1965). The grebes of Madagascar. *Ardea* **53**, 9–31.

Voute, C. (1962). Geological and morphological evolution of the Niger and Benue Valleys. *Annls Mus. r. Congo belge Sér. 8vo* **40** (1), 189–205.

Walter, H. (1964). Productivity of vegetation in arid countries, the savannah problem and bush encroachment after grazing. *In* "The Ecology of Man in this Tropical Environment", pp. 221–229. International Union for the Conservation of Nature, Series 4.

Ward, P. (1963). Lipid levels in birds preparing to cross the Sahara. *Ibis* **105**, 109–111.

Ward, P. (1965). The feeding biology of the Black-faced Dioch *Quelea quelea* in Nigeria. *Ibis* **107**, 173–214.

Weber, N. A. (1943). The ants of the Imatong Mountains, Anglo-Egyptian Sudan. *Bull. Mus. comp. Zool. Harv.* **93**, 261–389.

Weimarck, H. (1941). Phytogeographical groups, centres and intervals in the Cape flora. *Acta Univ. lund.* N.F. (2) **37** (5), 1–143.

Wetmore, A. (1959). Birds of the Pleistocene in North America. *Smithson. misc. Collns* **138** (4), 1–24.

Wetmore, A. (1960). A classification for the birds of the world. *Smithson. misc. Collns* **139** (11).

White, C. M. N. (1960). A check list of the Ethiopian Muscicapidae (Sylviinae). *Occ. Pap. natn. Mus. Sth. Rhod.* **24B**, 399–430.

White, C. M. N. (1961). A revised check list of African broadbills, pittas, larks, wagtails and pipits. Govt. Printer, Lusaka.

White, C. M. N. (1962a). A check list of the Ethiopian Muscicapidae (Sylviinae). *Occ. Pap. natn. Mus. Sth. Rhod.* **26B**, 653–738.

White, C. M. N. (1962b). A revised check list of African shrikes, orioles, drongos, starlings, crows, waxwings, cuckoo-shrikes, bulbuls, accentors, thrushes and babblers. Govt. Printer, Lusaka.

White, C. M. N. (1963). A revised check list of African flycatchers, tits, tree creepers, sunbirds, white-eyes, honey eaters, buntings, finches, weavers and waxbills. Govt. Printer, Lusaka.

Willett, H. C. (1953). Atmospheric and oceanic circulation as factors in glacial-interglacial climates. *In* "Climatic Change: Evidence, Causes and Effects" (H. Shapley, ed.), pp. 51–71. Harvard University Press, Cambridge, Mass.

Williams, G. R. (1953). The dispersal from New Zealand and Australia of some introduced European passerines. *Ibis* **95**, 676–692.

Williams, J. G. (1951). The birds of Bwamba: some additions. *Uganda J.* **15**, 107–111.

Winterbottom, J. M. (1936). Bird population studies. V. An analysis of the avifauna of the Jeanes School Station, Mazabuka, Northern Rhodesia. *J. Anim. Ecol.* **5**, 294–311.

Winterbottom, J. M. (1949). Mixed bird parties in the tropics, with special reference to Northern Rhodesia. *Auk* **66**, 258–264.

Winterbottom, J. M. and Skead, C. J. (1962). A preliminary classification of bird habitats for the Cape Province south of the Orange River. S. Afr. Avifauna Ser. (Percy Fitzpatrick Inst. Afr. Orn.) 3.

Woldstedt, P. (1960). Die letzte Eiszeit in Nordamerika und Europa. *Eiszeitälter Gegenw.* **11**, 148–165.

Wyatt-Smith, J. (1952). Malayan forest types. *Malay. Nat. J.* **7**, 45–55.

Yapp, W. B. (1962). "Birds and Woods." Oxford University Press, London.

Zeuner, F. E. (1958). "Dating the Past", 4th ed. Methuen, London.

Taxonomic Index

A

abyssinica, see Hirundo, Zosterops
abyssinicus, see Alcippe, Asio, Coracias, Turdus
Accipiter badius, 243
Accipitridae, 243
Acridotheres tristis, 333, 345
Acrocephalus,
    arundinaceus griseldis, 261
    baeticatus, 245, 316, 352
    brevipennis, 315, 316, 317
    gracilirostris, 315
    palustris, 266
    rufescens, 315
    scirpaceus, 63
Acryllium vulturinum, 280
acuta, see Anas
adametzi, see Cisticola
adsimilis, see Dicrurus
Aegypiidae, 83, 131, 148, 154, 305, 355
aegyptiaca, see Alopochen
aegyptiacus, see Pluvianus
aeneus, see Tangarius
Aepyornithidae, 327, 328, 335
Aerops albicollis, 244, 246
aethiopica, see Hirundo, Threskiornis
afer, see Eurystomus, Nilaus, Philostomus
affinis, see Apus
afra, see Nectarinia
africana, see Coturnix, Luscinia
africanus, see Phalacrocorax
Afropavo congensis, 166
Afroxyechus forbesi, 243
Agapornis, 340
Agelastes meleagridis, 161
    niger, 161
Alaemon alaudipes, 315
Alaudidae, 65, 87, 91, 94, 110, 133, 149, 155, 157, 177, 182, 244, 260, 277, 300, 307, 355, 367
alaudipes, see Alaemon, Certhilauda
alba, see Tyto
albicapilla, see Cossypha
albiceps, see Psalidoprocne, Xiphidopterus
albicollis, see Aerops, Corvultur, Corvus
albigularis, see Hirundo
albirostris, see Bubalornis, Onychognathus

albogularis, see Francolinus
albonotatus, see Trochocercus
alboterminatus, see Tockus
albus, see Casmerodius, Corvus
Alcedinidae, 86, 90, 132, 149, 154, 243, 276, 286, 306
alcinus, see Machaeramphus
Alcippe abyssinicus, 189, 193
Alectroenas, 340
Alethe,
    anomala, 214
    montana, 214
    see also Malacocincla
alius, see Malaconotus
Alopochen aegyptiaca, 74, 138, 243
alpestris, see Eremophila
Alseonax, see Muscicapa
Amaurocichla bocagei, 323
Ammomanes,
    cincturus, 65, 66
    deserti, 65, 66
    dunni, 65
Ammoperdix heyi, 64, 65
amurensis, see Falco
Anabathmis, see Nectarinia
Anas,
    acuta, 138
    angustirostris, 316
    capensis, 138, 243
    erythrophthalma, 238
    erythrorhyncha, 238
    punctata, 238
    querquedula, 138
    sparsa, 138, 225
Anastomus lamelligerus, 241, 243
Anatidae, 84, 130, 138, 148, 150, 243, 260, 304, 339
Andropadus,
    importunus, 93
    tephrolaema, 195
Anhinga rufa, 76
angolensis, see Gypohierax, Pitta, Ploceus
angustirostris, see Anas
Anhingidae, 84, 130, 148, 304
anomala, see Alethe
anselli, see Centropus
ansorgei, see Nesocharis
Anthoscopus, see Remiz

Meropidae, 86, 90, 94, 103, 132, 149, 154, 155, 176, 244, 260, 276, 286, 306
Meropogon, 102, 103
Merops,
    apiaster, 123, 137, 245
    malimbicus, 239, 244
    nubicus, 183, 185, 244, 245
    orientalis, 74, 116
    persicus, 251
    superciliosus, 251–252
merula, see *Turdus*
Mesitornithidae, 305, 333
Mesopicos griseocephalus, 93
metabates, see *Melierax*
metallica, see *Nectarinia*
metopias, see *Artisornis*
mfumbiri, see *Laniarius*
Micronisus gabar, 69, 74
Microscelis, see *Hypsipetes*
migrans, see *Milvus*
Milvus,
    migrans, 243, 316, 317, 324
    milvus, 316, 317
Mimidae, 155
minor, see *Indicator*, *Lanius*
minulla, see *Batis*
Mirafra,
    buckleyi, 174
    javanica, 116
    nigricans, 244
    ruddi, 183, 184
Modulatrix stictigula, 214
moesta, see *Oenanthe*
molleri, see *Prinia*
Molothrus ater, 158
monacha, see *Oenanthe*
monachus, see *Oriolus*
mongolus, see *Charadrius*
montana, see *Alethe*
montanus, see *Cercococcyx*
Monticola, 139
    explorator, 122
    rufocinerea, 122
    rupestris, 122
    saxatilis, 122, 259
    solitarius, 158
Moorhen, 63, 112, 324
moreaui, see *Apalis*
Motacilla,
    cinerea, 120, 122, 230, 263, 311
    clara, 120, 122
    flava, 259, 263, 272
Motacillidae, 87, 91, 112, 133, 149, 155, 177, 245, 260, 277, 307, 355
mouroniensis, see *Zosterops*

moussieri, see *Diplootocus*
Muscicapa adusta, 88, 194, 233, 241, 245
Muscicapidae, 87, 91, 94, 103, 104, 133, 149, 155, 177, 245, 260, 277, 287, 307
Musophaga,
    rossae, 180
    violacea, 180
Musophagidae, 86, 90, 104, 108, 132, 149, 154, 176, 276, 286, 306
musicus, see *Bias*

N

naevia, see *Locustella*
naevius, see *Coracias*
nana, see *Sylvia*
narina, see *Apaloderma*
nasutus, see *Tockus*
natalicus, see *Turdus*
naumanni, see *Falco*
Nectarinia,
    afra, 220
    balfouri, 355
    bifasciata, 233
    bocagei, 220
    famosa, 220, 228, 233
    johnstoni, 219, 225, 227
    kilimensis, 220, 228
    mediocris, 195
    metallica, 74, 245
    oseus, 245
    preussi, 233
    purpureiventris, 220
    reichenowi, 220, 228
    tacazze, 220, 228
    thomensis, 367
    veroxii, 357
    violacea, 181
Nectariniidae, 87, 91, 133, 149, 153, 155, 177, 219, 245, 277, 287, 300, 307, 336, 355
Neocossyphus rufus, 167
Neodrepanis, 110, 336, 343
Neomixis, 331
Neophron percnopterus, 317, 355
Neospiza concolor, 319, 323, 365
Neotis denhami, 64, 244
Nesillas,
    mariae, 364
    typicus, 364
Nesocharis ansorgei, 82
Netta, see *Anas*
Nettapus auritus, 138
newtoni, see *Cyanomitra*, *Lanius*
Newtonia brunneicauda, 341

Spatula clypeata, 138
Speirops, 222, 326, 367
  brunnea, 320
  melanocephala, 222
Spermophaga,
  cana, 167
  ruficapilla, 167
Sphenorhynchus, see *Ciconia*
spilodera, see *Petrochelidon*
spilonotos, see *Salpornis*
spinosus, see *Hoplopterus*
Spizocorys razae, 315
splendidus, see *Lamprocolius*, *Lamprotor-nis*
Spreo shelleyi, 245
Sprosser, 259
Starling,
  Amethyst, 245, 248
  Wattled, 231
stellata, see *Pogonocichla*
Stephanoaetus, 292, 339
stictigula, see *Modulatrix*
Stigmatopelia, see *Streptopelia*
Stonechat, 72, 116, 118, 119, 224, 233, 312, 337
Stork,
  Abdim's, 241, 246
  Black, 121, 122, 139
  Marabou, 115, 243
  Openbilled, 241, 243
  Whale-headed, 69, 74, 104, 109
  White, 123, 124, 266
Streptopelia,
  capicola, 98
  lugens, 114, 157, 195
  roseogrisea, 64
  semitorquata, 88
  senegalensis, 63, 74, 116, 355
  turtur, 63, 263
streptophorus, see *Francolinus*
striatus, see *Butorides*, *Colius*
Strigidae, 83, 95, 131, 148, 154, 260, 305, 355
striolata, see *Emberiza*, *Poliospiza*
Struthio camelus, 68, 104, 109, 183
Struthionidae, 83, 90, 104, 109, 131, 154, 276, 302
Sturnidae, 87, 91, 94, 104, 133, 149, 155, 157, 177, 182, 245, 277, 287, 307, 355
subbuteo, see *Falco*
subflava, see *Prinia*
Sugarbird, Cape, 105
sulphureopectus, see *Chlorophoneus*
superciliosus, see *Centropus*, *Merops*

Swallow,
  European, 258, 259, 264, 265
  Wire-tailed, 116
swierstrai, see *Francolinus*
Swift, Alpine, 122, 220
swynnertoni, see *Pogonocichla*
Sylvia,
  atricapilla, 257, 262, 316
  borin, 258, 263
  cantillans, 272
    communis, 263–264
  conspicillata, 316, 317
  melanocephala, 63, 70
  nana, 65, 66
  rüppellii, 263
Sylviidae, 65, 87, 91, 94, 102, 103, 112, 133, 149, 155, 158, 177, 182, 245, 260, 266, 277, 287, 307, 340, 355

T

tacazze, see *Nectarinia*
taeniolaema, see *Campethera*
tahapisi, see *Fringillaria*
Tangarius aeneus, 158
tarda, see *Otis*
Tauraco,
  bannermani, 199
  erythrolophus, 199
  fischeri, 353
  hartlaubi, 108, 194, 195
  ruspolii, 208
Tchagra senegalus, 73, 74
Teal, Hottentot, 238
temminckii, see *Cursorius*
tenuirostris, see *Onychognathus*
Tephrocorys, see *Calandrella*
tephrolaema, see *Andropadus*
terrestris, see *Phyllastrephus*
Terpsiphone, 324, 368
  paradisi, 116
  viridis, 92, 245
teydea, see *Fringilla*
Thalassornis leuconotos, 138
theklae, see *Galerida*
thomensis, see *Cyanomitra*, *Nectarinia*
thoracica, see *Apalis*
Threskiornis aethiopica, 69, 74, 244
Threskiornithidae, 84, 130, 148, 244, 304
Thrush,
  Blue Rock, 158
  Common Rock, 259
  Mistle, 158
  Song, 230
Timaliidae, 65, 87, 91, 94, 103, 104, 133, 149, 155, 177, 277, 287, 307

Subject Index

A

Abyssinia,
  afroalpine birds, 225
  biological affinities with Europe, 119
  montane forest birds, 207–9
  montane non-forest birds, 220–1
Acacia biome,
  extensions, 183–7
Acacia steppe,
  sample bird communities, 278–82
Adaptability,
  birds in South Africa, 93
  birds to artificial habitats, 89
  birds to untypical natural habitats, 92–93
  desert birds, 67
  Golden Eagle, 63
  Madagascar birds, 341
  reed-warblers, 315, 352
  see also individual islands
Adaptive radiation, *see* Radiation
Affinities of bird faunas,
  Ethiopian/European, 118–25
  Ethiopian/Indian, 113–18, 166
  Ethiopian/S. American, 112–13
  see also individual islands
Africa,
  rainfall, 15–16
  rivers and lakes, 15
  surface relief, 13, 14
  temperatures, past, 45–50
  temperatures, present, 17–19
  vegetation, general, 19–38
  see also individual vegetation types
Afroalpine birds, 224–8
Afroalpine environment, 223–4
Ages,
  islands, *see* individual islands
  species and subspecies, 9
Agriculture,
  effect on birds, 36
Ahaggar,
  bird fauna, 63
  in Pleistocene, 55
Algeria, *see* Maghreb
Altitude zones,
  specificity of birds to, 96–97

America,
  migrant birds, 257–8
  North, number of bird species, 143
  South, number of bird species, 143–4
Angola,
  faunistic importance of scarp, 187
  montane forest birds, 209–11
  Pleistocene climatic vicissitudes, 51
Anjouan Island, birds, 345–50
Annobon Island, birds, 324
Arabia and Afro-Indian affinities, 113–14
Area and number of species,
  Africa, 129
  islands, 359
Arizona,
  birds, compared with Somaliland, 153–8
Australia,
  number of bird species, 143
Avifaunas,
  composition of, *see* Composition
Avifaunal "districts", 173–4

B

Barbary, *see* Maghreb
Bee-eater, family and latitude, 139
Bird fauna, Ethiopian Region,
  analysis, 80–103
  ecological specificity, 83–92
Birds,
  eco-taxonomic categories, 5–6
Brachystegia,
  bird endemism, 187–8
  sample bird communities, 282–4
Breeding of individual birds,
  possibility in both South Africa and Palaearctic, 123
Breeding seasons,
  African birds, 39–41
  desert falcons, 67
Budongo forest,
  birds, 289–90
Butterflies,
  endemism low on Marsabit Mt, 194
  faunistics in Africa, 2
  "Malagasy sub-region", 334
  montane distribution, 189
  West African, on Kenya coast, 168–9

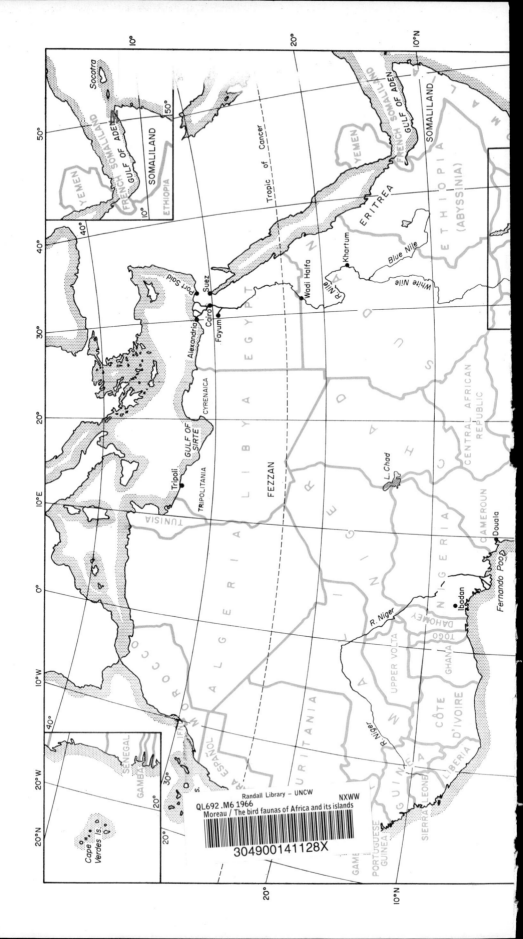